COMPOSITES MANUFACTURING
Materials, Product, and Process Engineering

Sanjay K. Mazumdar, Ph.D.

CRC PRESS

Boca Raton   London   New York   Washington, D.C.

Library of Congress Cataloging-in-Publication Data

Mazumdar, Sanjay K.
　　Composites manufacturing : materials, product, and process
engineering / by Sanjay K. Mazumdar.
　　　　p.　cm.
　　Includes bibliographical references and index.
　　ISBN 0-8493-0585-3
　　1. Composite materials.
TA418.9.C6 M34 2001
620.1'18--dc21　　　　　　　　　　　　　　　　　　　　　　　　　　　　2001004994

This book contains information obtained from authentic and highly regarded sources. Reprinted material is quoted with permission, and sources are indicated. A wide variety of references are listed. Reasonable efforts have been made to publish reliable data and information, but the author and the publisher cannot assume responsibility for the validity of all materials or for the consequences of their use.

Neither this book nor any part may be reproduced or transmitted in any form or by any means, electronic or mechanical, including photocopying, microfilming, and recording, or by any information storage or retrieval system, without prior permission in writing from the publisher.

The consent of CRC Press LLC does not extend to copying for general distribution, for promotion, for creating new works, or for resale. Specific permission must be obtained in writing from CRC Press LLC for such copying.

Direct all inquiries to CRC Press LLC, 2000 N.W. Corporate Blvd., Boca Raton, Florida 33431.

Trademark Notice: Product or corporate names may be trademarks or registered trademarks, and are used only for identification and explanation, without intent to infringe.

Visit the CRC Press Web site at www.crcpress.com

© 2002 by CRC Press LLC

No claim to original U.S. Government works
International Standard Book Number 0-8493-0585-3
Library of Congress Card Number 2001004994
Printed in the United States of America　　4 5 6 7 8 9 0
Printed on acid-free paper

Surrendered to the Lord of the Universe

Preface

Early large-scale commercial applications of composite materials started during World War II (late 1940s and early 1950s) with marine applications for the military; but today, composite products are manufactured by a diverse range of industries, including aerospace, automotive, marine, boating, sporting goods, consumer, infrastructure, and more. In recent years, the development of new and improved composites manufacturing processes has caused unlimited product development opportunities. New high-volume production methods such as compression molding (SMC) reveal a gained maturity level and are routinely used for making automotive, consumer, and industrial parts with a good confidence level. The use of composite materials is no longer limited to only naval and spacecraft applications. New material innovations, and a drop in pricing and development of improved manufacturing processes have given rise to the presence of composite materials in almost every industrial sector. In fact, because of styling detail possibilities and the high surface finish quality attainable by composites fabrication processes, composites are considered materials of choice for certain industry sectors (e.g., automotive).

Until recently, only a few universities offered courses on composites manufacturing, probably due to the lack of a suitable textbook. This book offers all the related materials to make composites manufacturing a part of the curriculum in the composite materials discipline. This book covers important aspects of composites product manufacturing, such as product manufacturability, product development, processing science, manufacturing processes, cost estimating, and more. These aspects of fabrication issues, which are crucial in the production of good composite parts, have not been covered in any of the books available in the composites industry. The most common courses offered at universities in the composite materials field are related to the introduction and design aspects of composite materials. Without processing and product development knowledge, successful composite products cannot be fabricated. This book bridges this gap and covers important elements of product manufacturing using composite materials. This book is suitable for students, engineers, and researchers working in the composite materials field. This book offers valuable insight into the production of cost-competitive and high-quality composite parts. Engineers and professionals working in the composites industry can significantly benefit from the content of this book.

This book discusses the subject of manufacturing within the framework of the fundamental classification of processes. This should help the reader understand where a particular manufacturing process fits within the overall fabrication scheme and what processes might be suitable for the manufacture of a particular component. The subject matters are adequately descriptive for those unfamiliar with the various fabrication techniques and yet sufficiently analytical for an academic course in composites manufacturing.

The book takes the reader step-by-step from raw material selection to final part fabrication and recycling. Chapter 2 details the raw materials available in the composites industry for the fabrication of various composite products. Methods of selecting the correct material from the thousands of materials available are discussed in Chapter 3. Chapter 4 discusses the six important phases of the product development process. It provides roadmaps to engineers and team members for the activities and deliverables required for the design, development, and fabrication of the part. Chapter 5 describes procedures to design a product, taking manufacturing into consideration. To be competitive in the current global marketplace, products must be designed in a minimum amount of time and with minimum resources and cost. Design for manufacturing (DFM) plays a key role in concept generation, concept approval, and concept improvement, and comes up with a better design in the shortest time. It integrates processing knowledge into the design of a part to get the maximum benefit and capabilities from the manufacturing method. As compared to metals, composite materials offer the higher potential of utilizing DFM and part integration, and therefore can significantly reduce the cost of production. Chapter 6 discusses various composite manufacturing techniques in terms of their advantages, disadvantages, raw materials requirements, applications, tooling and mold requirements, methods of applying heat and pressure, processing steps, and more. Process selection criteria and basic steps in composite manufacturing processes are discussed in this chapter. Process models for key manufacturing processes are described in Chapter 7. Process models are used to determine optimum processing conditions for making good-quality composite parts. This eliminates processing problems before a manufacturing process begins or before the part design is finalized. Preproduction guidelines and methods of writing manufacturing instructions and bill of materials are discussed in Chapter 8. Joining and machining of composite parts require a different approach than the joining and machining of metal parts; these are discussed in Chapters 9 and 10, respectively. Cost-estimating techniques are elaborated in Chapter 11. Tools for selecting a best technology/fabrication process to get a competitive advantage in the marketplace are included in this chapter. Finally, recycling aspects of composite materials, which are becoming growing concerns in industry and government sectors, are discussed in Chapter 12. Overall, this book provides professionals with valuable information related to composites product manufacturing as well as best state-of-the-art knowledge in this field.

Acknowledgments

With great devotion I acknowledge the grace of God for the successful completion of this book.

I am grateful to all the engineers, researchers, scientists, and professionals who contributed to the development of composites manufacturing processes and technologies. With their efforts, composites technology has gained maturity and has been used confidently for various applications. I am thankful to Professor Timothy Gutowski (M.I.T.), Gerald Sutton (Intellitec), John Marks (COI Materials), and John Taylor (Goodrich Corp.) for reviewing this book and for providing excellent comments.

I am thankful to my friends and relatives for their kindness and support. I am grateful to my wife Gargi Mazumdar for her patience and support during the writing of this book. Thanks to my 6-year-old daughter Ria Mazumdar for letting me work on my book. And I am thankful to my parents for their love and support.

Author

Dr. Sanjay K. Mazumdar is president and CEO of E-Composites.com, Inc., Grandville, Michigan, U.S.A., a leading service-oriented company providing market reports, job bank services, *CompositesWeek* newsletter, CompositesExchange for buy and sell trades, product marketing, product commercialization, and various other services for the composite materials industry. E-Composites.com, Inc. provides various platforms for connecting buyers to sellers, vendors to customers, employers to employees, and technical professionals to a wealth of information. E-Composites.com, Inc. is dedicated to rapid development of the composite materials industry and connects more than 20,000 composite users and suppliers from more than 50 countries using its weekly newsletter and Web site. E-Composites.com, Inc.'s clients range from small to Fortune 500 companies such as AOC, BFGoodrich, Bayer Corp., Dow Corning, General Electric, Hexcel, Johns Manville, Lockheed Martin, Owens Corning, Saint-Gobain, Zeon Chemical, and many more.

Dr. Mazumdar has published more than 25 professional papers on processing, joining, and testing of composite materials in reputed international journals and conference proceedings. He has designed and developed more than 100 composite products for a variety of applications, including automotive, aerospace, electronic, consumer, and industrial applications. He has two Society of Plastics Engineers (SPE) awards and two General Motors' Record of Innovation awards for his creativity and innovations. He has worked as adjunct faculty at the University of Michigan, Dearborn, and Concordia University, Montreal, and has taught composite materials-related courses to undergraduate and graduate students. He has given seminars and presentations at international conferences and reputed universities, including the University of California, Berkeley, and Fortune 500 companies.

Dr. Mazumdar can be contacted by e-mail at Sanjaym@e-composites.com *or* visit the Web site — www.e-composites.com — for details.

Contents

1 Introduction 1
1.1 Conventional Engineering Materials 1
    1.1.1 Metals 2
    1.1.2 Plastics 3
    1.1.3 Ceramics 3
    1.1.4 Composites 3
1.2 What Are Composites? 4
1.3 Functions of Fibers and Matrix 5
1.4 Special Features of Composites 6
1.5 Drawbacks of Composites 9
1.6 Composites Processing 10
1.7 Composites Product Fabrication 11
1.8 Composites Markets 13
    1.8.1 The Aerospace Industry 14
    1.8.2 The Automotive Industry 17
    1.8.3 The Sporting Goods Industry 18
    1.8.4 Marine Applications 18
    1.8.5 Consumer Goods 19
    1.8.6 Construction and Civil Structures 19
    1.8.7 Industrial Applications 19
1.9 Barriers in Composite Markets 20
References 20
Questions 20

2 Raw Materials for Part Fabrication 23
2.1 Introduction 23
2.2 Reinforcements 23
    2.2.1 Glass Fiber Manufacturing 27
    2.2.2 Carbon Fiber Manufacturing 27
    2.2.3 Aramid Fiber Manufacturing 28
2.3 Matrix Materials 28
    2.3.1 Thermoset Resins 29
        2.3.1.1 Epoxy 30
        2.3.1.2 Phenolics 31
        2.3.1.3 Polyesters 31
        2.3.1.4 Vinylesters 32
        2.3.1.5 Cyanate Esters 32
        2.3.1.6 Bismaleimide (BMI) and Polyimide 32
        2.3.1.7 Polyurethane 32

		2.3.2	Thermoplastic Resins..33
			2.3.2.1 Nylons..34
			2.3.2.2 Polypropylene (PP)..35
			2.3.2.3 Polyetheretherketone (PEEK)..................................35
			2.3.2.4 Polyphenylene Sulfide (PPS) 35

2.4 Fabrics..36
    2.4.1 Woven Fabrics..36
    2.4.2 Noncrimp Fabrics...37
2.5 Prepregs..40
    2.5.1 Thermoset Prepregs 43
    2.5.2 Thermoplastic Prepregs...43
2.6 Preforms 44
2.7 Molding Compound...47
    2.7.1 Sheet Molding Compound 48
    2.7.2 Thick Molding Compound (TMC)...49
    2.7.3 Bulk Molding Compound (BMC)..51
    2.7.4 Injection Moldable Compounds 52
2.8 Honeycomb and Other Core Materials...52
References 55
Questions...55

3 Material Selection Guidelines 57
3.1 Introduction...57
3.2 The Need for Material Selection 57
3.3 Reasons for Material Selection 58
3.4 Material Property Information 59
3.5 Steps in the Material Selection Process..60
    3.5.1 Understanding and Determining the Requirements...........60
    3.5.2 Selection of Possible Materials...60
    3.5.3 Determination of Candidate Materials.................................61
    3.5.4 Testing and Evaluation..62
3.6 Material Selection Methods 63
    3.6.1 Cost vs. Property Analysis 63
    3.6.2 Weighted Property Comparison Method.............................66
        3.6.2.1 Scaling for Maximum Property Requirement.............67
        3.6.2.2 Scaling for Minimum Property Requirement 67
        3.6.2.3 Scaling for Nonquantitative Property 67
    3.6.3 Expert System for Material Selection....................................68
Bibliography 69
Questions...69

4 Product Development 71
4.1 Introduction...71
4.2 What Is the Product Development Process....................................72
4.3 Reasons for Product Development..72
4.4 Importance of Product Development..73

4.5	Concurrent Engineering		74
4.6	Product Life Cycle		76
4.7	Phases of Product Development		77
	4.7.1	Concept Feasibility Phase	77
	4.7.2	Detailed Design Phase	78
	4.7.3	Prototype Development and Testing Phase	78
	4.7.4	Preproduction Demonstration, or Pilot-Scale Production	79
	4.7.5	Full-Scale Production and Distribution	80
	4.7.6	Continuous Improvement	80
4.8	Design Review		80
4.9	Failure Modes and Effects Analysis (FMEA)		81
Reference			83
Bibiliography			83
Questions			83

5 Design for Manufacturing 85

5.1	Introduction		85
5.2	Design Problems		86
5.3	What Is DFM?		87
5.4	DFM Implementation Guidelines		88
	5.4.1	Minimize Part Counts	88
	5.4.2	Eliminate Threaded Fasteners	89
	5.4.3	Minimize Variations	89
	5.4.4	Easy Serviceability and Maintainability	90
	5.4.5	Minimize Assembly Directions	90
	5.4.6	Provide Easy Insertion and Alignment	90
	5.4.7	Consider Ease for Handling	90
	5.4.8	Design for Multifunctionality	91
	5.4.9	Design for Ease of Fabrication	91
	5.4.10	Prefer Modular Design	91
5.5	Success Stories		92
	5.5.1	Composite Pickup Box	92
	5.5.2	Laser Printer	92
	5.5.3	Black & Decker Products	93
5.6	When to Apply DFM		93
5.7	Design Evaluation Method		93
5.8	Design for Assembly (DFA)		94
	5.8.1	Benefits of DFA	95
	5.8.2	Assembly-Related Defects	95
	5.8.3	Guidelines for Minimizing Assembly Defects	97
References			98
Questions			98

6 Manufacturing Techniques 99
6.1	Introduction	99

6.2	Manufacturing Process Selection Criteria		100
	6.2.1	Production Rate/Speed	100
	6.2.2	Cost	101
	6.2.3	Performance	101
	6.2.4	Size	101
	6.2.5	Shape	101
6.3	Product Fabrication Needs		103
6.4	Mold and Tool Making		104
	6.4.1	Mold Design Criteria	104
		6.4.1.1 Shrinkage Allowance	104
		6.4.1.2 Coefficient of Thermal Expansion of Tool Material and End Product	104
		6.4.1.3 Stiffness of the Mold	105
		6.4.1.4 Surface Finish Quality	105
		6.4.1.5 Draft and Corner Radii	105
	6.4.2	Methods of Making Tools	105
		6.4.2.1 Machining	105
		6.4.2.2 FRP Tooling for Open Molding Processes	106
	6.4.3	Tooling Guidelines for Closed Molding Operations	108
6.5	Basic Steps in a Composites Manufacturing Process		115
	6.5.1	Impregnation	116
	6.5.2	Lay-up	116
	6.5.3	Consolidation	116
	6.5.4	Solidification	117
6.6	Advantages and Disadvantages of Thermoset and Thermoplastic Composites Processing		117
	6.6.1	Advantages of Thermoset Composites Processing	117
	6.6.2	Disadvantages of Thermoset Composites Processing	118
	6.6.3	Advantages of Thermoplastic Composites Processing	118
	6.6.4	Disadvantages of Thermoplastic Composites Processing	118
6.7	Composites Manufacturing Processes		118
6.8	Manufacturing Processes for Thermoset Composites		119
	6.8.1	Prepreg Lay-Up Process	119
		6.8.1.1 Major Applications	120
		6.8.1.2 Basic Raw Materials	122
		6.8.1.3 Tooling Requirements	122
		6.8.1.4 Making of the Part	122
		6.8.1.5 Methods of Applying Heat and Pressure	126
		6.8.1.6 Basic Processing Steps	127
		6.8.1.7 Typical Manufacturing Challenges	127
		6.8.1.8 Advantages of the Prepreg Lay-Up Process	128
		6.8.1.9 Limitations of the Prepreg Lay-Up Process	128
	6.8.2	Wet Lay-Up Process	128
		6.8.2.1 Major Applications	128
		6.8.2.2 Basic Raw Materials	131
		6.8.2.3 Tooling Requirements	131

	6.8.2.4 Making of the Part	131
	6.8.2.5 Methods of Applying Heat and Pressure	134
	6.8.2.6 Basic Processing Steps	134
	6.8.2.7 Advantages of the Wet Lay-Up Process	135
	6.8.2.8 Limitations of the Wet Lay-Up Process	135
6.8.3	Spray-Up Process	135
	6.8.3.1 Major Applications	135
	6.8.3.2 Basic Raw Materials	136
	6.8.3.3 Tooling Requirements	136
	6.8.3.4 Making of the Part	139
	6.8.3.5 Methods of Applying Heat and Pressure	139
	6.8.3.6 Basic Processing Steps	139
	6.8.3.7 Advantages of the Spray-Up Process	140
	6.8.3.8 Limitations of the Spray-Up Process	140
6.8.4	Filament Winding Process	140
	6.8.4.1 Major Applications	141
	6.8.4.2 Basic Raw Materials	143
	6.8.4.3 Tooling	143
	6.8.4.4 Making of the Part	144
	6.8.4.5 Methods of Applying Heat and Pressure	146
	6.8.4.6 Methods of Generating the Desired Winding Angle	147
	6.8.4.7 Basic Processing Steps	148
	6.8.4.8 Advantages of the Filament Winding Process	149
	6.8.4.9 Limitations of the Filament Winding Process	150
6.8.5	Pultrusion Process	150
	6.8.5.1 Major Applications	150
	6.8.5.2 Basic Raw Materials	152
	6.8.5.3 Tooling	154
	6.8.5.4 Making of the Part	154
	6.8.5.4.1 Wall Thickness	156
	6.8.5.4.2 Corner Design	157
	6.8.5.4.3 Tolerances, Flatness, and Straightness	157
	6.8.5.4.4 Surface Texture	157
	6.8.5.5 Methods of Applying Heat and Pressure	157
	6.8.5.6 Basic Processing Steps	158
	6.8.5.7 Advantages of the Pultrusion Process	158
	6.8.5.8 Limitations of the Pultrusion Process	158
6.8.6	Resin Transfer Molding Process	159
	6.8.6.1 Major Applications	159
	6.8.6.2 Basic Raw Materials	161
	6.8.6.3 Tooling	162
	6.8.6.4 Making of the Part	164
	6.8.6.5 Methods of Applying Heat and Pressure	169
	6.8.6.6 Basic Processing Steps	170
	6.8.6.7 Advantages of the Resin Transfer Molding Process	173

		6.8.6.8	Limitations of the Resin Transfer Molding Process 174

		6.8.6.9	Variations of the RTM Process 175
			6.8.6.9.1 VARTM 175
			6.8.6.9.2 SCRIMP 175
	6.8.7	Structural Reaction Injection Molding (SRIM) Process 175
		6.8.7.1	Major Applications 176
		6.8.7.2	Basic Raw Materials 176
		6.8.7.3	Tooling 178
		6.8.7.4	Making of the Part 178
		6.8.7.5	Methods of Applying Heat and Pressure 178
		6.8.7.6	Basic Processing Steps 179
		6.8.7.7	Advantages of the SRIM Process 179
		6.8.7.8	Limitations of the SRIM Process 179
	6.8.8	Compression Molding Process 179
		6.8.8.1	Major Applications 180
		6.8.8.2	Basic Raw Materials 181
		6.8.8.3	Making of the Part 181
		6.8.8.4	Mold Design 186
		6.8.8.5	Methods of Applying Heat and Pressure 187
		6.8.8.6	Basic Processing Steps 187
		6.8.8.7	Advantages of the Compression Molding Process 188
		6.8.8.8	Limitations of the Compression Molding Process 188
	6.8.9	Roll Wrapping Process 188
		6.8.9.1	Major Applications 188
		6.8.9.2	Basic Raw Materials 189
		6.8.9.3	Tooling 189
		6.8.9.4	Making of the Part 189
		6.8.9.5	Methods of Applying Heat and Pressure 194
		6.8.9.6	Basic Processing Steps 194
		6.8.9.7	Advantages of the Roll Wrapping Process 196
		6.8.9.8	Limitations of the Roll Wrapping Process 196
		6.8.9.9	Common Problems with the Roll Wrapping Process 196
	6.8.10	Injection Molding of Thermoset Composites 197
		6.8.10.1 Major Applications 197
		6.8.10.2 Basic Raw Materials 197
		6.8.10.3 Tooling 198
		6.8.10.4 Making of the Part 198
6.9	Manufacturing Processes for Thermoplastic Composites 200
	6.9.1	Thermoplastic Tape Winding 201
		6.9.1.1	Major Applications 202
		6.9.1.2	Basic Raw Materials 202
		6.9.1.3	Tooling 202

	6.9.1.4	Making of the Part 203
	6.9.1.5	Methods of Applying Heat and Pressure 206
	6.9.1.6	Advantages of the Thermoplastic Tape Winding Process 207
	6.9.1.7	Limitations of the Thermoplastic Tape Winding Process 207
6.9.2	Thermoplastic Pultrusion Process 208	
	6.9.2.1	Major Applications 208
	6.9.2.2	Basic Raw Materials 208
	6.9.2.3	Tooling 208
	6.9.2.4	Making of the Part 209
	6.9.2.5	Methods of Applying Heat and Pressure 209
	6.9.2.6	Advantages of the Thermoplastic Pultrusion Process 210
	6.9.2.7	Limitations of the Thermoplastic Pultrusion Process 210
6.9.3	Compression Molding of GMT 210	
	6.9.3.1	Major Applications 210
	6.9.3.2	Basic Raw Materials 211
	6.9.3.3	Tooling 213
	6.9.3.4	Part Fabrication 213
	6.9.3.5	Methods of Applying Heat and Pressure 214
	6.9.3.6	Advantages of Compression Molding of GMT 214
	6.9.3.7	Limitations of Compression Molding of GMT 215
6.9.4	Hot Press Technique 215	
	6.9.4.1	Major Applications 215
	6.9.4.2	Basic Raw Materials 215
	6.9.4.3	Tooling 215
	6.9.4.4	Making of the Part 216
	6.9.4.5	Methods of Applying Heat and Pressure 217
	6.9.4.6	Basic Processing Steps 217
	6.9.4.7	Advantages of the Hot Press Technique 218
	6.9.4.8	Limitations of the Hot Press Technique 218
6.9.5	Autoclave Processing 218	
	6.9.5.1	Major Applications 219
	6.9.5.2	Basic Raw Materials 219
	6.9.5.3	Tooling 219
	6.9.5.4	Making the Part 219
	6.9.5.5	Methods of Applying Heat and Pressure 221
	6.9.5.6	Basic Processing Steps 221
	6.9.5.7	Advantages of Autoclave Processing 221
	6.9.5.8	Limitations of Autoclave Processing 222
6.9.6	Diaphragm Forming Process 222	
	6.9.6.1	Major Applications 222
	6.9.6.2	Basic Raw Materials 223
	6.9.6.3	Tooling 223

		6.9.6.4	Making of the Part..223
		6.9.6.5	Methods of Applying Heat and Pressure...................225
		6.9.6.6	Advantages of the Diaphragm Forming Process......225
		6.9.6.7	Limitations of the Diaphragm Forming Process.......225
	6.9.7	Injection Molding...226	
		6.9.7.1	Major Applications..226
		6.9.7.2	Basic Raw Materials..226
		6.9.7.3	Tooling...226
		6.9.7.4	Making of the Part...227
		6.9.7.5	Basic Processing Steps..228
		6.9.7.6	Methods of Applying Heat and Pressure...................229
		6.9.7.7	Advantages of the Injection Molding Process...........229
		6.9.7.8	Limitations of the Injection Molding Process...........229

References 230
Bibliography 233
Questions...233

7 Process Models 235
7.1 Introduction 235
7.2 The Importance of Models in Composites Manufacturing................235
7.3 Composites Processing 236
7.4 Process Models for Selected Thermosets and Thermoplastics Processing 237
    7.4.1 Thermochemical Sub-Model 239
        7.4.1.1 Autoclave or Hot Press Process for Thermoset Composites...241
        7.4.1.2 Filament Winding of Thermoset Composites...........242
        7.4.1.3 Tape Winding of Thermoplastic Composites............243
    7.4.2 Flow Sub-Model 252
        7.4.2.1 Compaction and Resin Flow during Autoclave Cure..253
            7.4.2.1.1 Resin Flow Normal to the Tool Plate........254
            7.4.2.1.2 Resin Flow Parallel to the Tool Plate 257
            7.4.2.1.3 Total Resin Flow 260
        7.4.2.2 Compaction and Resin Flow during Filament Winding...263
        7.4.2.3 Consolidation of Thermoplastic Composites during Autoclave or Hot Press Processing 270
        7.4.2.4 Consolidation and Bonding Models for Thermoplastic Tape Laying and Tape Winding 275
    7.4.3 Void Sub-Model...280
    7.4.4 Stress Sub-Model...283
7.5 Process Model for RTM 284
References 285
Questions...287

8 Production Planning and Manufacturing Instructions 289
8.1 Introduction 289
8.2 Objectives of Production Planning 290
8.3 Bill of Materials 290
8.4 Manufacturing Instructions 294
    8.4.1 Manufacturing Instructions for Making Tooling Panels 296
    8.4.2 Manufacturing Instructions for Making Flaps 296
8.5 Capacity Planning 304
    8.5.1 Problem Definition 305
    8.5.2 Assumptions 306
    8.5.3 Capacity Analysis 306
        8.5.3.1 Autoclave Capacity Analysis 306
        8.5.3.2 Freezer Storage Requirement 308
Questions 308

9 Joining of Composite Materials 309
9.1 Introduction 309
9.2 Adhesive Bonding 310
    9.2.1 Failure Modes in Adhesive Bonding 313
    9.2.2 Basic Science of Adhesive Bonding 313
        9.2.2.1 Adsorption Theory 314
        9.2.2.2 Mechanical Theory 314
        9.2.2.3 Electrostatic and Diffusion Theories 315
    9.2.3 Types of Adhesives 315
        9.2.3.1 Two-Component Mix Adhesives 315
            9.2.3.1.1 Epoxy Adhesives 315
            9.2.3.1.2 Polyurethane Adhesives 316
        9.2.3.2 Two-Component, No-Mix Adhesives 316
            9.2.3.2.1 Acrylic Adhesives 316
            9.2.3.2.2 Urethane Methacrylate Ester (Anaerobic) Adhesives 317
        9.2.3.3 One-Component, No-Mix Adhesives 317
            9.2.3.3.1 Epoxies 317
            9.2.3.3.2 Polyurethanes 317
            9.2.3.3.3 Cyanoacrylates 317
            9.2.3.3.4 Hot-Melt Adhesives 317
            9.2.3.3.5 Solvent- or Water-Based Adhesives 318
    9.2.4 Advantages of Adhesive Bonding over Mechanical Joints 318
    9.2.5 Disadvantages of Adhesive Bonding 319
    9.2.6 Adhesive Selection Guidelines 319
    9.2.7 Surface Preparation Guidelines 320
        9.2.7.1 Degreasing 321
        9.2.7.2 Mechanical Abrasion 321
        9.2.7.3 Chemical Treatment 322
    9.2.8 Design Guidelines for Adhesive Bonding 322

        9.2.9   Theoretical Stress Analysis for Bonded Joints 323
9.3   Mechanical Joints 323
        9.3.1   Advantages of Mechanical Joints 325
        9.3.2   Disadvantages of Mechanical Joints 325
        9.3.3   Failure Modes in a Bolted Joint 325
        9.3.4   Design Parameters for Bolted Joints 326
        9.3.5   Preparation for the Bolted Joint 327
References 328
Questions 329

10   Machining and Cutting of Composites 331
10.1   Introduction 331
10.2   Objectives/Purposes of Machining 331
10.3   Challenges during Machining of Composites 332
10.4   Failure Mode during Machining of Composites 333
10.5   Cutting Tools 334
10.6   Types of Machining Operations 336
        10.6.1   Cutting Operation 336
                10.6.1.1 Waterjet Cutting 337
                10.6.1.2 Laser Cutting 339
        10.6.2   Drilling Operation 341
References 343
Questions 344

11   Cost Estimation 345
11.1   Introduction 345
11.2   The Need for Cost Estimating 346
11.3   Cost Estimating Requirements 348
11.4   Types of Cost 348
        11.4.1   Nonrecurring (Fixed) Costs 348
        11.4.2   Recurring (Variable) Costs 349
11.5   Cost Estimating Techniques 350
        11.5.1   Industrial Engineering Approach (Methods Engineering) 351
        11.5.2   ACCEM Cost Model 351
        11.5.3   First-Order Model 352
        11.5.4   Cost Estimating by Analogy 361
11.6   Cost Analysis for Composite Manufacturing Processes 361
        11.6.1   Hand Lay-up Technique for Aerospace Parts 362
        11.6.2   Filament Winding for Consumer Goods 366
        11.6.3   Compression Molded SMC Parts for Automotive
                Applications 366
11.7   Learning Curve 370
11.8   Guidelines for Minimization of Production Cost 372
References 373
Bibliography 374
Questions 374

12 Recycling of Composites 375
12.1 Introduction 375
12.2 Categories of Dealing with Wastes 377
    12.2.1 Landfilling or Burying 377
    12.2.2 Incineration or Burning 377
    12.2.3 Recycling 377
12.3 Recycling Methods 378
    12.3.1 Regrinding 379
    12.3.2 Pyrolysis 379
12.4 Existing Infrastructure for Recycling 379
    12.4.1 Automotive Recycling Infrastructure 380
    12.4.2 Aerospace Recycling Infrastructure 381
References 382
Questions 383

Index 385

1
Introduction

1.1 Conventional Engineering Materials

There are more than 50,000 materials available to engineers for the design and manufacturing of products for various applications. These materials range from ordinary materials (e.g., copper, cast iron, brass), which have been available for several hundred years, to the more recently developed, advanced materials (e.g., composites, ceramics, and high-performance steels). Due to the wide choice of materials, today's engineers are posed with a big challenge for the right selection of a material and the right selection of a manufacturing process for an application. It is difficult to study all of these materials individually; therefore, a broad classification is necessary for simplification and characterization.

These materials, depending on their major characteristics (e.g., stiffness, strength, density, and melting temperature), can be broadly divided into four main categories: (1) metals, (2) plastics, (3) ceramics, and (4) composites. Each class contains large number of materials with a range of properties which to some extent results in an overlap of properties with other classes. For example, most common ceramic materials such as silicon carbide (SiC) and alumina (Al_2O_3) have densities in the range 3.2 to 3.5 g/cc and overlap with the densities of common metals such as iron (7.8 g/cc), copper (6.8 g/cc), and aluminum (2.7 g/cc). Table 1.1 depicts the properties of some selected materials in each class in terms of density (specific weight), stiffness, strength, and maximum continuous use temperature. The maximum operating temperature in metals does not degrade the material the way it degrades the plastics and composites. Metals generally tend to temper and age at high temperatures, thus altering the microstructure of the metals. Due to such microstructural changes, modulus and strength values generally drop. The maximum temperature cited in Table 1.1 is the temperature at which the material retains its strength and stiffness values to at least 90% of the original values shown in the table.

TABLE 1.1
Typical Properties of Some Engineering Materials

Material	Density (ρ) (g/cc)	Tensile Modulus (E) (GPa)	Tensile Strength (σ) (GPa)	Specific Modulus (E/ρ)	Specific Strength (σ/ρ)	Max. Service Temp. (°C)
Metals						
Cast iron, grade 20	7.0	100	0.14	14.3	0.02	230–300
Steel, AISI 1045 hot rolled	7.8	205	0.57	26.3	0.073	500–650
Aluminum 2024-T4	2.7	73	0.45	27.0	0.17	150–250
Aluminum 6061-T6	2.7	69	0.27	25.5	0.10	150–250
Plastics						
Nylon 6/6	1.15	2.9	0.082	2.52	0.071	75–100
Polypropylene	0.9	1.4	0.033	1.55	0.037	50–80
Epoxy	1.25	3.5	0.069	2.8	0.055	80–215
Phenolic	1.35	3.0	0.006	2.22	0.004	70–120
Ceramics						
Alumina	3.8	350	0.17	92.1	0.045	1425–1540
MgO	3.6	205	0.06	56.9	0.017	900–1000
Short fiber composites						
Glass-filled epoxy (35%)	1.90	25	0.30	8.26	0.16	80–200
Glass-filled polyester (35%)	2.00	15.7	0.13	7.25	0.065	80–125
Glass-filled nylon (35%)	1.62	14.5	0.20	8.95	0.12	75–110
Glass-filled nylon (60%)	1.95	21.8	0.29	11.18	0.149	75–110
Unidirectional composites						
S-glass/epoxy (45%)	1.81	39.5	0.87	21.8	0.48	80–215
Carbon/epoxy (61%)	1.59	142	1.73	89.3	1.08	80–215
Kevlar/epoxy (53%)	1.35	63.6	1.1	47.1	0.81	80–215

1.1.1 Metals

Metals have been the dominating materials in the past for structural applications. They provide the largest design and processing history to the engineers. The common metals are iron, aluminum, copper, magnesium, zinc, lead, nickel, and titanium. In structural applications, alloys are more frequently used than pure metals. Alloys are formed by mixing different materials, sometimes including nonmetallic elements. Alloys offer better properties than pure metals. For example, cast iron is brittle and easy to corrode, but the addition of less than 1% carbon in iron makes it tougher, and the addition of chromium makes it corrosion-resistant. Through the principle of alloying, thousands of new metals are created.

Metals are, in general, heavy as compared to plastics and composites. Only aluminum, magnesium, and beryllium provide densities close to plastics. Steel is 4 to 7 times heavier than plastic materials; aluminum is 1.2 to 2 times heavier than plastics. Metals generally require several machining operations to obtain the final product.

Metals have high stiffness, strength, thermal stability, and thermal and electrical conductivity. Due to their higher temperature resistance than plastics, they can be used for applications with higher service temperature requirements.

1.1.2 Plastics

Plastics have become the most common engineering materials over the past decade. In the past 5 years, the production of plastics on a volume basis has exceeded steel production. Due to their light weight, easy processability, and corrosion resistance, plastics are widely used for automobile parts, aerospace components, and consumer goods. Plastics can be purchased in the form of sheets, rods, bars, powders, pellets, and granules. With the help of a manufacturing process, plastics can be formed into near-net-shape or net-shape parts. They can provide high surface finish and therefore eliminate several machining operations. This feature provides the production of low-cost parts.

Plastics are not used for high-temperature applications because of their poor thermal stability. In general, the operating temperature for plastics is less than 100°C. Some plastics can take service temperature in the range of 100 to 200°C without a significant decrease in the performance. Plastics have lower melting temperatures than metals and therefore they are easy to process.

1.1.3 Ceramics

Ceramics have strong covalent bonds and therefore provide great thermal stability and high hardness. They are the most rigid of all materials. The major distinguishing characteristic of ceramics as compared to metals is that they possess almost no ductility. They fail in brittle fashion. Ceramics have the highest melting points of engineering materials. They are generally used for high-temperature and high-wear applications and are resistant to most forms of chemical attack. Ceramics cannot be processed by common metallurgical techniques and require high-temperature equipment for fabrication. Due to their high hardness, ceramics are difficult to machine and therefore require net-shape forming to final shape. Ceramics require expensive cutting tools, such as carbide and diamond tools.

1.1.4 Composites

Composite materials have been utilized to solve technological problems for a long time but only in the 1960s did these materials start capturing the attention of industries with the introduction of polymeric-based composites. Since then, composite materials have become common engineering materials and are designed and manufactured for various applications including automotive components, sporting goods, aerospace parts, consumer goods, and in the marine and oil industries. The growth in composite usage also came about because of increased awareness regarding product performance and increased competition in the global market for lightweight components. Among all materials, composite materials have the potential to replace widely used steel and aluminum, and many times with better performance. Replacing steel components with composite components can save 60 to 80% in component weight, and 20 to 50% weight by replacing aluminum parts. Today, it appears that composites are the materials of choice for many engineering applications.

1.2 What Are Composites?

A composite material is made by combining two or more materials to give a unique combination of properties. The above definition is more general and can include metals alloys, plastic co-polymers, minerals, and wood. Fiber-reinforced composite materials differ from the above materials in that the constituent materials are different at the molecular level and are mechanically separable. In bulk form, the constituent materials work together but remain in their original forms. The final properties of composite materials are better than constituent material properties.

The concept of composites was not invented by human beings; it is found in nature. An example is wood, which is a composite of cellulose fibers in a matrix of natural glue called lignin. The shell of invertebrates, such as snails and oysters, is an example of a composite. Such shells are stronger and tougher than man-made advanced composites. Scientists have found that the fibers taken from a spider's web are stronger than synthetic fibers. In India, Greece, and other countries, husks or straws mixed with clay have been used to build houses for several hundred years. Mixing husk or sawdust in a clay is an example of a particulate composite and mixing straws in a clay is an example of a short-fiber composite. These reinforcements are done to improve performance.

The main concept of a composite is that it contains matrix materials. Typically, composite material is formed by reinforcing fibers in a matrix resin as shown in Figure 1.1. The reinforcements can be fibers, particulates, or whiskers, and the matrix materials can be metals, plastics, or ceramics.

The reinforcements can be made from polymers, ceramics, and metals. The fibers can be continuous, long, or short. Composites made with a polymer matrix have become more common and are widely used in various industries. This book focuses on composite materials in which the matrix materials are polymer-based resins. They can be thermoset or thermoplastic resins.

The reinforcing fiber or fabric provides strength and stiffness to the composite, whereas the matrix gives rigidity and environmental resistance. Reinforcing fibers are found in different forms, from long continuous fibers to

Introduction

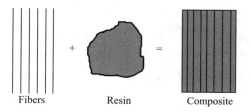

FIGURE 1.1
Formation of a composite material using fibers and resin.

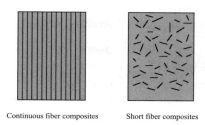

FIGURE 1.2
Continuous fiber and short fiber composites.

woven fabric to short chopped fibers and mat. Each configuration results in different properties. The properties strongly depend on the way the fibers are laid in the composites. All of the above combinations or only one form can be used in a composite. The important thing to remember about composites is that the fiber carries the load and its strength is greatest along the axis of the fiber. Long continuous fibers in the direction of the load result in a composite with properties far exceeding the matrix resin itself. The same material chopped into short lengths yields lower properties than continuous fibers, as illustrated in Figure 1.2. Depending on the type of application (structural or nonstructural) and manufacturing method, the fiber form is selected. For structural applications, continuous fibers or long fibers are recommended; whereas for nonstructural applications, short fibers are recommended. Injection and compression molding utilize short fibers, whereas filament winding, pultrusion, and roll wrapping use continuous fibers.

1.3 Functions of Fibers and Matrix

A composite material is formed by reinforcing plastics with fibers. To develop a good understanding of composite behavior, one should have a good knowledge of the roles of fibers and matrix materials in a composite. The important functions of fibers and matrix materials are discussed below.

The main functions of the fibers in a composite are:

- To carry the load. In a structural composite, 70 to 90% of the load is carried by fibers.
- To provide stiffness, strength, thermal stability, and other structural properties in the composites.
- To provide electrical conductivity or insulation, depending on the type of fiber used.

A matrix material fulfills several functions in a composite structure, most of which are vital to the satisfactory performance of the structure. Fibers in and of themselves are of little use without the presence of a matrix material or binder. The important functions of a matrix material include the following:

- The matrix material binds the fibers together and transfers the load to the fibers. It provides rigidity and shape to the structure.
- The matrix isolates the fibers so that individual fibers can act separately. This stops or slows the propagation of a crack.
- The matrix provides a good surface finish quality and aids in the production of net-shape or near-net-shape parts.
- The matrix provides protection to reinforcing fibers against chemical attack and mechanical damage (wear).
- Depending on the matrix material selected, performance characteristics such as ductility, impact strength, etc. are also influenced. A ductile matrix will increase the toughness of the structure. For higher toughness requirements, thermoplastic-based composites are selected.
- The failure mode is strongly affected by the type of matrix material used in the composite as well as its compatibility with the fiber.

1.4 Special Features of Composites

Composites have been routinely designed and manufactured for applications in which high performance and light weight are needed. They offer several advantages over traditional engineering materials as discussed below.

1. Composite materials provide capabilities for part integration. Several metallic components can be replaced by a single composite component.
2. Composite structures provide in-service monitoring or online process monitoring with the help of embedded sensors. This feature is used to monitor fatigue damage in aircraft structures or can be

utilized to monitor the resin flow in an RTM (resin transfer molding) process. Materials with embedded sensors are known as "smart" materials.

3. Composite materials have a high specific stiffness (stiffness-to-density ratio), as shown in Table 1.1. Composites offer the stiffness of steel at one fifth the weight and equal the stiffness of aluminum at one half the weight.
4. The specific strength (strength-to-density ratio) of a composite material is very high. Due to this, airplanes and automobiles move faster and with better fuel efficiency. The specific strength is typically in the range of 3 to 5 times that of steel and aluminum alloys. Due to this higher specific stiffness and strength, composite parts are lighter than their counterparts.
5. The fatigue strength (endurance limit) is much higher for composite materials. Steel and aluminum alloys exhibit good fatigue strength up to about 50% of their static strength. Unidirectional carbon/epoxy composites have good fatigue strength up to almost 90% of their static strength.
6. Composite materials offer high corrosion resistance. Iron and aluminum corrode in the presence of water and air and require special coatings and alloying. Because the outer surface of composites is formed by plastics, corrosion and chemical resistance are very good.
7. Composite materials offer increased amounts of design flexibility. For example, the coefficient of thermal expansion (CTE) of composite structures can be made zero by selecting suitable materials and lay-up sequence. Because the CTE for composites is much lower than for metals, composite structures provide good dimensional stability.
8. Net-shape or near-net-shape parts can be produced with composite materials. This feature eliminates several machining operations and thus reduces process cycle time and cost.
9. Complex parts, appearance, and special contours, which are sometimes not possible with metals, can be fabricated using composite materials without welding or riveting the separate pieces. This increases reliability and reduces production times. It offers greater manufacturing feasibility.
10. Composite materials offer greater feasibility for employing design for manufacturing (DFM) and design for assembly (DFA) techniques. These techniques help minimize the number of parts in a product and thus reduce assembly and joining time. By eliminating joints, high-strength structural parts can be manufactured at lower cost. Cost benefit comes by reducing the assembly time and cost. DFM and DFA techniques are discussed in Chapter 5.

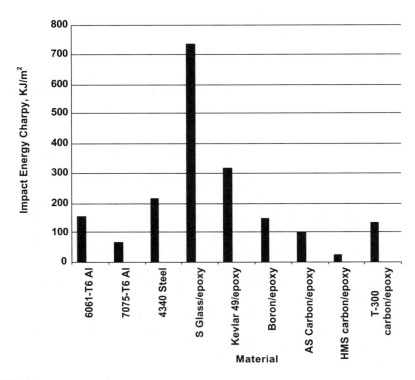

FIGURE 1.3
Impact properties of various engineering materials. Unidirectional composite materials with about 60% fiber volume fraction are used. (*Source:* Data adapted from Mallick.[1])

11. Composites offer good impact properties, as shown in Figures 1.3 and 1.4. Figure 1.3 shows impact properties of aluminum, steel, glass/epoxy, kevlar/epoxy, and carbon/epoxy continuous fiber composites. Glass and Kevlar composites provide higher impact strength than steel and aluminum. Figure 1.4 compares impact properties of short and long glass fiber thermoplastic composites with aluminum and magnesium. Among thermoplastic composites, impact properties of long glass fiber nylon 66 composite (NylonLG60) with 60% fiber content, short glass fiber nylon 66 composite (NylonSG40) with 40% fiber content, long glass fiber polypropylene composite (PPLG40) with 40% fiber content, short glass fiber polypropylene composite (PPSG40) with 40% fiber content, long glass fiber PPS composite (PPSLG50) with 50% fiber content, and long glass fiber polyurethane composite (PULG60) with 60% fiber content are described. Long glass fiber provides three to four times improved impact properties than short glass fiber composites.

12. Noise, vibration, and harshness (NVH) characteristics are better for composite materials than metals. Composite materials dampen

Introduction

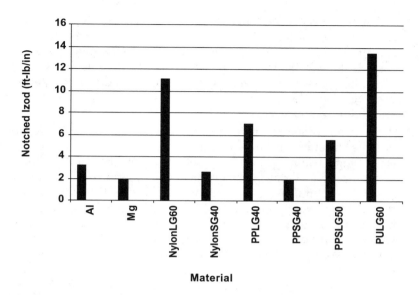

FIGURE 1.4
Impact properties of long glass (LG) and short glass (SG) fibers reinforced thermoplastic composites. Fiber weight percent is written at the end in two digits.

vibrations an order of magnitude better than metals. These characteristics are used in a variety of applications, from the leading edge of an airplane to golf clubs.

13. By utilizing proper design and manufacturing techniques, cost-effective composite parts can be manufactured. Composites offer design freedom by tailoring material properties to meet performance specifications, thus avoiding the over-design of products. This is achieved by changing the fiber orientation, fiber type, and/or resin systems.

14. Glass-reinforced and aramid-reinforced phenolic composites meet FAA and JAR requirements for low smoke and toxicity. This feature is required for aircraft interior panels, stowbins, and galley walls.

15. The cost of tooling required for composites processing is much lower than that for metals processing because of lower pressure and temperature requirements. This offers greater flexibility for design changes in this competitive market where product lifetime is continuously reducing.

1.5 Drawbacks of Composites

Although composite materials offer many benefits, they suffer from the following disadvantages:

1. The materials cost for composite materials is very high compared to that of steel and aluminum. It is almost 5 to 20 times more than aluminum and steel on a weight basis. For example, glass fiber costs $1.00 to $8.00/lb; carbon fiber costs $8 to $40/lb; epoxy costs $1.50/lb; glass/epoxy prepreg costs $12/lb; and carbon/epoxy prepreg costs $12 to $60/lb. The cost of steel is $0.20 to $1.00/lb and that of aluminum is $0.60 to $1.00/lb.
2. In the past, composite materials have been used for the fabrication of large structures at low volume (one to three parts per day). The lack of high-volume production methods limits the widespread use of composite materials. Recently, pultrusion, resin transfer molding (RTM), structural reaction injection molding (SRIM), compression molding of sheet molding compound (SMC), and filament winding have been automated for higher production rates. Automotive parts require the production of 100 to 20,000 parts per day. For example, Corvette volume is 100 vehicles per day, and Ford-Taurus volume is 2000 vehicles per day. Steering system companies such as Delphi Saginaw Steering Systems and TRW produce more than 20,000 steering systems per day for various models. Sporting good items such as golf shafts are produced on the order of 10,000 pieces per day.
3. Classical ways of designing products with metals depend on the use of machinery and metals handbooks, and design and data handbooks. Large design databases are available for metals. Designing parts with composites lacks such books because of the lack of a database.
4. The temperature resistance of composite parts depends on the temperature resistance of the matrix materials. Because a large proportion of composites uses polymer-based matrices, temperature resistance is limited by the plastics' properties. Average composites work in the temperature range −40 to +100°C. The upper temperature limit can range between +150 and +200°C for high-temperature plastics such as epoxies, bismaleimides, and PEEK. Table 1.2 shows the maximum continuous-use temperature for various polymers.
5. Solvent resistance, chemical resistance, and environmental stress cracking of composites depend on the properties of polymers. Some polymers have low resistance to solvents and environmental stress cracking.
6. Composites absorb moisture, which affects the properties and dimensional stability of the composites.

1.6 Composites Processing

Processing is the science of transforming materials from one shape to the other. Because composite materials involve two or more different materials,

TABLE 1.2

Maximum Continuous-Use Temperatures for Various Thermosets and Thermoplastics

Materials	Maximum Continuous-Use Temperature (°C)
Thermosets	
Vinylester	60–150
Polyester	60–150
Phenolics	70–150
Epoxy	80–215
Cyanate esters	150–250
Bismaleimide	230–320
Thermoplastics	
Polyethylene	50–80
Polypropylene	50–75
Acetal	70–95
Nylon	75–100
Polyester	70–120
PPS	120–220
PEEK	120–250
Teflon	200–260

the processing techniques used with composites are quite different than those for metals processing. There are various types of composites processing techniques available to process the various types of reinforcements and resin systems. It is the job of a manufacturing engineer to select the correct processing technique and processing conditions to meet the performance, production rate, and cost requirements of an application. The engineer must make informed judgments regarding the selection of a process that can accomplish the most for the least resources. For this, engineers should have a good knowledge of the benefits and limitations of each process. This book discusses the various manufacturing processes frequently used in the fabrication of thermoset and thermoplastic composites, as well as the processing conditions, fabrication steps, limitations, and advantages of each manufacturing method. Figure 1.5 classifies the frequently used composites processing techniques in the composites industry. These methods are discussed in Chapter 6.

1.7 Composites Product Fabrication

Composite products are fabricated by transforming the raw material into final shape using one of the manufacturing process discussed in Section 1.6.

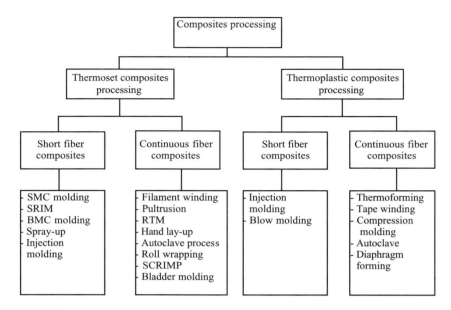

FIGURE 1.5
Classification of composites processing techniques.

The products thus fabricated are machined and then joined with other members as required for the application. The complete product fabrication is divided into the following four steps:

1. **Forming.** In this step, feedstock is changed into the desired shape and size, usually under the action of pressure and heat. All the composites processing techniques described in Section 1.6 are in this category. Composite-forming operations are discussed in detail in Chapter 6.
2. **Machining.** Machining operations are used to remove extra or undesired material. Drilling, turning, cutting, and grinding come in this category. Composites machining operations require different tools and operating conditions than that required by metals. The machining of composites is discussed in Chapter 10.
3. **Joining and assembly.** Joining and assembly is performed to attach different components in a manner so that it can perform a desired task. Adhesive bonding, fusion bonding, mechanical fastening, etc. are commonly used for assmbling two components. These operations are time consuming and cost money. Joining and assembly should be avoided as much as possible to reduce product costs. This is achieved by part integration as discussed in detail in Chapter 5. Joining and assembly operations used for composite product formations are discussed in Chapter 9.

4. **Finishing.** Finishing operations are performed for several reasons, such as to improve outside appearance, to protect the product against environmental degradation, to provide a wear-resistant coating, and/or to provide a metal coating that resembles that of a metal. Golf shaft companies apply coating and paints on outer composite shafts to improve appearance and look.

It is not necessary that all of the above operations be performed at one manufacturing company. Sometimes a product made in one company is sent to another company for further operations. For example, an automotive driveshaft made in a filament winding company is sent to automakers (tier 1 or tier 2) for assembly with their final product, which is then sold to OEMs (original equipment manufacturers). In some cases, products such as golf clubs, tennis rackets, fishing rods, etc. are manufactured in one company and then sent directly to the distributor for consumer use.

1.8 Composites Markets

There are many reasons for the growth in composite applications, but the primary impetus is that the products fabricated by composites are stronger and lighter. Today, it is difficult to find any industry that does not utilize the benefits of composite materials. The largest user of composite materials today is the transportation industry, having consumed 1.3 billion pounds of composites in 2000. Composite materials have become the materials of choice for several industries.

In the past three to four decades, there have been substantial changes in technology and its requirement. This changing environment created many new needs and opportunities, which are only possible with the advances in new materials and their associated manufacturing technology.

In the past decade, several advanced manufacturing technology and material systems have been developed to meet the requirements of the various market segments. Several industries have capitalized on the benefits of composite materials. The vast expansion of composite usage can be attributed to the decrease in the cost of fibers, as well as the development of automation techniques and high-volume production methods. For example, the price of carbon fiber decreased from \$150.00/lb in 1970 to about \$8.00/lb in 2000. This decrease in cost was due to the development of low-cost production methods and increased industrial use.

Broadly speaking, the composites market can be divided into the following industry categories: aerospace, automotive, construction, marine, corrosion-resistant equipment, consumer products, appliance/business equipment, and others. U.S. composite shipments in the above markets are shown in Figure 1.6 for the years 1999 and 2000 (projected).

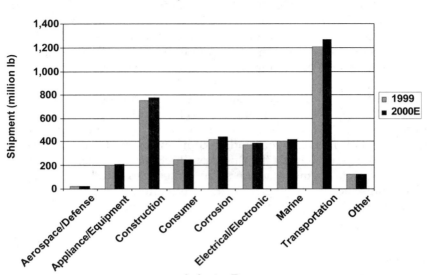

FIGURE 1.6
Composite shipments in various industries in 1999 and those projected for 2000. (*Source:* Data adapted from the Composites Fabricators Association.[2])

1.8.1 The Aerospace Industry

The aerospace industry was among the first to realize the benefits of composite materials. Airplanes, rockets, and missiles all fly higher, faster, and farther with the help of composites. Glass, carbon, and Kevlar fiber composites have been routinely designed and manufactured for aerospace parts. The aerospace industry primarily uses carbon fiber composites because of their high-performance characteristics. The hand lay-up technique is a common manufacturing method for the fabrication of aerospace parts; RTM and filament winding are also being used.

In 1999, the aerospace industry consumed 23 million pounds of composites, as shown in Figure 1.6. Military aircrafts, such as the F-11, F-14, F-15, and F-16, use composite materials to lower the weight of the structure. The composite components used in the above-mentioned fighter planes are horizontal and vertical stabilizers, wing skins, fin boxes, flaps, and various other structural components as shown in Table 1.3. Typical mass reductions achieved for the above components are in the range of 20 to 35%. The mass saving in fighter planes increases the payload capacity as well as the missile range.

Figure 1.7 shows the typical composite structures used in commercial aircraft and Figure 1.8 shows the typical composite structures used in military aircraft. Composite components used in engine and satellite applications are shown in Figures 1.9 and 1.10, respectively.

Introduction

TABLE 1.3

Composite Components in Aircraft Applications

	Composite Components
F-14	Doors, horizontal tails, fairings, stabilizer skins
F-15	Fins, rudders, vertical tails, horizontal tails, speed brakes, stabilizer skins
F-16	Vertical and horizontal tails, fin leading edge, skins on vertical fin box
B-1	Doors, vertical and horizontal tails, flaps, slats, inlets
AV-8B	Doors, rudders, vertical and horizontal tails, ailerons, flaps, fin box, fairings
Boeing 737	Spoilers, horizontal stabilizers, wings
Boeing 757	Doors, rudders, elevators, ailerons, spoilers, flaps, fairings
Boeing 767	Doors, rudders, elevators, ailerons, spoilers, fairings

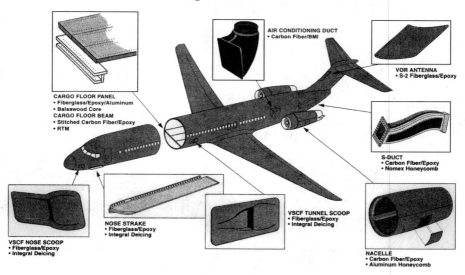

FIGURE 1.7
Typical composite structures used in commercial aircraft. (Courtesy of Composites Horizon, Inc.)

The major reasons for the use of composite materials in spacecraft applications include weight savings as well as dimensional stability. In low Earth orbit (LEO), where temperature variation is from −100 to +100°C, it is important to maintain dimensional stability in support structures as well as in reflecting members. Carbon epoxy composite laminates can be designed to give a zero coefficient of thermal expansion. Typical space structures are tubular truss structures, facesheets for the payload baydoor, antenna reflectors, etc. In space shuttle composite materials provide weight savings of 2688 lb per vehicle.

16   Composites Manufacturing: Materials, Product, and Process Engineering

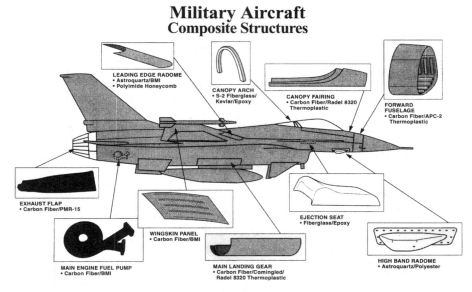

FIGURE 1.8
Typical composite structures used in military aircraft. (Courtesy of Composites Horizon, Inc.)

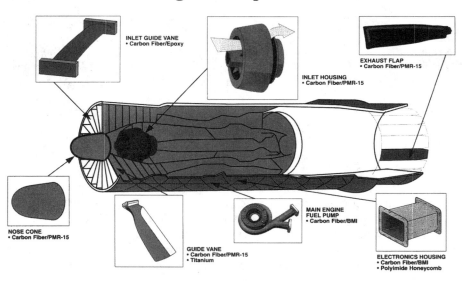

FIGURE 1.9
Composite components used in engine applications. (Courtesy of Composites Horizon, Inc.)

Introduction 17

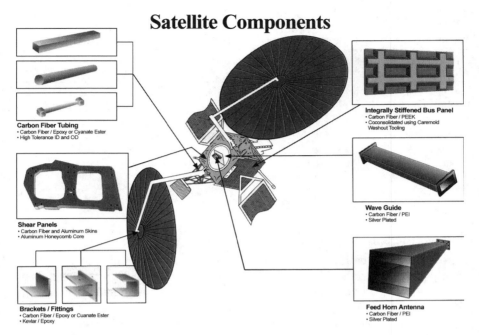

FIGURE 1.10
Composite components used in satellite applications. (Courtesy of Composites Horizon, Inc.)

Passenger aircrafts such as the Boeing 747 and 767 use composite parts to lower the weight, increase the payload, and increase the fuel efficiency. The components made out of composites for such aircrafts are shown in Table 1.3.

1.8.2 The Automotive Industry

Composite materials have been considered the "material of choice" in some applications of the automotive industry by delivering high-quality surface finish, styling details, and processing options. Manufacturers are able to meet automotive requirements of cost, appearance, and performance utilizing composites. Today, composite body panels have a successful track record in all categories — from exotic sports cars to passenger cars to small, medium, and heavy truck applications. In 2000, the automotive industry used 318 million pounds of composites.

Because the automotive market is very cost-sensitive, carbon fiber composites are not yet accepted due to their higher material costs. Automotive composites utilize glass fibers as main reinforcements. Table 1.4 provides a breakdown of automotive composite usage by applications, matrix materials, and manufacturing methods.

TABLE 1.4
Average Use of Composites in Automobiles per Year, 1988–1993

Applications	Usage (kg × 10⁶)	Matrix Material	Usage (kg × 10⁶)	Manufacturing Process	Usage (kg × 10⁶)
Bumper beam	42	Polyester (TS)	42	SMC (comp. mold)	40
Seat/load floor	14	Polypropylene	22	GMT (comp. mold)	20
Hood	13	Polycarbonate/PBT	10	Injection molding	13
Radiator support	4	Polyethylene	4	Ext. blow mold	5
Roof panel	4	Epoxy	4	Filament wound	3
Other	11	Other	7	Other	8
Total	89	Total	89	Total	89

Source: The Automotive Composites Consortium.[3]

1.8.3 The Sporting Goods Industry

Sports and recreation equipment suppliers are becoming major users of composite materials. The growth in structural composite usage has been greatest in high-performance sporting goods and racing boats. Anyone who has visited a sporting goods store can see products such as golf shafts, tennis rackets, snow skis, fishing rods, etc. made of composite materials. These products are light in weight and provide higher performance, which helps the user in easy handling and increased comfort.

Total 1999 U.S. sports equipment shipment cost (including golf, hockey, basketball, baseball, tennis, etc.) was estimated to be $17.33 billion, as reported by the Sporting Goods Manufacturers Association (North Palm Beach, Florida). The market for recreational transport (bicycles, motorcycles, pleasure boats, RVs, snowmobiles, and water scooters) was estimated at $17.37 billion, up from 1998 sales of $15.39 billion. The total shipment for golf was $2.66 billion for 1999, including balls, clubs, and others, with a third of that amount attributed to golf clubs. The ice skates and hockey are estimated to $225 million, snowboards to $183 million, and snow skiing to about $303 million wholesale values in 1999. There are no statistics available that describe the amount of composites usage in the above sporting segments. In North America, 6 million hockey sticks are manufactured every year, with composites capturing 1 to 3% of this market (shafts retail for $60 to $150).[4] The Kite Trade Association, San Francisco, estimated a total sale of $215 million in 1990 worldwide in kites, which are generally made by roll wrapping composite tubes or pultruded tubes. Composite bicycle frames and components represent half a million of these parts, or 600,000 lb of material worldwide in top-of-the-line bicycles, which sell in the range of $3000 to $5000 per unit.[4]

1.8.4 Marine Applications

Composite materials are used in a variety of marine applications such as passenger ferries, power boats, buoys, etc. because of their corrosion resistance and light weight, which gets translated into fuel efficiency, higher cruising speed, and portability. The majority of components are made of glass-reinforced plastics (GRP) with foam and honeycomb as core materials. About 70% of all recreational boats are made of composite materials according to a 361-page market report on the marine industry.[5] According to this report total annual domestic boat shipments in the United States was $8.85 billion and total composite shipments in the boating industry worldwide is estimated as 620 million lbs in 2000.

Composites are also used in offshore pipelines for oil and gas extractions. The motivation for the use of GRP materials for such applications includes reduced handling and installation costs as well as better corrosion resistance and mechanical performance. Another benefit comes from the use of adhesive bonding, which minimizes the need for a hot work permit if welding is employed.

1.8.5 Consumer Goods

Composite materials are used for a wide variety of consumer good applications, such as sewing machines, doors, bathtubs, tables, chairs, computers, printers, etc. The majority of these components are short fiber composites made by molding technology such as compression molding, injection molding, RTM, and SRIM.

1.8.6 Construction and Civil Structures

The construction and civil structure industries are the second major users of composite materials. Construction engineering experts and engineers agree that the U.S. infrastructure is in bad shape, particularly the highway bridges. Some 42% of this nation's bridges need repair and are considered obsolete, according to Federal Highway Administration officials. The federal government has budgeted approximately $78 billion over the next 20 years for major infrastructure rehabilitation. The driving force for the use of glass- and carbon-reinforced plastics for bridge applications is reduced installation, handling, repair, and life-cycle costs as well as improved corrosion and durability. It also saves a significant amount of time for repair and installation and thus minimizes the blockage of traffic.

Composite usage in earthquake and seismic retrofit activities is also booming. The columns wrapped by glass/epoxy, carbon/epoxy, and aramid/epoxy show good potential for these applications.

1.8.7 Industrial Applications

The use of composite materials in various industrial applications is growing. Composites are being used in making industrial rollers and shafts for the printing industry and industrial driveshafts for cooling-tower applications.

Filament winding shows good potential for the above applications. Injection-molded, short fiber composites are used in bushings, pump and roller bearings, and pistons. Composites are also used for making robot arms and provide improved stiffness, damping, and response time.

1.9 Barriers in Composite Markets

The primary barrier to the use of composite materials is their high initial costs in some cases, as compared to traditional materials. Regardless of how effective the material will be over its life cycle, industry considers high up-front costs, particularly when the life-cycle cost is relatively uncertain. This cost barrier inhibits research into new materials.

In general, the cost of processing composites is high, especially in the hand lay-up process. Here, raw material costs represent a small fraction of the total cost of a finished product. There is already evidence of work moving to Asia, Mexico, and Korea for the cases where labor costs are a significant portion of the total product costs.

The recycling of composite materials presents a problem when penetrating a high-volume market such as the automotive industry, where volume production is in the millions of parts per year. With the new goverment regulations and environmental awareness, the use of composites has become a concern and poses a big challenge for recycling. Recycling of composites is discussed in Chapter 12.

References

1. Mallick, P.K., *Fiber Reinforced Composites: Materials, Manufacturing and Design*, Marcel Dekker, New York, 1993.
2. Composites Fabricators Association, U.S.A., 2000.
3. Automotive Composites Consortium, U.S.A., 1994.
4. McConnel, V.P., Sports applications — composites at play, *High Performance Composites*, January/February 1994.
5. Market report on "Composites in Marine Industry — 2001: Market Analysis, Opportunities and Trend," Publisher: E-Composites.com, Total 362 pages.

Questions

1. What are the different categories of materials? Rank these materials based on density, specific stiffness, and specific strength.
2. What are the benefits of using composite materials?

3. What is the function of a matrix in a composite material?
4. What are the processing techniques for short fiber thermoset composites?
5. What are the four major steps typically taken in the making of composite products?
6. What are the manufacturing techniques available for continuous thermoplastic composites?

2

Raw Materials for Part Fabrication

2.1 Introduction

Each manufacturing method utilizes a specific type of material system for part fabrication. One material system may be suitable for one manufacturing method, whereas the same material system may not be suitable for another fabrication method. For example, the injection molding process utilizes molding compounds in pellet form, which cannot be used for filament winding or pultrusion process. Filament winding and pultrusion utilize continuous fibers and wet resin systems in most cases. The roll wrapping process requires prepreg systems for making golf club shafts, bicycle tubes, and other products. Therefore, it is important to have a good knowledge of the various raw materials available for the manufacture of good composite products.

Figure 2.1 depicts the various types of composite manufacturing techniques and their corresponding material systems. In general, raw materials for composite manufacturing processes can be divided into two categories: thermoset-based and thermoplastic-based composite materials. Thermoset plastics are those that once solidified (cured) cannot be remelted. Thermoplastics can be remelted and reshaped once they have solidified. Thermosets and thermoplastics have their own advantages and disadvantages in terms of processing, cost, recyclability, storage, and performance. In all these composite systems, there are two major ingredients: reinforcements and resins.

2.2 Reinforcements

Reinforcements are important constituents of a composite material and give all the necessary stiffness and strength to the composite. These are thin rod-like structures. The most common reinforcements are glass, carbon, aramid and boron fibers. Typical fiber diameters range from 5 µm (0.0002 in.) to 20 µm (0.0008 in.). The diameter of a glass fiber is in the range of 5 to 25 µm,

24 *Composites Manufacturing: Materials, Product, and Process Engineering*

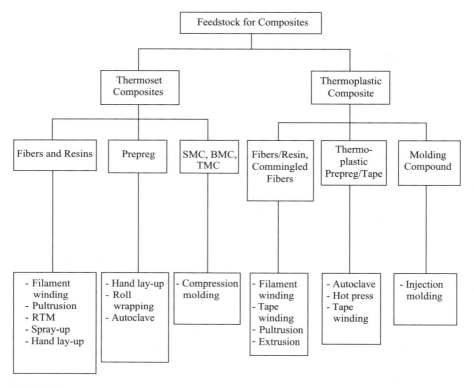

FIGURE 2.1
Classification of raw materials.

a carbon fiber is 5 to 8 μm, an aramid fiber is 12.5 μm, and a boron fiber is 100 μm. Because of this thin diameter, the fiber is flexible and easily conforms to various shapes. In general, fibers are made into strands for weaving or winding operations. For delivery purposes, fibers are wound around a bobbin and collectively called a "roving." An untwisted bundle of carbon fibers is called "tow." In composites, the strength and stiffness are provided by the fibers. The matrix gives rigidity to the structure and transfers the load to fibers.

Fibers for composite materials can come in many forms, from continuous fibers to discontinuous fibers, long fibers to short fibers, organic fibers to inorganic fibers. The most widely used fiber materials in fiber-reinforced plastics (FRP) are glass, carbon, aramid, and boron. Glass is found in abundance and glass fibers are the cheapest among all other types of fibers. There are three major types of glass fibers: E-glass, S-glass, and S2-glass. The properties of these fibers are given in Table 2.1. The cost of E-glass is around $1.00/lb, S-glass is around $8.00/lb, and S-2 glass is $5.00/lb. Carbon fibers range from low to high modulus and low to high strength. Cost of carbon fibers

TABLE 2.1
Properties of Fibers and Conventional Bulk Materials

Material	Diameter (μm)	Density (ρ) (g/cm³)	Tensile Modulus (E) (GPa)	Tensile Strength (σ) (GPa)	Specific Modulus (E/ρ)	Specific Strength	Melting Point (°C)	% Elongation at Break	Relative Cost
Fibers									
E-glass	7	2.54	70	3.45	27	1.35	1540+	4.8	Low
S-glass	15	2.50	86	4.50	34.5	1.8	1540+	5.7	Moderate
Graphite, high modulus	7.5	1.9	400	1.8	200	0.9	>3500	1.5	High
Graphite, high strength	7.5	1.7	240	2.6	140	1.5	>3500	0.8	High
Boron	130	2.6	400	3.5	155	1.3	2300	—	High
Kevlar 29	12	1.45	80	2.8	55.5	1.9	500(D)	3.5	Moderate
Kevlar 49	12	1.45	130	2.8	89.5	1.9	500(D)	2.5	Moderate
Bulk materials									
Steel		7.8	208	0.34–2.1	27	0.04–0.27	1480	5–25	<Low
Aluminum alloys		2.7	69	0.14–0.62	26	0.05–0.23	600	8–16	Low

FIGURE 2.2
Glass rovings and yarns. Three spools of glass yarns are shown at the center and four roving spools are shown at edges. (Courtesy of Saint-Gobain Vetrotex America.)

fall in a wide range from $8.00 to $60.00/lb. Aramid fibers cost approximately $15.00 to $20.00/lb. Some of the common types of reinforcements include:

- Continuous carbon tow, glass roving, aramid yarn
- Discontinuous chopped fibers
- Woven fabric
- Multidirectional fabric (stitch bonded for three-dimensional properties)
- Stapled
- Woven or knitted three-dimensional preforms

Continuous fibers are used for filament winding, pultrusion, braiding, weaving, and prepregging applications. Glass rovings and yarns are shown in Figure 2.2 and individual rovings in Figure 2.3. Continuous fibers are used with most thermoset and thermoplastic resin systems. Chopped fibers are used for making injection molding and compression molding compounds. Chopped fibers are made by cutting the continuous fibers. In spray-up and other processes, continuous fibers are used but are chopped by machine into small pieces before the application. Woven fabrics are used for making prepregs as well as for making laminates for a variety of applications (e.g., boating, marine, and sporting). Preforms are made by braiding and other processes and used as reinforcements for RTM and other molding operations.

The next section provides a brief description of manufacturing techniques for glass, carbon, and Kevlar fibers.

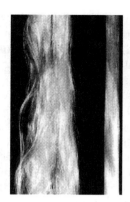

FIGURE 2.3
Photograph of individual rovings. (Courtesy of GDP-DFC, France.)

2.2.1 Glass Fiber Manufacturing

The properties of fibers depend on how the fibers are manufactured. The raw materials used for making E-glass fibers are silica sand, limestone, fluorspar, boric acid, and clay. Silica accounts for more than 50% of the total ingredients. By varying the amounts of raw materials and the processing parameters, other glass types are produced. The raw materials are mixed thoroughly and melted in a furnace at 2,500 to 3,000°F. The melt flows into one or more bushings containing hundreds of small orifices. The glass filaments are formed as the molten glass passes through these orifices and successively goes through a quench area where water and/or air quickly cool the filaments below the glass transition temperature. The filaments are then pulled over a roller at a speed around 50 miles per hour. The roller coats them with sizing. The amount of sizing used ranges from 0.25 to 6% of the original fiber weight. All the filaments are then pulled into a single strand and wound onto a tube.

Sizing is applied to the filaments to serve several purposes; it promotes easy fiber wetting and processing, provides better resin and fiber bonding, and protects fibers from breakage during handling and processing. The sizing formulation depends on the type of application; for example, sizing used for epoxy would be different than that used for polyester.

2.2.2 Carbon Fiber Manufacturing

Carbon and graphite fibers are produced using PAN-based or pitch-based precursors. The precursor undergoes a series of operations. In the first step, the precursors are oxidized by exposing them to extremely high temperatures. Later, they go through carbonization and graphitization processes. During these processes, precursors go through chemical changes that yield high stiffness-to-weight and stength-to-weight properties. The successive surface treatment and sizing process improves its resin compatibility and handleability.

PAN refers to polyacrylonitrile, a polymer fiber of textile origin. Pitch fiber is obtained by spinning purified petroleum or coal tar pitch. PAN-based fibers are most widely used for the fabrication of carbon fibers. Pitch-based fibers tend to be stiffer and more brittle. The cost of carbon fiber depends on the cost of the raw material and process. The PAN-based precursor costs around $1.50 to $2.00/lb. During oxidation and carbonization processes, the weight reduces to almost 50% of the original weight. Considering the weight loss, the cost of fibers based on raw material alone becomes $3.00 to $4.00/lb. The fabrication method for the production of carbon fibers is slow and capital intensive. Therefore, higher tow count is produced to lower the cost of the fibers. There is a limitation on increasing the tow size. For example, a tow size more than 12K creates processing and handling difficulties during filament winding and braiding operations.

Pitch-based carbon fibers are produced in the same way as PAN-based fibers but pitch is more difficult to spin and the resultant fiber is more difficult to handle. Pitch itself costs pennies a kilogram, but processing and purifying it to the fiber form are very expensive. Generally, pitch-based fibers are more expensive than PAN-based fibers.

The cost of carbon fibers depends on the strength and stiffness properties as well as on the tow size (number of filaments in a fiber bundle). Fibers with high stiffness and strength properties cost more. The higher the tow size, the lower the cost will be. For example, 12K tow (12,000 filaments per fiber bundle) costs less than 6K tow.

2.2.3 Aramid Fiber Manufacturing

Aramid fibers provide the highest tensile strength-to-weight ratio among reinforcing fibers. They provide good impact strength. Like carbon fibers, they provide a negative coefficient of thermal expansion. The disadvantage of aramid fibers is that they are difficult to cut and machine. Aramid fibers are produced by extruding an acidic solution (a proprietary polycondensation product of terephthaloyol chloride and *p*-phenylenediamine) through a spinneret. The filaments are drawn through several orifices. During the drawing operation, aramid molecules beome highly oriented in the longitudinal direction.

2.3 Matrix Materials

As discussed, composites are made of reinforcing fibers and matrix materials. Matrix surrounds the fibers and thus protects those fibers against chemical and environmental attack. For fibers to carry maximum load, the matrix must have a lower modulus and greater elongation than the reinforcement.

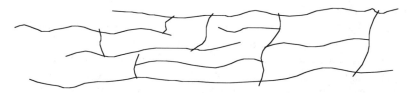

FIGURE 2.4
Cross-linking of thermoset molecules during curing.

Matrix selection is performed based on chemical, thermal, electrical, flammability, environmental, cost, performance, and manufacturing requirements. The matrix determines the service operating temperature of a composite as well as processing parameters for part manufacturing. Maximum continuous-use temperatures of the various types of thermoset and thermoplastic resins are shown in Table 1.2 in Chapter 1.

2.3.1 Thermoset Resins

Thermoset materials once cured cannot be remelted or reformed. During curing, they form three-dimensional molecular chains, called cross-linking, as shown in Figure 2.4. Due to these cross-linkings, the molecules are not flexible and cannot be remelted and reshaped. The higher the number of cross-linkings, the more rigid and thermally stable the material will be. In rubbers and other elastomers, the densities of cross-links are much less and therefore they are flexible. Thermosets may soften to some extent at elevated temperatures. This characteristic is sometimes used to create a bend or curve in tubular structures, such as filament-wound tubes. Thermosets are brittle in nature and are generally used with some form of filler and reinforcement. Thermoset resins provide easy processability and better fiber impregnation because the liquid resin is used at room temperature for various processes such as filament winding, pultrusion, and RTM. Thermosets offer greater thermal and dimensional stability, better rigidity, and higher electrical, chemical, and solvent resistance. The most common resin materials used in thermoset composites are epoxy, polyester, vinylester, phenolics, cyanate esters, bismaleimides, and polyimides. Some of the basic properties of selected thermoset resins are shown in Table 2.2.

TABLE 2.2

Typical Unfilled Thermosetting Resin Properties

Resin Material	Density (g/cm³)	Tensile Modulus GPa (10⁶ psi)	Tensile Strength MPa (10³ psi)
Epoxy	1.2–1.4	2.5–5.0 (0.36–0.72)	50–110 (7.2–16)
Phenolic	1.2–1.4	2.7–4.1 (0.4–0.6)	35–60 (5–9)
Polyester	1.1–1.4	1.6–4.1 (0.23–0.6)	35–95 (5.0–13.8)

2.3.1.1 *Epoxy*

Epoxy is a very versatile resin system, allowing for a broad range of properties and processing capabilities. It exhibits low shrinkage as well as excellent adhesion to a variety of substrate materials. Epoxies are the most widely used resin materials and are used in many applications, from aerospace to sporting goods. There are varying grades of epoxies with varying levels of performance to meet different application needs. They can be formulated with other materials or can be mixed with other epoxies to meet a specific performance need. By changing the formulation, properties of epoxies can be changed; the cure rate can be modified, the processing temperature requirement can be changed, the cycle time can be changed, the drape and tack can be varied, the toughness can be changed, the temperature resistance can be improved, etc. Epoxies are cured by chemical reaction with amines, anhydrides, phenols, carboxylic acids, and alcohols. An epoxy is a liquid resin containing several epoxide groups, such as diglycidyl ether of bisphenol A (DGEBA), which has two epoxide groups. In an epoxide group, there is a three-membered ring of two carbon atoms and one oxygen atom. In addition to this starting material, other liquids such as diluents to reduce its viscosity and flexibilizers to increase toughness are mixed. The curing (cross-linking) reaction takes place by adding a hardener or curing agent (e.g., diethylenetriamine [DETA]). During curing, DGEBA molecules form cross-links with each other as shown in Figure 2.4. These cross-links grow in a three-dimensional network and finally form a solid epoxy resin. Cure rates can be controlled through proper selection of hardeners and/or catalysts. Each hardener provides different cure characteristics and different properties to the final product. The higher the cure rate, the lower the process cycle time and thus higher production volume rates.

Epoxy-based composites provide good performance at room and elevated temperatures. Epoxies can operate well up to temperatures of 200 to 250°F, and there are epoxies that can perform well up to 400°F. For high-temperature and high-performance epoxies, the cost increases, but they offer good chemical and corrosion resistance.

Epoxies come in liquid, solid, and semi-solid forms. Liquid epoxies are used in RTM, filament winding, pultrusion, hand lay-up, and other processes with various reinforcing fibers such as glass, carbon, aramid, boron, etc. Semi-solid epoxies are used in prepreg for vacuum bagging and autoclave processes. Solid epoxy capsules are used for bonding purposes. Epoxies are more costly than polyester and vinylesters and are therefore not used in cost-sensitive markets (e.g., automotive and marine) unless specific performance is needed.

Epoxies are generally brittle, but to meet various application needs, toughened epoxies have been developed that combine the excellent thermal properties of a thermoset with the toughness of a thermoplastic. Toughened epoxies are made by adding thermoplastics to the epoxy resin by various patented processes.

2.3.1.2 Phenolics

Phenolics meet FAA (and JAR) requirements for low smoke and toxicity. They are used for aircraft interiors, stowbins, and galley walls, as well as other commercial markets that require low-cost, flame-resistant, and low-smoke products.

Phenolics are formed by the reaction of phenol (carbolic acid) and formaldehyde, and catalyzed by an acid or base. Urea, resorcinol, or melamine can be used instead of phenol to obtain different properties. Their cure characteristics are different than other thermosetting resins such as epoxies, due to the fact that water is generated during cure reaction. The water is removed during processing. In the compression molding process, water can be removed by bumping the press. Phenolics are generally dark in color and therefore used for applications in which color does not matter. The phenolic products are usually red, blue, brown, or black in color. To obtain light-colored products, urea formaldehyde and melamine formaldehyde are used.

Other than flame-resistant parts, phenolic products have demonstrated their capabilities in various other applications where:

- High temperature resistance is required.
- Electrical properties are needed.
- Wear resistance is important.
- Good chemical resistance and dimensional stability are essential.

Phenolics are used for various composite manufacturing processes such as filament winding, RTM, injection molding, and compression molding. Phenolics provide easy processability, tight tolerances, reduced machining, and high strength. Because of their high temperature resistance, phenolics are used in exhaust components, missile parts, manifold spacers, commutators, and disc brakes.

2.3.1.3 Polyesters

Polyesters are low-cost resin systems and offer excellent corrosion resistance. The operating service temperatures for polyesters are lower than for epoxies. Polyesters are widely used for pultrusion, filament winding, SMC, and RTM operations. Polyesters can be a thermosetting resin or a thermoplastic resin.

Unsaturated polyesters are obtained by the reaction of unsaturated difunctional organic acids with a difunctional alcohol. The acids used include maleic, fumaric, phthalic, and terephthalic. The alcohols include ethylene glycol, propylene glycol, and halogenated glycol. For the curing or cross-linking process, a reactive monomer such as styrene is added in the 30 to 50 wt% range. The carbon-carbon double bonds in unsaturated polyester molecules and styrene molecules function as the cross-linking site.

With the growing health concerns over styrene emissions, the use of styrene is being reduced for polyester-based composite productions. In recent methods, catalysts are used for curing polyesters with reduced styrene.

2.3.1.4 Vinylesters

Vinylesters are widely used for pultrusion, filament winding, SMC, and RTM processes. They offer good chemical and corrosion resistance and are used for FRP pipes and tanks in the chemical industry. They are cheaper than epoxies and are used in the automotive and other high-volume applications where cost is critical in making material selection.

Vinylesters are formed by the chemical reaction of an unsaturated organic acid with an epoxide-terminated molecule. In vinylester molecules, there are fewer unsaturated sites for cross-linking than in polyesters or epoxies and, therefore, a cured vinylester provides increased ductility and toughness.

2.3.1.5 Cyanate Esters

Cyanate esters offer excellent strength and toughness, better electrical properties, and lower moisture absorption compared to other resins. If they are formulated correctly, their high-temperature properties are similar to bismaleimide and polyimide resins. They are used for a variety of applications, including spacecrafts, aircrafts, missiles, antennae, radomes, microelectronics, and microwave products.

Cyanate esters are formed via the reaction of bisphenol esters and cyanic acid that cyclotrimerize to produce triazine rings during a second cure. Cyanate esters are more easily cured than epoxies. The toughness of cyanate esters can be increased by adding thermoplastics or spherical rubber particles.

2.3.1.6 Bismaleimide (BMI) and Polyimide

BMI and polyimide are used for high-temperature applications in aircrafts, missiles, and circuit boards. The glass transition temperature (T_g) of BMIs is in the range of 550 to 600°F, whereas some polyimides offer T_g greater than 700°F. These values are much higher than for epoxies and polyesters. The lack of use of BMIs and polyimides is attributed to their processing difficulty. They emit volatiles and moisture during imidization and curing. Therefore, proper venting is necessary during the curing of these resins; otherwise, it may cause process-related defects such as voids and delaminations. Other drawbacks of these resins include the fact that their toughness values are lower than epoxies and cyanate esters, and they have a higher moisture absorption ability.

2.3.1.7 Polyurethane

Polyurethane is widely used for structural reaction injection molding (SRIM) processes and reinforced reaction injection molding (RRIM) processes, in which isocyanate and polyol are generally mixed in a ratio of 1:1 in a reaction chamber and then rapidly injected into a closed mold containing short or long fiber reinforcements. RRIM and SRIM processes are low-cost and high-volume production methods. The automotive industry is a big market for these processes. Polyurethane is currently used for automotive applications such as bumper beams, hoods, body panels, etc. Unfilled polyurethane is used for various applications, including truck wheels, seat and furniture cushions, mattress foam, etc. Polyurethane is also used for wear and impact resistance coatings.

Polyurethane can be a thermosetting or thermoplastic resin, depending on the functionality of the selected polyols. Thermoplastic-based polyurethane contains linear molecules, whereas thermoset-based resin contains cross-linked molecules.

Polyurethane is obtained by the reaction between polyisocyanate and a polyhydroxyl group. There are a variety of polyurethanes available by selecting various types of polyisocyanate and polyhydroxyl ingredients. Polyurethane offers excellent wear, tear, and chemical resistance, good toughness, and high resilience.

2.3.2 Thermoplastic Resins

Thermoplastic materials are, in general, ductile and tougher than thermoset materials and are used for a wide variety of nonstructural applications without fillers and reinforcements. Thermoplastics can be melted by heating and solidified by cooling, which render them capable of repeated reshaping and reforming. Thermoplastic molecules do not cross-link and therefore they are flexible and reformable. Thermoplastics can be either amorphous or semi-crystalline, as shown in Figure 2.5. In amorphous thermoplastics, molecules are randomly arranged; whereas in the crystalline region of semi-crystalline plastics, molecules are arranged in an orderly fashion. It is not possible to have 100% crystallinity in plastics because of the complex nature of the molecules. Some of the properties of themoplastics are given in Table 2.3. Their lower stiffness and strength values require the use of fillers and reinforcements for structural applications. Thermoplastics generally exhibit poor creep resistance, especially at elevated temperatures, as compared to thermosets. They are more susceptible to solvents than thermosets. Thermoplastic resins can be welded together, making repair and joining of parts more simple than for thermosets. Repair of thermoset composites is a complicated process, requiring adhesives and careful surface preparation. Thermoplastic composites typically require higher forming temperatures and pressures than comparable thermoset systems. Thermoplastic composites do not enjoy as high a level of integration as is currently obtained with thermosetting systems. The higher viscosity of thermoplastic resins makes some manufacturing processes, such as hand lay-up and tape winding operations, more difficult. As a consequence of this, the fabrication of thermoplastic composite parts have drawn a lot of attention from researchers to overcome these problems.

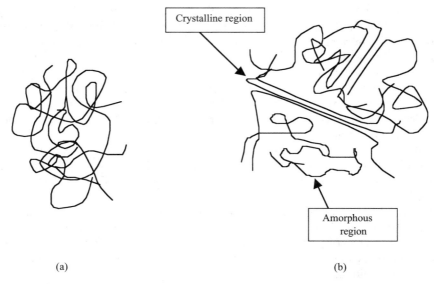

FIGURE 2.5
Molecular arrangements in (a) amorphous and (b) semi-crystalline polymers.

TABLE 2.3
Typical Unfilled Thermoplastic Resin Properties

Resin Material	Density (g/cm³)	Tensile Modulus GPa (10⁶ psi)	Tensile Strength MPa (10³ psi)
Nylon	1.1	1.3–3.5 (0.2–0.5)	55–90 (8–13)
PEEK	1.3–1.35	3.5–4.4 (0.5–0.6)	100 (14.5)
PPS	1.3–1.4	3.4 (0.49)	80 (11.6)
Polyester	1.3–1.4	2.1–2.8 (0.3–0.4)	55–60 (8–8.7)
Polycarbonate	1.2	2.1–3.5 (0.3–0.5)	55–70 (8–10)
Acetal	1.4	3.5 (0.5)	70 (10)
Polyethylene	0.9–1.0	0.7–1.4 (0.1–0.2)	20–35 (2.9–5)
Teflon	2.1–2.3	—	10–35 (1.5–5.0)

2.3.2.1 Nylons

Nylons are used for making intake manifolds, housings, gears, bearings, bushings, sprockets, etc. Glass-filled and carbon-filled nylons in pellet form are available for injection molding purposes. Nylons are most widely used for injection molding purposes, but are also available as prepregs with various reinforcements. Nylons have been used for various pultruded components.

Nylons are also called polyamides. There are several types of nylon, including nylon 6, nylon 66, nylon 11, etc., each offering a variety of mechanical and physical properties; but as a whole, they are considered engineering plastics. Nylons provide a good surface appearance and good lubricity. The important design consideration with nylons is that they absorb moisture, which affects the properties and dimensional stability of the part. Glass reinforcement minimizes this problem and produces a strong, impact-resistant material. Impact resistance of long glass-filled nylon is higher than conventional engineering materials such as aluminim and magnesium, as shown in Figure 1.4 in Chapter 1.

2.3.2.2 Polypropylene (PP)

Polypropylene (PP) is a low-cost, low-density, versatile plastic and is available in many grades and as a co-polymer (ethylene/propylene). It has the lowest density (0.9 g/cm^3) of all thermoplastics and offers good strength, stiffness, chemical resistance, and fatigue resistance. PP is used for machine parts, car components (fans, fascia panels, etc.), and other household items, and has also been pultruded with various reinforcements.

2.3.2.3 Polyetheretherketone (PEEK)

PEEK is a new-generation thermoplastic that offers the possibility of use at high service temperatures. Carbon-reinforced PEEK composites (APC-2) have already demostrated their usefulness in fuselage, satellite parts, and other aerospace structures; they can be used continuously at 250°C. The glass transition temperature (T_g) of PEEK is 143°C and crystalline melting temperature is ~336°C. PEEK/carbon thermoplastic composites (APC-2, aromatic polymer composites) have generated significant interest among researchers and in the aircraft industry because of their greater damage tolerance, better solvent resistance, and high-temperature usage. As well, PEEK has the advantage of almost 10 times lower water absorption than epoxies. The water absorption of PEEK is 0.5% at room temperature, whereas aerospace-grade epoxies have 4 to 5% water absorption. The drawback of PEEK-based composites is that the materials cost is very high, more than $50.00/lb.

PEEK/carbon is processed in the range of 380 to 400°C for autoclave, hot press, and diaphragm molding processes whereas for tape winding operation, more than 500°C is suggested for better interply consolidation.[1] It is a semi-crystalline material with a maximum crystallinity of 48%. In general, the crystallinity of PEEK is 30 to 35%. The toughness offered by PEEK is 50 to 100 times higher than that of epoxies.

2.3.2.4 Polyphenylene Sulfide (PPS)

PPS is an engineering thermoplastic with a maximum crystallinity of 65%. It provides high operating temperatures and can be used continuously at 225°C. The T_g of PPS is 85°C and crystalline melt temperature is 285°C.

Prepreg tape of PPS with several reinforcements is available. The trade names of PPS-based prepreg systems are Ryton and Techtron. It is processed in the temperature range of 300 to 345°C. PPS-based composites are used for applications where great strength and chemical resistance are required at elevated temperature.

2.4 Fabrics

There are two major types of fabrics available in composites industry: woven fabrics and nonwoven (noncrimp) fabrics.

2.4.1 Woven Fabrics

Woven fabrics are used in trailers, containers, barge covers, and water tower blades, and in other marine wet lay-up applications. These fabrics are woven yarns, rovings, or tows in mat form in a single layer. Common weave styles are shown in Figure 2.6. Figure 2.7 shows carbon fabrics in a variety of weave styles. The amount of fiber in different directions is controlled by the weave pattern. For example, in unidirectional woven fabrics, fibers are woven in such a way that the fibers in 0° are up to 95% of the total weight of the fabric. In a plain-weave pattern, fibers in 0° and 90° directions are equally distributed. Hybrid fabrics in various combinations, such as glass/carbon and

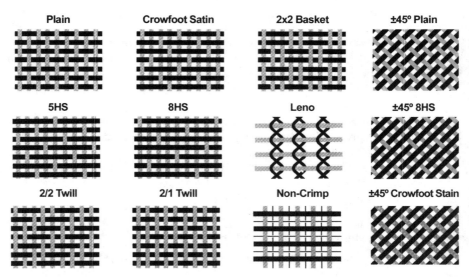

FIGURE 2.6
Various weave styles for fabrics. (Courtesy of Cytec Fiberite.)

Raw Materials for Part Fabrication

FIGURE 2.7
Carbon fabrics with a variety of weave styles. (Courtesy of Cytec Fiberite.)

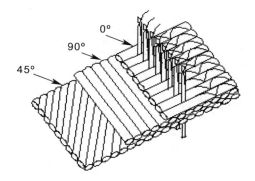

FIGURE 2.8
Schematic of noncrimp fabrics.

aramid/carbon, are also available. For lightning strike purposes, conductive wires are woven into fabric forms to distribute the energy imparted by lightning, thus minimizing damage to the structure.

Woven fabrics are also used to make prepregs, as well as in RTM and SRIM processes as feedstock. Woven fabrics have the advantage of being inexpensive.

2.4.2 Noncrimp Fabrics

In noncrimp fabrics, yarns are placed parallel to each other as shown in Figure 2.8 and then stitched together using polyester thread. Warp unidirectional fabric is used when fibers are needed in one direction only, for example, in stiffness-critical applications such as water ski applications where the fabric is laid along the length of the ski to improve resistance to bending. In warp fabrics, reinforcements are laid at 0° (or warp direction) only as shown in Figure 2.9; whereas in weft unidirectional fabrics, reinforcements are laid at 90° (or weft direction) only as shown in Figure 2.10. Weft fabrics are typically used in filament wound tubes and pipes and also pultruded components

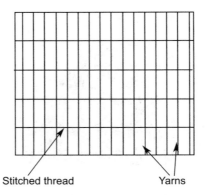

FIGURE 2.9
Warp unidirectional fabrics.

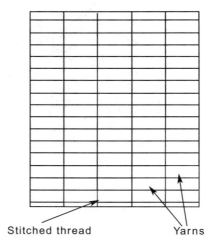

FIGURE 2.10
Weft unidirectional fabrics.

where reinforcement in the weft direction is necessary. Warp triaxial (0°, ±45°) fabric is used to increase longitudinal stiffness and torsional rigidity, whereas weft triaxial (90°, ±45°) fabric is used to increase transverse stiffness and torsional rigidity. Quadraxial fabrics are quasi-isotropic, providing strength in all four fiber axial directions. Fabrics typically come in the weight range of 9 to 200 oz/yd^2. Noncrimp fabrics offer greater flexibility compared to woven fabrics. For example, fibers can be laid at almost any angle from 0° to 90°, including 45°, 90°, 30°, 60°, and 22°, and then stitched to make multiaxial stitched plies, whereas woven fabrics are made from rovings mostly on the 0° and 90° axes. Noncrimp fabrics offer greater

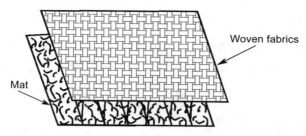

FIGURE 2.11
Illustration of bi-ply fabric.

FIGURE 2.12
Photograph of a mat. (Courtesy of GDP-DFC, France.)

strength because fibers remain straight; whereas in woven fabrics, fibers bend over each other. Noncrimp fabrics are available in a thick layer and thus an entire laminate could be achieved in a single-layer fabric. This is useful in making thicker laminates such as boat hulls and reduces the number of fabrication steps.

To make noncrimp glass fabrics, input rovings are selected by yield numbers in combinations of 113, 218, 450, 675, 1200, and 1800 yd/lb. A larger yield number denotes a finer roving and, therefore, more yards are required to achieve a given weight. The selection of yield number is determined by the physical, mechanical, and aesthetic requirements of the laminate. The finer filaments mean higher fiber content and less resin. This improves strength and can reduce weight.

To meet the market need for heavier fabrics, stitched fabrics with various combinations of plies are produced. Figure 2.11 shows a bi-ply fabric in which woven fabrics and a chopped strand mat are stitched together to form the fabric. Figure 2.12 shows a mat made using continuous fibers or random cut fibers 5 to 10 cm in length.

FIGURE 2.13
Prepreg types: unidirectional tape, woven fabric prepregs, and rovings. (Courtesy of Cytec Fiberite.)

2.5 Prepregs

A prepreg is a resin-impregnated fiber, fabric, or mat in flat form, which is stored for later use in hand lay-up or molding operations. Fibers laid at 0° orientation and preimpregnated with resin are called unidirectional tape. Figure 2.13 shows the various types of prepregs available as unidirectional tape, woven fabric tape, and rovings. Figure 2.14 shows the making of unidirectional prepreg tape. Unidirectional tape provides the ability to tailor the composite properties in the desired direction. Woven fabric prepregs are used to make highly contoured parts in which material flexibility is key. It is also used to make sandwich panels using honeycomb as a core material. Preimpregnated rovings are primarily used in filament winding applications.

Epoxy-based prepregs are very common in industry and come in flat sheet form in a thickness range of 0.127 mm (0.005 in.) to 0.254 mm (0.01 in.). Prepregs can be broadly classified as thermoset-based prepregs and thermoplastic-based prepregs/tapes, the difference between the two being the type of resin used. Reinforcements in a prepreg can be glass, carbon, or aramid, and are used in filament or woven fabric or mat form in either type of prepreg. Table 2.4 provides an overview of the basic mechanical and physical properties of advanced composite prepreg systems. Properties of thermoset- and thermoplastic-based prepreg tapes with unidirectional fibers, as well as plain weave fabric forms, are also shown in the table. The table presents a wide range of material possibilities and performance potentials, which provides easy and quick assessment of the various prepregs. There are more

FIGURE 2.14
Making unidirectional prepreg tape. (Courtesy of Cytec Fiberite.)

than 100 prepreg types available on the market to meet various application needs.

Prepregs provide consistent properties as well as consistent fiber/resin mix and complete wet-out. They eliminate the need for weighing and mixing resin and catalyst. Various types of drape and tack are provided with prepregs to meet various application needs. Drape is the ability of prepreg to take the shape of a contoured surface. For example, thermoplastic prepregs are not easy to drape, whereas thermoset prepregs are easy to drape. Tack is the stickiness of uncured prepregs. A certain amount of tack is required for easy laying and processing. Because thermoplastic tapes do not have any tack, they are welded with another layer while laying up. Thermoset prepregs have a limited shelf life and require refrigeration for storage. Refrigeration minimizes the degree of cure in the prepreg materials because heat causes thermoset prepregs to cure.

Unidirectional prepreg tapes are available in a wide variety of widths, ranging from 0.5 to 60 in. By laying unidirectional tapes at desired ply orientations, the stiffness, strength, and coefficient of thermal expansion properties of the structure can be controlled. Fabric prepregs are available in a number of weave patterns in standard widths ranging from 39 to 60 in. Fabric prepregs are made by preimpregnating woven fabrics with resin via a hot melt process or by solution treatment. They provide a good amount of flexibility in highly contoured and complex parts. They are used in making aircraft parts, sandwich panels, sporting goods, industrial products, and

TABLE 2.4
Properties of Various Prepreg Materials

Prepreg Material	Fiber Volume Fraction (%)	Processing Temp. (°F)	Tensile Modulus (Msi)	Tensile Strength (ksi)	Compressive Modulus (Msi)	Compressive Strength (ksi)	Maximum Service Temp. (0°F, Dry)	Shelf Life (0°F, Months)	Out Time at Room Temp. (Days)
Unidirectional thermoset									
Carbon (AS4, T-300)/epoxy	55–65	250	15–22	180–320	15–20	160–250	180–250	6–12	14–30
Carbon (IM7)/epoxy	55–60	250	20–25	320–440	18.5–20	170–237	250	12	30
S-2 glass/epoxy	55–63	250–350	6.0–8.0	120–230	6–8.0	100–160	180–250	6	10–30
Kevlar/epoxy	55–60	250–285	10	140	9	33	180	6	10–30
Carbon (AS4)/bismaleimide	55–62	350–475	15–22	200–320	15–20	245	450–600	6	25
Carbon (IM7)/bismaleimide	60–66	350–440	20–25	380–400	22–23	235–255	450–600	6–12	25
Carbon (IM7)/cyanate ester	55–63	350–450	20–25	100–395	18.5–23	205–230	450	6	10
S-2 glass/cyanate ester	55–60	250–350	7	180	9	130	400	6	10
Unidirectional thermoplastic									
Carbon (IM7)/PEEK	57–63	550	26	410	22	206	350	Indefinite	Indefinite
Carbon (G34/700)/Nylon 6	55–62	450–500	16	216	14	90	200	Indefinite	Indefinite
Aramid/Nylon 12	52	400	6.8	205	6.5	—	—	Indefinite	Indefinite
Carbon (AS4)/PPS	64	450–520	17.5	285	16.5	155	—	Indefinite	Indefinite
Carbon (IM7)/polyimide	62	610–665	25	380	22	156	400	Indefinite	Indefinite
Fabric (plain weave) thermoset									
Carbon (AS4)/epoxy	57–63	250	8–9	75–124	7–9.5	65–95	—	6	10
S-2 glass/epoxy	55	250	5	80	4.5	55	180	6	10
Fabric (plain weave) thermoplastic tape									
Carbon HM (T650-35)/polyimide	58–62	660–730	10–18	130–155	15.5	130	500–600	12	Indefinite

printed circuit boards. Prepregs in roving form are also available for filament winding purposes.

Prepregs are used in a wide variety of applications, including aerospace parts, sporting goods, printed circuit boards, medical components, and industrial products. The advantages of prepreg materials over metals are their higher specific stiffness, specific strength, corrosion resistance, and faster manufacturing. The major disadvantage of prepreg materials is their higher cost. Products made with prepreg materials provide a higher fiber volume fraction than those made by filament winding and pultrusion. Prepregs also provide more controlled properties and higher stiffness and strength properties than other composite products.

2.5.1 Thermoset Prepregs

The most common resin used in thermoset prepreg materials is epoxy. These prepregs are generally stored in a low-temperature environment and have a limited shelf life. Room-temperature prepregs are also becoming available. Usually, the resin is partially cured to a tack-free state called B-staging. Several additives (e.g., flame retardants, catalysts, and inhibitors) are added to meet various end-use properties and processing and handling needs. Thermoset prepregs require a longer process cycle time, typically in the range of 1 to 8 hr due to their slower kinetic reactions. Due to higher production needs, rapid-curing thermoset prepregs are being developed.

Thermoset prepregs are more common and more widely used than thermoplastic prepregs. They are generally made by solvent impregnation and hot melt technology. In the solvent impregnation method, the resin is dissolved by a chemical agent, creating a low-viscosity liquid into which fibers are dipped. Due to growing environmental awareness, disposal of the solvent resulting from this process is becoming a concern. The hot melt technology eliminates the use of solvents. In this process, the matrix resin is applied in viscous form. The drawback of this process is that fiber wetting is not easily achievable due to the higher viscosity of the resin.

Prepregs are generally used for hand lay-up, roll wrapping, compression molding, and automatic lay-up processes. Once the prepregs are laid on a tool, it is cured in the presence of pressure and temperature to obtain the final product.

2.5.2 Thermoplastic Prepregs

Thermoplastic prepregs have an unlimited shelf life at room temperature and are generally processed at the melting temperature of the resin. The most common resins are nylon, polyetheretherketone (PEEK), polyphenylene sulfide, polyimide, etc. The process cycle time for thermoplastic composites is much faster than thermoset composites, in the range of a few minutes. It is a relatively new technology and provides several processing and design advantages over thermoset prepregs. The benefits of thermoplastic prepregs are:

- Recyclability
- Good solvent and chemical resistance
- Reduced process cycle time
- Higher toughness and impact resistance
- Indefinite shelf life with no refrigeration
- Reshaping and reforming flexibility
- Greater flexibility for joining and assembly by fusion bonding and *in situ* consolidation
- Better repairability potential

The disadvantages of thermoplastic prepregs are that they require higher temperatures and pressures for processing. They provide some processing difficulties because of their poor drape capabilities.

Thermoplastic prepregs are manufactured by solvent impregnation and hot melt coating techniques similar to thermoset prepreg manufacturing. Solvent impregnation becomes difficult because thermoplastics offer more chemical resistance. The hot melt coating technique is similar to an extrusion process, wherein fibers and resins are extruded simultaneously in sheet form. There are other manufacturing methods such as film stacking and dry powder deposition methods for prepreg fabrication. In the film stacking process, the thermoplastic resin film is stacked together with the reinforcements and consolidated under heat and pressure to fully impregnate the fibers. This process is clean and solvent-free but requires proper care for the production of a void-free prepreg. In the dry powder deposition technique, the resin must be in powder form as a starting material. The powder is fluidized and charged to form a resin cloud. The fibers are passed through the cloud and coated with the charged resin as they get attracted to the fibers. The coated fibers are then passed through a heat source to fully melt the resin and form a continuous sheet of material.

2.6 Preforms

Preforms are feedstock for the RTM and SRIM processes, where a reinforcement in the form of a thick two- or three-dimensional fiber architecture is put in the mold cavity and then resin is injected into the cavity to obtain the composite part. Preforms are made in several ways. To make a preform by braiding and filament winding, dry fibers are laid over a mandrel as shown in Figure 2.15.

Raw Materials for Part Fabrication 45

FIGURE 2.15
Braiding of fiberglass preform for an airfoil application. (Courtesy of Fiber Innovations Inc.)

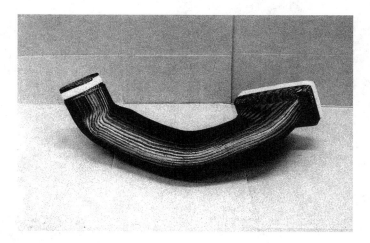

FIGURE 2.16
Braided carbon fiber duct preform for an aircraft application. (Courtesy of Fiber Innovations, Inc.)

The braided preform is becoming common and is widely used for RTM processes. Braiding can be done over a mandrel of nearly any shape or size. Helical and longitudinal fiber yarns are interlaced to give a single or multiple layers of braided preform. Each pass by the machine produces a layer of fiber reinforcements that conforms to the shape of the mandrel. Biaxial and triaxial preforms can be produced by this method. Three-dimensional fiber architecture provides better interlaminar properties. Figure 2.16 shows a braided carbon fiber duct preform for an aircraft application and Figure 2.17 shows a braided carbon/epoxy composite spar after resin transfer molding.

FIGURE 2.17
Braided carbon/epoxy composite spar after resin transfer molding. (Courtesy of Fiber Innovations, Inc.)

Preforms can be of any shape, depending on the requirements and size of the component. Preforms are stable and offer a good strength-to-weight ratio.

One method of making a short fiber preform is shown in Figure 2.18. The chopped fibers and binder materials are sprayed into a perforated preform

Raw Materials for Part Fabrication

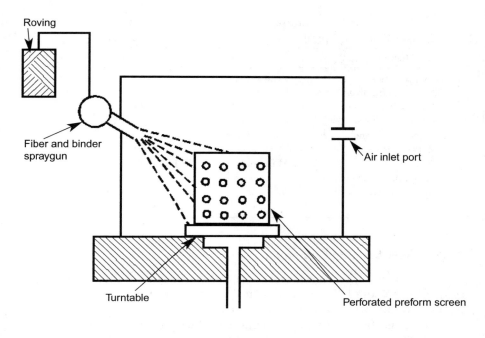

FIGURE 2.18
Schematic of short fiber preform fabrication.

screen mold. The mold rotates but the spraygun remains stationary as shown in Figure 2.18. With gradual application of chopped fibers and binders, a suitable preform thickness builds up. The binder keeps the fibers together and maintains the shape of the preform. In another version, a robot is used to apply the chopped fibers and binders. When a robot is used, the mold is kept stationary and the robot moves around the mold.

Preforms can also be made by stitching woven fabrics, by stitching mats with woven fabrics, or by stitching braided preforms with woven fabrics. This process creates z-direction reinforcements and allows the manufacture of complex preforms.

2.7 Molding Compound

There are several types of molding compounds available to meet various needs. Molding compounds are made of short or long fibers impregnated with resins. In general, they are used for compression molding and injection molding processes.

TABLE 2.5
Composition of a Typical SMC

Ingredients	Purpose	Weight %
Chopped glass fibers	Reinforcement	30.00
Unsaturated polyester resin	Base resin	10.50
Calcium carbonate	Filler	40.70
Styrene monomer	Co-monomer	13.40
Polyvinyl acetate	Low shrink additive	3.45
Magnesium oxide	Thickener	0.70
Zinc stearate	Mold release (lubricant)	1.00
t-Butyl perbenzoate	Catalyst (initiator)	0.25
Hydroquinone	Inhibitor	Trace amount, <0.005 g
		Total = 100%

2.7.1 Sheet Molding Compound

SMC (sheet molding compound) is a sheet of ready-to-mold composites containing uncured thermosetting resins and uniformly distributed short fibers and fillers. It primarily consists of polyester or vinylester resin, chopped glass fibers, inorganic fillers, additives, and other materials. Normally, SMC contains 30% by weight short glass fibers. The typical composition of various ingredients in SMC is shown in Table 2.5. The purpose of adding filler is to reduce the overall cost, increase dimensional stability, and reduce shrinkage during molding. Thickener is used to increase resin viscosity. Styrene is added for the curing or cross-linking process. An inhibitor prevents premature curing of the resin mix. A release agent (mold release) is added to the compound for easy release of compression molded parts from the mold. SMC containing 50 to 60 weight fraction glass fibers is called HMC. SMC is a low-cost technology and used for high-volume production of composite components requiring moderate strength. Currently, SMC dominates the automotive market because of its low-cost, high-volume production capabilities. SMC comes in various thicknesses, up to a maximum thickness of 6 mm. SMC is cut into a rectangular strip and then kept in the mold cavity for compression molding. During molding, suitable temperature and pressure are applied to spread the charge into the cavity and then to cure it. The process cycle time for compression molding is in the range of 1 to 4 min.

The manufacture of sheet molding compound is shown in Figure 2.19. In this process, a resin paste (resin, inhibitor, thickener, filler, etc., except fiber) as described in Table 2.5 is placed on a polyethylene moving film through a metering device. Before placing, all the ingredients of the resin paste are thoroughly mixed. Continuous strands of glass fibers are chopped through a chopping machine and evenly dispersed over the moving resin paste. Another layer of resin paste is placed over dispersed fibers for good fiber impregnation. Another moving polyethylene film is placed on top of the above compound. These top and bottom polyethylene films remain until it

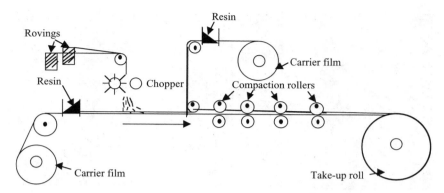

FIGURE 2.19
Schematic of SMC manufacturing.

is placed in a compression mold. These carrier films help in packaging and handling. The thickness of the sheet is controlled by a mechanical adjuster. Instead of chopped glass fibers alone, continuous glass fiber or any other fiber (e.g., carbon fiber) can be added. The complete sheet then passes through a heated compaction roller and is then rolled up or cut into rectangular shapes for shipping. A polyester-based SMC requires 1 to 7 days of exposure at around 30°C prior to use in compression molding. This period is called the maturation period, wherein resin viscosity increases to a level satisfactory for its use for the molding operation.

There are three types of SMC commonly available, depending on the fiber form used:

1. SMC-R, for randomly oriented short fibers. The weight percent of the fiber is written after R. For example, SMC-R25 has 25 wt% short fibers (Figure 2.20a).
2. SMC-CR contains continuous (C) unidirectional fibers in addition to random (R) short fibers, as shown in Figure 2.8. The percentage amounts of C and R are denoted after the letters C and R as SMC-C30R20 (Figure 2.20b).
3. XMC represents a mixture of random short fibers with continuous fibers in an X pattern. The angle between cross fibers is in the range of 5 to 7° (Figure 2.20c).

Table 2.6 provides mechanical properties of selected SMC and Table 2.7 provides fatigue data for selected SMC parts.

2.7.2 Thick Molding Compound (TMC)

Thick molding compound (TMC) is a thicker form of SMC. The thickness of TMC goes up to 50 mm whereas the maximum thickness of SMC is 6 mm.

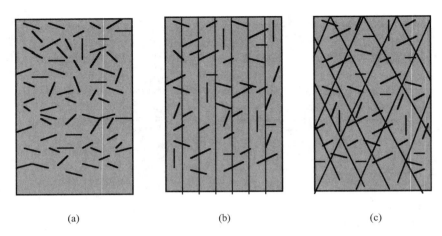

FIGURE 2.20
Common types of SMC: (a) SMC-R; (b) SMC-CR; and (c) XMC.

TABLE 2.6
Mechanical Properties of Some Selected SMC (Glass Fibers in Polyester Resin)

Properties	SMC-R25	SMC-R50	SMC-R65	SMC-C20R30	XMC-31
Specific gravity	1.83	1.87	1.82	1.81	1.97
Tensile modulus, GPa	13.2	15.8	14.8	21(L), 12(T)	36(L), 12(T)
Tensile strength, MPa	82.4	164	227	289(L), 84(T)	561(L), 70(T)
Poisson's ratio	0.25	0.31	0.26	0.30(LT), 0.18(TL)	0.31(LT), 0.12(TL)
Strain to failure (%)	1.34	1.73	1.63	1.7 (L), 1.6(T)	1.7(L), 1.5(T)
Compressive modulus (GPa)	11.7	15.9	17.9	20(L), 12(T)	37(L), 14(T)
Compressive strength (MPa)	183	225	241	306(L), 166(T)	480(L), 160(T)
Flexural strength (MPa)	220	314	403	640(L), 160(T)	970(L), 140(T)
In-plane shear modulus (GPa)	4.5	5.9	5.4	4.1	4.5
In-plane shear strength (MPa)	79	62	128	85	91
Interlaminar shear strength (MPa)	30	25	45	41	55

Source: From Reigner, D. A. and Sanders, B. A., *Proc. Natl. Tech. Conf.*, Society of Plastics Engineers, 1979.

TMC is used for making thicker molded parts. TMC eliminates having to use several SMC plies. Due to its greater thickness, TMC provides reduced pliability. In TMC, fibers are randomly distributed in three dimensions, whereas in SMC fibers are in two dimensions.

Raw Materials for Part Fabrication

TABLE 2.7

Fatigue Data for Selected SMC Parts

Material	Fatigue Strength at 23°C (MPa)	Fatigue Strength at 90°C (MPa)	Ratio of Fatigue to Ultimate Static Strength (23°C)	Ratio of Fatigue to Ultimate Static Strength (90°C)
SMC-R25	40	25	0.49	0.55
SMC-R50	63	53	0.38	0.41
SMC-R65	70	35	0.31	0.28
SMC-C20R30(L)	130	110	0.45	0.44
SMC-C20R30(T)	44	35	0.52	0.43
XMC-31(L)	130	85	0.23	0.18
XMC-31(T)	19	27	0.15	0.26

Source: From Reigner, D. A. and Sanders, B. A., *Proc. Natl. Tech. Conf.*, Society of Plastics Engineers, 1979.

FIGURE 2.21
Various forms of molding compounds. (Courtesy of Cytec Fiberite.)

2.7.3 Bulk Molding Compound (BMC)

Bulk molding compound (BMC) is a compound that is in log or rope form. It is also known as dough molding compound (DMC). BMC is obtained by mixing the resin paste with fibers and then extruding the compound in log or rope form. The extruded part is cut to length, depending on the requirement. BMC generally contains 15 to 20% fiber in a polyester or vinylester resin. The fiber length ranges from 6 to 12 mm. Due to the lower fiber volume fraction and shorter fiber length, BMC composites provide lower mechanical properties than SMC composites.

2.7.4 Injection Moldable Compounds

Injection molding is a widely used manufacturing method for thermoplastic product fabrication. This method provides increased production rates and low-cost fabrication of parts. Because unfilled thermoplastics have lower stiffness and strength, reinforcements are made by adding short fibers.

Injection molding compound (Figure 2.21) can be made with thermoset or thermoplastic resins. Thermoplastic molding compound is made by passing the fiber strand through a die similar to the pultrusion process. The counter-flow die induces shear that lowers the resin viscosity and enhances fiber wet-out and dispersion. The rod-like structure is then pulled and cut into lengths of around 10 mm in pellet form. The molded part contains fibers in the range 0.2 to 6 mm long. The reduction in fiber length takes place during the molding operation. Glass, carbon, and aramid fibers can be reinforced to make pellets. A wide variety of resins (e.g., nylon, PPS, PP, PE, etc.) can be used to make injection moldable thermoplastic parts. A long fiber-molded compound (15 mm fiber length) is also available.

Thermoset molding compounds are also used for the injection molding process. The fiber length is generally 1.5 mm for short fiber composites. The typical resins used are epoxy, phenolics, polyester, etc. The fiber length reduces to 0.1 to 0.5 mm after the molding process. Some of the molding compounds are shown in Figure 2.21.

2.8 Honeycomb and Other Core Materials

These materials are generally used for sandwich structures as cores between two thin high-strength facings. These materials are joined with facings using an adhesive strong enough to transfer the loads from one face to another. The honeycomb material acts like a web of I-beams, taking the shear loads as well as providing structural rigidity by keeping high-strength materials away from the neutral axis where tensile and compressive stresses are high. The difference between sandwich structure and I-beam is that in sandwich structure, the web is spread over the entire cross section, providing high torsional rigidity; whereas in I-beam, the web is only in the middle, thus providing less torsional rigidity. The sandwich construction provides the highest stiffness-to-weight ratio and strength-to-weight ratio.

Aluminum, Nomex, and thermoplastic honeycombs are shown in Figure 2.22. These honeycombs can be cut in any shape, however, flat sheets of honeycomb cores are mostly used. A typical flat sheet comes in a 4 × 8 ft size having 0.125 to 12 in. thickness. The cell sizes most commonly used are 0.125 to 1.0 in. diameter. The shape of a cell in an aluminum and Nomex core is mostly hexagonal, whereas thermoplastic honeycombs have a circular shape.

Raw Materials for Part Fabrication 53

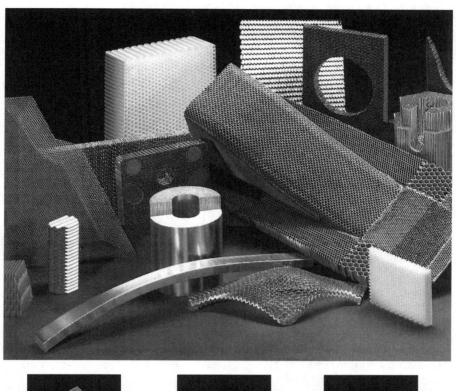

FIGURE 2.22
Commonly used aluminum, Nomex, and thermoplastic honeycombs. (Courtesy of Alcore, Inc.)

Honeycomb materials are used in aircraft, transportation, marine, communication, sporting goods, and many other industries. Honeycomb materials provide predictable crash behavior and are used for the design of crash-resistant parts. These materials are used for the design of computer and communications rooms because of their radiation shielding characteristics. The repetitive cellular structure acts as a myriad of waveguides, attenuating signals across a wide frequency range.[3] Metallic and nonmetallic materials are used for the manufacture of honeycomb cores (Figure 2.22). Some of the properties of selected honeycomb cores are given in Table 2.8. Most prevalent honeycomb materials are made of aluminum, Nomex, polycarbonate, and polypropylene (Figure 2.22).

TABLE 2.8
Physical Properties of Some Selected Honeycomb Materials

Material	Cell Diameter (in.)	Density (lb/ft³)	Max. Service Temp. (°F)	Stabilized Compressive Modulus (ksi)	Stabilized Compressive Strength (psi)
5052 Aluminum	0.125	8.1	350	350	1470
5052 Aluminum	0.125	3.1	350	75	275
3003 Aluminum	0.25	5.2	350	148	625
3003 Aluminum	0.5	2.5	350	40	165
Polycarbonate	0.125	7.9	200	55	695
Polycarbonate	0.25	3.0	200	15	110
Polyetherimide	0.125	5.1	—	32	580
Aramid (Nomex)	0.125	8.0	350	80	1900
Aramid (Nomex)	0.125	4.0	350	28	560
Aramid (Nomex)	0.25	3.1	350	21	285
Glass-reinforced polyimide	0.1875	8.0	500	126	1300
Glass-reinforced phenolic	0.1875	5.5	350	95	940

There are two major methods for the manufacture of honeycomb cores: expansion and corrugation. The expansion method is more common and is used for making aluminum and Nomex honeycombs. In the expansion process, sheets of material are stacked together in a block form. Before stacking, adhesive nodelines are printed on the sheets to obtain interrupted adhesive bonding. The stack of sheets is then cured. Slices of appropriate thickness are cut from the block and then expanded to obtain the desired cell size and shape.

In the corrugation method, the sheet of material is transformed into corrugation form using corrugating rolls. The corrugated sheets are stacked together, bonded, and cured. Honeycomb panels are cut from the block into the desired shape and size without any expansion.

Other than honeycomb materials, core materials such as balsa wood, plywood, and foam cores are available to obtain sandwich structures. There are a wide variety of open and closed cell foams available that are rigid, flexible, or rubbery. Foams can be made using thermosetting resin or thermoplastic resin by various techniques, including gas injection, blowing agent, expandable bead process, etc. All these processes supply gas or blowing agent to the resin to decrease its density by forming closed or open gaseous cells. The purpose of the foam is to increase the bending stiffness and thickness of the structural members without proportionately increasing the weight of the member.

References

1. Mazumdar, S.K. and Hoa, S.V., Determination of manufacturing conditions for hot gas aided thermoplastic tape winding technique, *J. Thermoplastic Composite Mater.*, 9, January 1996.
2. Reigner, D.A. and Sanders, B.A., A characterization study of automotive continuous and random glass fiber composites, *Proc. Natl. Tech. Conf.*, Society of Plastics Engineers, 1979.
3. English, L.K., Honeycomb: million-year-old material of the future, *Mat. Eng.*, p. 29, January 1985.

Questions

1. Why do thermoplastics have shorter processing times than thermosets?
2. Why is it easier to process with thermosets than with thermoplastics?
3. Compare the percentage of elongation at break for glass, graphite, and aramid fibers with aluminum and steel. What material properties are improved with increase in percentage elongation at break?
4. What are the commonly used fibers and resins in the thermoset composites industry?
5. Which thermoset and thermoplastic resins are used for high temperature composite applications?
6. What are the differences between woven and non-woven fabrics? Write the pros and cons of these fabrics?
7. Write down the composite manufacturing techniques that use prepregs?
8. How are short fiber preforms manufactured?
9. Why are honeycomb and core materials used?
10. List different types of molding compounds?

3
Material Selection Guidelines

3.1 Introduction

The behavior and performance of a product depend on the types of materials used in making the part. There are more than 50,000 materials available for the design and manufacture of a product. Every material cannot be a right choice for a given application; therefore, there is a need for suitable material selection. Depending on the selection of a material, the design, processing, cost, quality, and performance of the product change. Material selection becomes vital for civil and mechanical structures where material cost is almost 50% of the total product cost. For microelectronic applications such as computers, material cost is almost 5% of the product cost. Because the volume used in civil and mechanical structures is very high, there are greater opportunities for material innovations.

This chapter illustrates how material properties and systematic selection methods are important to quick and effective selection of a suitable material. Cost vs. property analysis, a weighted property comparison method, and expert systems are described as tools for material selections. Once a suitable material is selected for an application, design and manufacturing considerations begin.

3.2 The Need for Material Selection

With the technological advancements, new material systems and processes are being developed and are growing faster than ever before. In the past, steel and aluminum were more dominant for product design. This is no longer the case. With growing awareness and customer needs, ignorance of opportunities offered by advanced material systems such as composite materials can cause decreased competitiveness and can lead to loss of market.

With the increase in customer demand for higher performance and quality, competition among companies has increased. To capture the global market and be at the cutting edge of the technology, companies are utilizing new and advanced materials for increased performance. Government is passing new laws for a better environment. The Federal Government encouraged automakers to develop vehicles with fuel efficiency of 80 miles per gallon (almost 3 times that of existing mileage). In light of global needs and environmental awareness, lightweight materials are gaining importance in various industry sectors.

Material selection for the automotive market is driven by cost and energy efficiency. The aerospace market emphasizes high performance and lightweight materials. Consumer products, where performance is not a critical factor, are driven by cost and handling. The sporting goods market demands high performance and lightweight materials. The marine industry emphasizes lightweight and corrosion-resistant materials. This interest in lightweight and high-performance materials has given rise to several opportunities for composites technology. Composite materials can provide solutions to fulfill the needs of various industrial sectors.

3.3 Reasons for Material Selection

There are two major reasons why an engineer becomes involved in the material selection process.

1. To redesign an existing product for better performance, lower cost, increased reliability, decreased weight, etc.
2. To select a material for a new product or application.

In either case, mere material substitution is not sufficient. The product must be redesigned for the selected material to utilize the maximum benefits of the material's properties and processing characteristics. When a composite material is substituted for a steel or aluminum product, the part needs to be redesigned to obtain the cost and weight benefits. Many times, a composite material is not selected for an application because of higher material cost, compared to steel and aluminum on a weight basis. Such a comparison of materials costs should be avoided in the selection of a new material. For actual comparison, the product's final cost should be compared. For example, parts made by injection molding should be compared with the parts made by the die casting or sand casting process, including the machining operations to get the final part. Similarly, the roll wrapping process for making golf shafts should be compared with the filament winding process; this

Material Selection Guidelines 59

comparison should be based on final product cost, not on the cost of initial materials. For the filament winding process, the raw material cost is 3 to 8 times cheaper than the graphite/epoxy prepreg used in the roll wrapping process; but for large production volume and higher performance, the roll wrapping process is selected for making golf shafts.

3.4 Material Property Information

Materials data enters at various stages of the design process, but the level of accuracy required on material property information differs at each stage. In the initial stage of the design process (conceptual stage), approximate data for wide range of materials is gathered and design options are kept open. Table 3.1 and Tables 2.1 and 2.2 in Chapter 2 provide basic information about the various types of composite materials. This information can be helpful in the preliminary assessment of these materials. In the preliminary stage, a designer identifies which matrix material and fibers are more suitable. For example, for a certain application, polyester resin might be a suitable choice; and for another application, epoxy may be a more suitable choice. Similarly, for a manufacturing process, one specific type of material system may be more appropriate than the other. For a manufacturing process, if the engineer selects a particular raw material, the engineer needs to clearly write down key characteristics of the selected raw material. For example, if prepreg material is selected for an application, then the designer has to identify the following parameters:

- Type of prepreg: unidirectional or woven fabric prepregs
- Fiber type (carbon, glass, Kevlar, etc.)
- Woven pattern (plain weave, satin weave, etc.)
- Width of the tape
- Resin content
- Areal weight
- Ply physical and mechanical properties

Similarly the manufacturing engineer has to specify tack, drape, cure cycle, processing temperature, gel time, flow characteristics, and more for the application.

Materials property information can be obtained from published literature, suppliers' brochures, and handbooks. In the published literature, single-value data is given on the property without any standard deviation. During the final design process, more detailed information on material property with high precisions is required.

TABLE 3.1
General Properties of Thermoset and Thermoplastic Composites

Property	Thermoset Composites	Thermoplastic Composites
Fiber volume	Medium to high	Low to medium
Fiber length	Continuous and discontinuous	Continuous and discontinuous
Molding time	Slow: 0.5 to 4 h	Fast: less than 5 min
Molding pressure	Low: 1 to 7 bars	High: greater than 14 bars
Material cost	Low to high	Low to medium
Safety/handling	Good	Excellent
Solvent resistance	High	Low
Heat resistance	Low to high	Low to medium
Storage life	Good (6 to 24 months with refrigeration)	Indefinite

3.5 Steps in the Material Selection Process

There are four steps involved in narrowing the list of materials to a suitable material of choice.

3.5.1 Understanding and Determining the Requirements

The first step in identifying a material is to define the requirements (e.g., cost, weight, service, performance, etc.) of a product. There may be several benefits a material can offer but some requirements are critical to the application. For example, weight might not be critical for a consumer product, whereas weight is critical for an aerospace part. Similarly, for one application, the wear resistance requirement may be very high, whereas for another application, wear resistance may be of no concern. Therefore, it is important to prioritize the requirements of a product in the beginning of a material selection process. Some of the requirements that need to be considered during material selection process are:

1. Strength
2. Impact resistance
3. Temperature resistance
4. Humidity, chemical and electrical resistance
5. Process
6. Production rate
7. Cost

3.5.2 Selection of Possible Materials

Based on the requirements of an application, possible materials and manufacturing processes that meet minimum or maximum requirements of the

Material Selection Guidelines

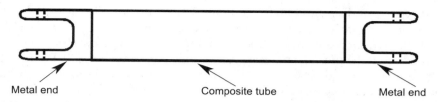

FIGURE 3.1
Composite driveshaft.

application are determined. Materials and manufacturing processes are discussed simultaneously because they go hand in hand with composite material systems. For example, if SMC is selected, then the designer is left with a choice of compression molding. To narrow the choice, one should set the minimum or maximum requirements that the material and the process must possess and should result in a positive "yes" or "no" answer. The purpose of this screening phase is to obtain a definite answer as to whether a particular material and the process should be considered for the application. Selection of potential materials is done from material databases obtained from material suppliers and handbooks.

3.5.3 Determination of Candidate Materials

Once a list of materials based on the above guidelines is created, the next task is to determine the candidate materials best suited for the application. To do this, methods discussed in Section 3.6 should be consulted. These techniques can greatly help the designer to narrow down the choices.

In the conceptual design phase, more than one material system and manufacturing method are selected to provide a wide choice of creative and innovative designs. Design options are kept open. For example, to make a composite driveshaft (Figure 3.1) for a truck or racing car application, the designer can look at pultrusion, filament winding, RTM, and roll wrapping processes for weight, cost, and performance comparisons. Steel and aluminum can also be considered a suitable choice; but for the present analysis, composite material choices are discussed. Composite driveshafts offer superior dampening characteristics, reduced rotating weight, greater fatigue life, and eliminate harmonic whipping. The manufacturing processes discussed above utilize different types of material systems, and one type of material system may not be suitable for all of the manufacturing processes. For example, glass fiber and vinylester resin systems are more suitable for a pultrusion process, a glass and polyester combination is common in the filament winding industry, graphite/epoxy prepreg is the basic raw material for the roll wrapping process, and preform and vinylester systems are common for the RTM process. Instead of glass fibers in the above filament winding or RTM or pultrusion process, an option of carbon fibers can be considered for greater weight saving and better vibration dampening characteristics. The costs per pound of these material systems are different. The pros and cons of these material systems and manufacturing processes must be checked with the requirements discussed in Sections 3.5.1 and 3.5.2. Moreover, the driveshaft in the present analysis has two metal ends as shown in Figure 3.1. These metal ends can be incorporated differently for these manufacturing processes. For example, adhesive bonding is most suitable for end connection in the pultrusion process, while mechanical locking may be more suitable in the filament winding process and insert molding may be the best in RTM. The cost, weight, and performance characteristics of each of these design concepts will vary and the engineer must look at all of these options to make the right selection of material systems and manufacturing processes. Stress and other analyses must be performed to evaluate each design concept. Finite element analysis (FEA) software and other tools can greatly reduce the cost and time associated with the product development phase. A make-and-break approach must be avoided. A good understanding of the product requirements will greatly help in making the right material and processing choices.

3.5.4 Testing and Evaluation

After selecting the candidate materials for the various types of feasible manufacturing processes, prototype parts are made and then tested to validate the design. Depending on the seriousness of an application, the number of parts to be tested should be decided. Aerospace and automobile parts generally require more tests to ensure that the part functions safely and reliably under various service conditions. For some consumer products, for which breaking of the part may not result in physical damage or injury, a minimum number of tests are required.

For critical applications, a large property database is created to understand the behavior of the part under various service conditions. For example, the driveshaft needs to be tested for automotive fluid exposures, water and salt exposures, and temperature extremes of –40 and +150°C. Static and dynamic tests are conducted under these conditions. The behavior of the adhesive (if used) is investigated for these service conditions. The effect of thermal cycling and stone impingement on the performance should be understood. All of the above tests are performed separately; but in some cases, a new test procedure is created to simulate the worst-case scenario. For this, a sequence of the above exposures are determined in such a way that the test results obtained provide a worst performance. For example, if adhesive bond strength has to be measured for the assembly, then to get the worst result, thermal cycling of the assembly is done first, then stone impingement, and then bond strength — instead of bond strength first. The purpose of all these tests is to generate statistically reliable test results under the conditions expected during the service life of the product.

3.6 Material Selection Methods

There is no standard technique used by the designer to select a right material for an application. Sometimes, a material is selected based on what worked before in similar conditions or what a competitor is using in their product. In the current competitive market, this short-cut method may cause one to overlook new emerging technologies and may put the product in a less competitive position. The job of materials and design engineers is to consider all possible opportunities to utilize new material systems and technologies for the reduction of manufacturing cost and weight for the same or increased performance. This section presents some material selection methods for the evaluation of a suitable material.

3.6.1 Cost vs. Property Analysis

Cost is a crucial factor in a material selection process. The present method compares the material cost for equivalent material property requirements. The property could be tensile strength, compressive strength, flexural strength, fatigue, creep, impact, or any other property that is critical to the application. The method determines the weight required by different materials to meet the desired property. From the weight, it determines the cost of the material. For structural applications, the volume of the material required to carry the load is determined and then, by multiplying with the specific gravity, the weight is determined. For equal volumes, steel is 3 to 4 times heavier than composites; thus, composites can offer significant weight savings as compared to steel if properly designed and manufactured.

Because cost is so important in any decision-making process, proper care should be taken to determine the actual cost of the product. Cost-estimating techniques are discussed in detail in Chapter 11. Many times, it is appropriate to consider total life-cycle cost as the basis for decision-making. Life-cycle cost includes not only material cost and manufacturing cost, but also the cost for serviceability, maintainability, and recycling.

In this section, only material cost is compared for equal performance requirement. For an application, the most critical performance requirement might be tensile strength, compressive strength, fatigue strength, impact, or creep, or any combination of these. If tensile strength is the determining factor for the selection of a material, then the weight required by different materials of choice are determined to have the same tensile strength values. Suppose materials A and B are selected for comparison purposes and their tensile strengths are σ_A and σ_B, densities are ρ_A and ρ_B, and cross-sectional areas are A_A and A_B, respectively. If the member is designed to carry an axial load P, then the cross-sectional area required by beam A will be:

$$A_A = \frac{P}{\sigma_A} \tag{3.1}$$

and beam B will be:

$$A_B = \frac{P}{\sigma_B} \tag{3.2}$$

If the member is a solid circular rod, then the corresponding diameters will be:

$$D_A^2 = \frac{4P}{\pi \sigma_A} \tag{3.3}$$

and

$$D_B^2 = \frac{4P}{\pi \sigma_B} \tag{3.4}$$

or

$$P = \frac{\pi}{4} D_A^2 \sigma_A = \frac{\pi}{4} D_B^2 \sigma_B \tag{3.5}$$

The ratio of the diameters will be:

$$\frac{D_A}{D_B} = \left(\frac{\sigma_B}{\sigma_A}\right)^{\frac{1}{2}} \tag{3.6}$$

For equal lengths of rod L, the volume ratio of the rod will be same as above. The weight ratio will be:

$$\frac{W_A}{W_B} = \frac{\rho_A (\pi/4) D_A^2 L}{\rho_B (\pi/4) D_B^2 L} = \frac{\rho_A}{\rho_B} \frac{\sigma_B}{\sigma_A} \tag{3.7}$$

If the cost per unit weight of materials A and B is C_A and C_B, respectively, then the ratio of total material costs T_A and T_B will be:

$$\frac{T_A}{T_B} = \frac{C_A}{C_B} \frac{W_A}{W_B} = \frac{C_A}{C_B} \frac{\rho_A}{\rho_B} \frac{\sigma_B}{\sigma_A} \tag{3.8}$$

In this example, a cost comparison is done for the tensile strength of the material. Similarly, it can be done for design strength, compressive strength, fatigue strength, etc.

Material Selection Guidelines

Similarly a beam can be designed for bending stiffness or torsional stiffness. For example, deflection of a simply supported beam loaded at the center is given by:

$$\delta = \frac{PL^3}{48EI} \qquad (3.9)$$

where P is the load applied at the center, L is the total length of the beam, E is the modulus of the composite, and I is the moment of inertia of the cross-section. The determination of E for a composite beam is given in design and introductory textbooks on composite materials. For example, if a beam is constructed of 16 plies with ±45° fiber orientations, then the equivalent modulus along the longitudinal direction (x-direction) is determined by the following relation:

$$\frac{1}{E_x} = \frac{\cos^4\theta}{E_L} + \frac{\sin^4\theta}{E_T} + \frac{1}{4}\left(\frac{1}{G_{LT}} - \frac{2v_{LT}}{E_L}\right)\sin^2 2\theta \qquad (3.10)$$

where E_L is the modulus of the composite along the longitudinal (fiber) direction and E_T is the modulus of the composite along the transverse (perpendicular to fiber) direction, G_{LT} is shear modulus of the composite, v_{LT} is Poisson ratio, and θ is the fiber orientation along the x-axis (tube length direction).

The stiffness of the beam may be given by:

$$\frac{P}{\delta} = \frac{48EI}{L^3} \qquad (3.11)$$

For the same length of beam, stiffness is proportional to EI as shown in Equation (3.11). Therefore, for two material systems A and B,

$$E_A I_A = E_B I_B \qquad (3.12)$$

The cross-section of the beam could be rectangular, I-shaped, circular, or any other shape. I for various cross-sections are given in any strength of materials book or machinery handbook. For a rectangular shape,

$$I = \frac{bh^3}{12} \qquad (3.13)$$

where b is the width and h is the height or thickness of the cross-section. For the same width, thickness h can be compared as:

$$h_B = h_A \left(\frac{E_A}{E_B}\right)^{1/3} \qquad (3.14)$$

Relative weight can be written as:

$$\frac{W_A}{W_B} = \frac{\rho_A h_A}{\rho_B h_B} \tag{3.15}$$

$$\frac{W_A}{W_B} = \frac{\rho_A}{\rho_B}\left(\frac{E_B}{E_A}\right)^{1/3} \tag{3.16}$$

Total material cost can be compared as:

$$\frac{T_A}{T_B} = \frac{C_A}{C_B}\frac{W_A}{W_B} = \frac{C_A}{C_B}\frac{\rho_A}{\rho_B}\left(\frac{E_B}{E_A}\right)^{1/3} \tag{3.17}$$

For various other properties, similar relationships can be determined. It is noted here that specific stiffness and specific strength are important parameters in tension or compression for comparison purposes, and not just the stiffness or strength. Other than strength and stiffness, there are several other properties (e.g., chemical resistance, corrosion resistance, wear resistance, durability) that could be important to an application. These properties need to be given sufficient consideration in material selection. For many structural applications, strength, stiffness, weight, and cost are important features in material selection. Other secondary features such as corrosion resistance or wear resistance can be incorporated by providing a coating to the structure. This secondary operation on the surface of the structure requires additional cost. It is beneficial to have these features without performing secondary operations. Unlike many metals, composites can provide good corrosion and chemical resistance without any coating. Composites are coated for wear resistance. Chrome plating, ceramic coating, and teflon coating can be applied on composite surfaces for additional surface characteristic requirements.

3.6.2 Weighted Property Comparison Method

In many circumstances, there are several factors (e.g., weight, performance, cost, serviceability, and machinability) that may be important for an application. The level of importance of each factor is different for different applications. For example, product cost is given more weight in automobile applications and weight is given higher consideration in aircraft applications. In commercial planes, a pound of weight saving translates into a $100 to $1000 cost savings. In space applications such as a satellite, a pound of weight saving translates into about $10,000 cost savings. In the automotive industry, a pound of weight saving translates into $5 to $10 cost savings. This method is suitable for cases in which more than one factor is used for material selection purposes.

TABLE 3.2
Scaling of Nonquantitative Property

Property	Materials under Consideration				
	A	B	C	D	E
Chemical resistance	Poor	Good	Excellent	Satisfactory	Very Good
Subjective rating	1	3	5	2	4
Scaled property	20	60	100	40	80

According to this method, each property is assigned a certain weight, depending on its importance during service. Because properties are measured in different units, each property is normalized to get the same numerical range. This is done by a scaling method in the following ways, depending on the type of property requirement.

3.6.2.1 Scaling for Maximum Property Requirement

There are material properties, such as strength, stiffness, and percentage elongation, that are desired in a structure to be a maximum value. Such properties are scaled in the range of 0 to 100 in the following way:

$$\alpha = \text{scaled property} = \frac{\text{Numerical value of a property}}{\text{Highest value in the same category}} \times 100 \qquad (3.18)$$

3.6.2.2 Scaling for Minimum Property Requirement

There are properties, such as cost, density, and friction, that are required as a low value in a design. Such properties are scaled as follows:

$$\alpha = \text{scaled property} = \frac{\text{Lowest value in the same category}}{\text{Numerical value of a property}} \times 100 \qquad (3.19)$$

3.6.2.3 Scaling for Nonquantitative Property

There are properties, such as wear resistance, corrosion resistance, repairability, machinability, recyclability, that cannot be quantified as a numerical value. Such properties are given subjective ratings as shown in Table 3.2.

Once material properties are scaled, the performance index of a material is determined as follows:

$$\gamma = \sum w_i \alpha_i \qquad (3.20)$$

where w is weighting factor, α is a scaled property, and i is the summation of all the properties under consideration.

TABLE 3.3

Material Evaluation for Automotive Leaf Spring

Material	Go/No-Go Screening		Scaled Property				Performance Index
	Corrosion Resistance[a]	Temp. Resistance	Flexural Strength (0.15)	Fatigue Strength (0.20)	Cost (0.25)	Mass (0.40)	
Gr/Ep	S	S	100	100	30	100	82.5
Glass/Ep	S	S	85	90	50	80	75.25
304 Steel	S	S	80	60	100	40	65
6061 Al	S	S	60	40	60	60	56

[a] S = satisfactory.

Example 3.1

Evaluate alternative materials for a leaf spring to be used in a vehicle as a suspension system on the basis of flexural strength, fatigue strength, cost, and mass. A management team gives weight to various properties as follows: flexural strength (0.15), fatigue strength (0.2), cost (0.25), and mass (0.4).

SOLUTION:

Various material systems such as graphite/epoxy (Gr/Ep), glass/epoxy (glass/Ep), aluminum (Al), and steel are considered for this application, as shown in Table 3.3. The analysis shows that graphite/epoxy is the best choice for this application. The result can vary, depending on the weight assigned to each property. Before evaluation, go/no-go screening is done for these materials to meet some other requirements such as corrosion resistance and temperature resistance of –40 to +120°C. In this case, all the above materials meet these requirements satisfactorily. For the present example, graphite/epoxy has the highest performance index and is thus the most suitable candidate for the leaf spring.

3.6.3 Expert System for Material Selection

Thousands of material choices are available to an engineer to assist in right material selection. This reveals the need for an expert system for the selection of alternative materials for a given application. Few expert systems are being developed for composite material selection. In an expert system, the user feeds in the service condition requirements (e.g., operating temperature range, chemical resistance, fluid exposure, percent elongation, fracture toughness, strength, etc.) and based on the available material database, the expert system provides material systems that are suitable for the application. Unlike metals, a large database for the performance of composite materials for various conditions is not available. Raw material suppliers provide designers and fabricators with a list of basic material properties. The datasheet is typically generated by testing standard coupons manufactured in their laboratory. These datasheets are useful for initial screening of materials but not for final selection.

Bibliography

1. Ashby, M.F., Materials selection in conceptual design, *Mater. Sci. Technol.*, 5, June 1989.
2. Dieter, G.E., *Engineering Design: A Materials and Processing Approach*, McGraw-Hill, New York, 1983.
3. Crane, F.A.A. and Charles, J.A., *Selection and Use of Engineering Materials*, Butterworths, London, 1984.
4. Smithells, C.J., Ed., *Metals Reference Book*, 6 ed., Butterworths, London, 1984.
5. Harper, C.A., Ed., *Handbook of Plastics and Elastomers*, McGraw-Hill, New York, 1975.
6. Morrell, R., *Handbook of Properties of Technical and Engineering Ceramics*, Part 1, Her Majesty's Stationery Office, London, 1985.

Questions

1. Why do materials and composites manufacturing processes go hand in hand?
2. Under what circumstances does a search for new materials start?
3. What are the steps involved in selecting a best material for an application?
4. An engineer has three materials — A, B, and C — from which to select for an application. The selection criteria are specific strength and machinability. How would you quantify the selection process and select the best material from among these three?

4
Product Development

4.1 Introduction

Every year, hundreds of new materials and products are launched into the marketplace. Material suppliers are coming up with new reinforcements, resins, and prepreg materials to meet various customer needs. The materials not available in the past have become the material of choice today. New epoxy-based prepregs with a shelf life of 1 year at room temperature are available. Preform manufacturers are coming up with new braiding technologies and preform materials for increased manufacturing feasibility. Honeycomb and core suppliers are introducing new core materials to meet new market needs. Sporting goods manufacturers, boat builders, and consumer industries are launching new products into the marketplace to maintain their competitive position. The automotive industry is replacing steel and aluminum parts with composite materials for fuel and weight savings. The aerospace industry is utilizing more and more composite materials to increase payload capacities and fuel savings.

The need for new material/product development has increased rapidly with the improvement in technology and globalization of the market. Companies are struggling to keep their competitive position in the marketplace by improving their product quality and performance. Customers are demanding better products at lower cost. Product life has decreased significantly. There has been greater demand for reducing the product development cycle time. It is said that the companies that introduce a new product first onto the market get more than 50% of the market share.

Understanding the product development process is very important in the fabrication of a good quality part. Major automotive, aerospace, and computer companies follow a systematic approach for the development of a new product/model or a specific part. Product fabrication is one element of a product development process. This chapter explains all the important elements of a product development process.

4.2 What Is the Product Development Process

Product development is a process for translating customer needs into product design and manufacturing. A broader view of product development involves managing mutual dependencies between all stages of the product life cycle, including the design, manufacturing, distribution, technical support, and disposal or recycle stages. The product development process provides a roadmap to team members for the activities and deliverables required to design, develop, and manufacture the product. It is a systematic approach for the successful launching of a product, from concept initiation to marketing. Product development consists of material selection, product design, selection of the right manufacturing and assembly techniques, prototyping, testing and validation, and successful launching of the product into the marketplace. In a typical industrial scenario, the sales or marketing department identifies a market need for a product. They study an application and pass the application requirement to the product development team (PDT). The PDT collects information about the product from the marketing team or by talking directly to the customer. Depending on the product need, suitable materials are then selected for the application. Engineers have many choices from which to select the right resin and reinforcing materials. Refer to Chapters 2 and 3 for various types of raw materials and material selection guidelines. The selection of a material is also affected by the selection of a manufacturing process because different manufacturing processes require different initial raw materials. The product is then designed by taking into consideration the pros and cons of a manufacturing process. Once design, material, and manufacturing processes are selected, prototype parts can be made using the selected manufacturing processes. Prototype parts are then tested and validated to meet the identified service and consumer needs. The various phases of product development are described in detail in Section 4.7.

The goals of a product development activity are to minimize life-cycle costs, maximize product quality, maximize customer satisfaction, maximize flexibility, and minimize lead times. Today's hot topic is to compress product development cycle time. A compressed product development process incorporates specialists from diverse groups such as design, manufacturing, marketing, purchasing, product engineering, reliability, and sometimes customers. This strategy drastically reduces the product development cycle time. This chapter focuses on the managerial and engineering steps needed during product development to meet the functional, performance, and customer requirements of the product.

4.3 Reasons for Product Development

Product development activities are undertaken to increase market share and create growth in the company. Following are the main reasons for launching a new product.

Product Development

1. To find new business opportunities and market shares
2. To add features and benefits over and above a competitor's product to increase market share
3. To retain customers, continuous product improvements are made
4. To attract more customers, new features are added

There are so many features that can be added to a product, but all such features cannot be incorporated because each feature costs money. Moreover, every customer's requirements are different and they are not willing to pay for all the features. Therefore, every supplier needs to supply a different mix of features for different groups of customers. The following list will help the reader get a good combination of product features to meet customer's need.

- List all buying concerns of customers.
- Separate the list into three categories: (1) must have, (2) important to have, (3) nice to have.
- Rate each feature on a scale of 1 to 10 and quantify the information.

This analysis will help to get a blend of features that will be best for various groups of customers.

4.4 Importance of Product Development

In today's customer-driven global marketplace, it is important to launch a high-quality product at low cost with quick turn-around time. In general, the company that enters into the market first with a new high-quality product captures the largest share of the market and makes a greater profit. To achieve this, one must simultaneously consider the requirements of manufacturing and assembly during the product design phase in order to reduce product cost and improve product quality. By employing concurrent engineering during the product development process, major design problems can be solved in the very early stages of the design phase. The cost of design changes at a later stage is very significant compared to an earlier stage. For example, a company decides to go with design option A and starts making prototype parts and testing them. Once it finds out at this stage that design B will be better than design A and tries to convert everything for design B, then it loses all the money and time spent thus far in design A. If the company decides to change the design after the product is manufactured and reaches the market, then it incurs additional costs for such design changes. A recall of the product could be for faulty design or a major drawback in the product. There is ample evidence available in which fitness-for-use problems can be traced to the design of the product. For example, in a study of seven space programs, 35.2% of component failures were due to design or specification errors.[1] During a typical period of 11 months at a chemical plant, 42% of the rework dollars were traced to research and development. For mechanical products of moderate complexity, it is estimated that the errors made during the product development phase caused about 40% of the fitness-for-use problems.

4.5 Concurrent Engineering

Concurrent engineering implies the simultaneous use of various engineering disciplines, such as design, manufacturing, marketing, packaging, reliability, etc., during product development activity. This branch of engineering is becoming popular because it has a significant effect on reducing development cycle time and cost. The conventional way of developing a product involves serial activities from various groups such as design, manufacturing, packaging, service and maintainability, reliability, and marketing as shown in Figure 4.1. In this approach, the product design group designs the product without interacting with manufacturing and/or other groups. Many times, the part designed by the design team may not be manufacturable or may not be production efficient. For example, the design group may specify a 0.002-in. tolerance for the outer diameter of a pultruded part and then passes it to the manufacturing group. The manufacturing team looks up the design and rejects it because the part cannot be fabricated. The design team again works on the design and makes related changes. Similarly, the design engineer may specify sharp corner, high flatness, high surface quality, and other features that may not be production feasible. Therefore, it is important to discuss the design with the manufacturing engineer early in the design phase to avoid rework and delay. Similarly, design needs to be discussed with the packaging department to ensure that the product size fits into the packaging space allocated for the product. This is critical in the automative industry. There is

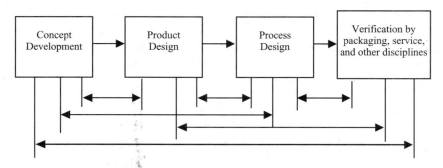

FIGURE 4.1
Serial approach to product design.

Product Development

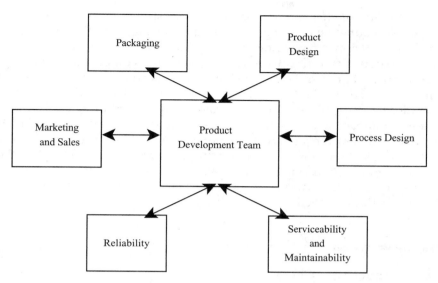

FIGURE 4.2
Schematic diagram of concurrent engineering.

a significant space constraint in the design of an automobile, in which thousands of parts are assembled together in compact form. This space constraint is increasing day by day because customers are asking for more and more features and a compact car for fuel efficiency. The service engineer should look at the design to ensure that the various parts are easily accessible during servicing. All these requirements may force the designer to change the design several times, and sometimes to restart the entire design process from scratch.

The current market demand is to perform product development activities in a concurrent engineering environment as shown in Figure 4.2. There is a greater need for strong interaction between the manufacturing engineer and the design engineer because manufacturing decisions are directly related to product design. It is required to simultaneously consider the requirements of manufacturing, assembly, packaging, service, distribution, and disposal right in the design phase to reduce the per-unit cost of production, reduce the lead time, and improve quality. The reason various disciplines need to work together is to avoid changes in the design late in the product development cycle. It is estimated that there is a tenfold increase in cost for making a change in the production phase as compared to changes made early in the design phase. Changing the design in the production phase may result in changes in tool, material, equipment, planning, and labor requirements. Therefore, there is greater emphasis on moving the engineering changes to the early design stage by utilizing concurrent engineering.

There are many ways a product can be designed to meet the functional and performance requirements of the application. Thus, early design decisions have a significant effect on product life-cycle costs. For example, a designer might select adhesive bonding, mechanical fastening, snap fitting, mechanical locking, or insert molding for the joining of two components. Each process requires a different production planning, equipment set-up, and production cost process. Depending on the production rate, application, and other requirements, an informed decision is made to select the right assembly operations. Clearly, the ability to improve the quality of design will have a significant impact on overall product cost. Design for manufacturing (DFM) and design for assembly (DFA) strategies can be used to improve the quality of design. Therefore, it is a good idea to invest more time and effort into the early design phase to come up with a design that meets (or exceeds) the requirements of the various disciplines. This will result in fewer or no changes during production.

4.6 Product Life Cycle

Every product, depending on its value and need, has a life from birth to death. In general, product inception takes place in the product design phase, where market needs and requirements are transformed into a three-dimensional shape and size. In most cases, the idea for a new product comes from the customer. For example, when customers complained to 3M Inc. about the dust problem in their sandpaper, 3M came up with the idea of wet sandpaper, in which the dust problem is removed. After its birth in a manufacturing company, the product is introduced into the marketplace at a smaller scale, called the test sample. During this stage, product sales remain low until the product starts to gain familiarity and acceptance. This phase critically affects the management to react on customer's complaints and feedback to improve the performance and its capabilities for greater acceptance in the marketplace. After the introductory phase, product enters into the growth phase where product sales increase through word of mouth, recognition, and advertisement. After the growth phase, product enters into its maturity phase where annual sales remain almost the same or increase at a rate economy increases. During this phase, the product experiences competition due to the introduction of other similar products. In this phase, continuous improvement is done to lower the cost and increase the quality of the product. Additional features are added to the product and sometimes the product is divided into different groups to meet the need(s) of various customer groups. After some period of time, product sales decline because of the introduction of a better product in the market to meet similar needs using better technology. For example, composite golf shafts, fishing rods, and composite boats replaced metal or wood counterparts. During this decline phase, management may try to revive the product by innovation or by incorporating new technology.

Launching a new product onto the market costs a significant amount of time and money, along with market uncertainities; these are the biggest barriers to product innovation. The high cost of investment for research, development, advertisement, and sampling increases the initial product cost. As the demand and production volume of the product increases, the cost decreases. The higher cost of investment is a big barrier for small companies to compete with big companies. Business acquisitions and mergers are taking place around the world to overcome this initial barrier to successfully compete in the market.

4.7 Phases of Product Development

The complete product development process (PDP) is divided into several stages (or phases) for the successful design and fabrication of the product. Each phase has its goals and associated activities to meet those goals. Each phase is divided into several activities or tasks. Different industries practice different types of product development processes. Some practice three-phase PDP, some practice four-phase PDP, and some practice seven-phase PDP. The goal of each of these is to minimize potential errors, reduce product costs, minimize lead time, and improve product quality. Six important phases of a PDP are discussed below.

4.7.1 Concept Feasibility Phase

In this phase, market need of a product with associated functional and performance requirements is identified. The ideas are then reviewed by a team of experts from various departments such as engineering, manufacturing, materials, marketing, sales, finance, and sometimes customers. The team members are not only experts in their areas but also knowledgeable in the discipline of other team members. This helps avoid communication problems and cultural barriers. The panel transforms customer needs into a three-dimensional product shape and size without much attention to specific geometries, manufacturing, or technical details. The purpose of this phase is to create a preliminary design and production scenerios and then evaluate the technical feasibility of designing and manufacturing the product. The important element of this feasibility study is to estimate the expected cost to the company and to the customer. At this stage, a list of competitors' products that compete with the proposed product are identified. Competitor prices are compared with the expected cost of the product, and the need for the proposed product is examined. Once studies of the market need, preliminary product and process designs, project cost, and return on investment (ROI) are reviewed, a go/no-go decision is made regarding the project.

4.7.2 Detailed Design Phase

This phase utilizes the information generated in the concept feasibilty phase to come up with multiple concepts about the product. A product development team (PDT) is formed, consisting of experts from the design engineering, manufacturing, materials, testing, packaging, reliability, service, marketing, and purchasing disciplines. This team is led by a product design engineering representative. This team meets on a regular basis for the design of the product. Various functional, performance, service condition, packaging, servicing, and other requirements are listed. Several brain-storming sessions are conducted among team members for the design of the product. The purpose of brain-storming is to generate as many ideas about components design, joint design, assembly methods, material selection, and manufacturing process selection. In a brain-storming session, criticism of an idea is not allowed. Significant amounts of time and effort are dedicated in this phase to come up with best ideas for the design of the product. Tools such as DFM, DFA, finite element analysis (FEA), and failure mode effective analysis (FMEA) are used to generate the best ideas.

Once there are many design options available for the product, the various design options are compared. Tools such as Pugh analysis, DFA software (Boothroyd and Dewhurst, Inc.), etc. are used for narrowing down the choices or for selecting the most promising design. A complete drawing of the product is made for the best design, and product and material specifications are written down. This information is then used to make prototype parts.

It may be possible that the design selected may need to be modified or may be discarded. In that case, a new design is selected from narrowed choices, or the above process is repeated again.

4.7.3 Prototype Development and Testing Phase

After the product design is complete, it is important to test the design for its functionality, performance, and other requirements. The product and material specifications developed in the previous phase are used to develop prototype parts. Making prototype parts sometimes involves extensive investment in tools. To avoid expensive tooling and other investments, rapid prototyping is used in some cases to make sure the design meets fitness and assembly needs. Rapid prototyping, such as stereolithography, is an inexpensive way of making parts for visual inspection. A stereolithography apparatus slices the three-dimensional CAD files (solid models) data into cross sections, and then constructs the pysical part by depositing plastics or wax layer by layer from bottom to top until the desired part is completed. Depending on the size of the part, a prototype can be made in a few hours. The resulting part provides a quick conceptual model for visualization and review. It can be used by marketing personnel for demonstration purposes or by purchasing personnel for increasing the accuracy of bids. With the help of rapid prototyping systems, flaws in the design can be determined early in the product development cycle before expensive investment in tooling and fabrication. Once the part meets the dimensional requirement, prototype parts are made using the manufacturing process specified for the product.

Following the fabrication of prototype parts, component- and assembly-based tests are performed. Before all the components are assembled and tested, it is a good idea to first test the individual component for its performance and other requirements. By doing this, the cost of assembly is avoided and a major source of failure is recognized. Once parts are assembled and tested, it is sometimes difficult to figure out which component is the real cause of failure. Component-based tests avoid such confusion. After parts meet the component-based requirement, then various joints and interfaces are tested for their design requirement. Once that is done, complete product assembly is created and tested.

The purpose of prototype testing is to determine the design capability of the product; the effects of service conditions such as fluid exposure, temperature extreme exposure, etc. on product performance, and the reliability of product operation over time. This phase concerns the three F's of engineering: function, fit, and form. The units built in this phase may not be complete but should be representative of the actual product. Ideally, the prototype parts should be made using the equipment and manufacturing processes selected for the full-scale fabrication of the product, but sometimes this is not possible. For example, prototype parts made by SRIM or compression molding of SMC are avoided because of higher tooling and equipment costs. Prototype testing provides a good indication of the technology and design feasibility, and is valuable in determining the adequacy of design process before committing resources to subsequent stages. Prototype testing also provides guidelines and directions for future efforts. In some cases, prototype parts are shown and demonstrated to potential customers to obtain their feedback on the design. It expedites the marketing effort and eases customer concerns about the product.

4.7.4 Preproduction Demonstration, or Pilot-Scale Production

After the product has passed the prototype and testing phase, the lessons learned are listed. The effort in this phase focuses on defining and simplifying production processes; cutting the cost of material and production; identifying machine, tools, and fixturing needed; standardizing the product and processes for interchangeability with existing product line; evaluating in-house resources and production facility; making decisions on outsourcing vs. in-house production; etc.

In this phase, the integrated discipline team is led by the manufacturing engineering representative. This helps to document the manufacturing efficiency of the product by the most qualified person on the team. By keeping the manufacturing engineer as a leader of the team, fabrication knowledge of machine, tools, fixturing, material handling, process flow, process cost, and product cost is utilized. The design engineer remains responsible for making sure that the product is in the product specification envelope.

The next step is to demostrate the capability of the production process for the production rate requirement, tolerance, and other requirements by producing a small batch of products. This is also called pilot-scale production. Quality and dimensional inspections are performed to make sure the product is made to specifications and is of the desired quality. The product thus manufactured is tested for service conditions in a test lab or in the proving ground, or under actual conditions. Based on the manufacturing experience, quality measurements, and test results, the design is either released or sent back to rework.

4.7.5 Full-Scale Production and Distribution

After successful demonstration of pilot-scale production, the product is ready for sale and distribution. In this phase, the focus is on packaging decisions, mode of distribution, quality assurance procedures, order entry systems, warehousing, and marketing strategy. This is the phase when the product is formally introduced to the market. In this phase, the product development team transfers the product knowledge and its value to other company departments such as sales, advertising, accounting, quality assurance, etc.

4.7.6 Continuous Improvement

Once the product is on the market, the company begins to receive feedback from customers, sales, marketing, and other groups. The plant employees may complain about health and safety features of a process or might suggest a better operation for performing various manufacturing steps. Based on experience gained in marketing, sales, production, and use of the product, design changes are made to improve product quality and performance, and to eliminate field failures that were caused during product use. These design changes can go through one or more of the previous phases of the product development. The company continuously strives to improve the quality and lower the cost of product to make it fit for competition.

During the product development process, various design reviews are set to make sure that the goals and objectives of the PDP are met. The following section briefly describes the design review process.

4.8 Design Review

Design review is an important element in the product development activity. In general, there are four to six design reviews such as preliminary design review (PDR), interim design review (IDR), critical design review (CDR), and final design review (FDR) in a product development cycle. These design reviews are held at different phases of the product development cycle to make sure the activity is going in the right direction. The design review is conducted by the leader of the PDT and attended by managers from various departments (e.g., manufacturing, application, marketing, purchasing, and materials) who are experts in the company's product lines and business, and not directly associated with the development of a design. These specialists are highly experienced professionals and are aware of the objectives of the project. The specialists can come from outside the company, such as a customer or a university professor. The PDT leader presents the results of developmental work performed up to that point in time, and committee members examine the results and provide future direction for the project. The purpose of the review is to get new ideas for the design, correct potential problems with the design, screen design choices, and provide future courses of action to meet the objectives of the project. The review ensures that the product will function successfully during use. It provides an opportunity for experts to ask critical questions about the product and process. A successful design review not only provides constructive criticism about the product and process design, but also provides solutions to various design problems and challenges.

Design review is a formal activity, typically completed in a few hours. The PDT leader schedules a meeting date and sends the agenda for the meeting. Minutes of the meeting are taken and circulated among committee and PDT members. Courses of action are noted and followed during the PDT meetings.

4.9 Failure Modes and Effects Analysis (FMEA)

FMEA is performed during PDP to determine the potential failure modes a product can experience during operation and to evaluate the effect and consequences of those failure modes on the functions and performance of the total system. It provides a systematic method for analyzing the potential causes and effects of failure before design is finalized. The aim of this analysis is to redesign the product for maximum reliability. FMEA identifies and addresses the causes and mechanisms of failure modes. For each failure mode, effects on total system, its seriousness, and its probability of occurrence are examined. The purpose of FMEA is to improve quality, reliability, safety, and durabilty. The net effect of this is to reduce product life-cycle cost. FMEA can be applied during product design as well as process design. FMEA should be incorporated during product development activities. FMEA helps the product development process by the following means:

- It establishes a list of potential failure modes in the product. The PDT goes through the list and develops a design and manufacturing plan to avoid these failure modes. Thus, the quality of the product improves significantly.
- It ranks the failure modes according to their effects on the overall system. PDT develops priority lists and test plans for design improvements according to failure mode rankings.
- It provides future reference for analyzing field failures and establishes guidelines for design changes.
- It serves as a tool in the selection of process and product design alternatives.

In performing the FMEA, one should ask the following questions:

1. What are the functions of the product or assembly?
2. How can the assembly, sub-assemblies, and individual components fail?
3. What are the probabilities of those failures and their severity?
4. What are the causes of failure?
5. What are the effects of each failure mode on the functions and performance of the product?
6. How can these failures be prioritized?
7. How will the failure be detected?
8. How can failure be avoided?
9. What are the design options to eliminate failure?
10. If failure takes place, then what corrective actions should be taken to reduce its severity?

Failure is the loss of function or ability to perform a prescribed task in a predetermined manner. Examples of failures include a product becoming noisy, or unable to do a task, or not meeting desired performance requirement. Failure modes are those that cause the failure to occur. Typical failure modes in a product include part broken, loose, tight, bent, leakage, joint failure, etc. For each failure mode, there are several possible causes of failure. It is very important to identify the real cause of a failure. Once the cause is known, the solution becomes very evident in many cases. Knowing the cause provides corrective action and design solutions.

To perform FMEA, the product is divided into assemblies. Each assembly is divided into sub-assemblies and then the sub-assemblies are divided into components. The functions of each assembly, sub-assembly, and component are determined and relationships among each other are established. Potential failure modes caused by operation and service conditions are evaluated for each component and its effect on the next higher item or on the total system

Product Development

is analyzed. The reliability group poses possible failure modes to the PDT about their design, and PDT takes corrective actions or preventive measures to reduce the effect of failure or totally eliminate the failure. The reliability group has a record of the probability of a certain type of failure in a product from the company's past experience.

Reference

1. Juran, J.M. and Gryna, F.M., Eds., *Juran's Quality Control Handbook*, 1970.

Bibiliography

1. Boothroyd, G., Poli, C.R., and Murch, L.E., *Automatic Assembly*, Marcel Dekker, New York, 1982.
2. Gryna, F.M. and Juran, J.M., *Quality Planning and Analysis: From Product Development Through Use*, McGraw-Hill, New York, 1993.

Questions

1. How can the manufacturing engineer assist the design engineer in developing a manufacturable product?
2. Why is it important for the manufacturing engineer to begin working with the design team very early in the product development stage?
3. What are the benefits of concurrent engineering?
4. What are the important elements of the product development phase?
5. What is FMEA? How does it help in coming up with better product design?
6. Why is it necessary to follow a systematic product development approach when making a product?

5
Design for Manufacturing

5.1 Introduction

Companies are constantly being challenged to find means to do things better, faster, and cheaper. Companies can no longer overdesign the product, nor can they afford a lengthy product development cycle time. The products can no longer be viewed individually, and designers can no longer pass the engineering concept to the manufacturing engineer for finding the ways to make it. The design engineer and manufacturing engineer need to work together to come up with a best design and manufacturing solutions for fabricating the products cost-effectively. For example, if design and manufacturing engineers work separately to create the design of the outer body panels of automobiles, the manufacturing engineer will come up with a flat or square box-like product that is cheaper and quicker to make, but no one would buy it. On the other hand, the design engineer will come up with a design that is creative, eye-catching, and satisfies all customer needs and requirements, but it would be unaffordable. In either case, the product will not sell.

To be competitive, the product needs to be designed in a minimum amount of time, with minimum resources and costs. To meet current market needs, several philosophies, such as design for manufacturing, design for assembly, design for quality, design for life cycle, and concurrent design, are being developed. The primary aim of these philosophies is to think about the manufacturing, assembly, quality, or life-cycle needs *during* the design process. This is achieved by working concurrently in a concurrent engineering environment to avoid later changes in the design.

A product can be designed in many ways to meet functional, performance, and other requirements. Therefore, different organizations come up with different design concepts to meet the same application needs. The solution for an application depends on how the problem is defined to the designer as well as the knowledge and creativity of the designer. Because there are many design solutions to a problem, the question arises as to how to know which design is the best solution. It is also possible that there may be other designs that may be better than the realm of the designer. Design for manufacture is

85

a tool that guides the designer in coming up with better design choices and then provides the optimum design. It is a tool for concept generation, concept approval, and concept improvement. It integrates processing knowledge into the design of a part to obtain maximum benefits and capabilities of the manufacturing method. To come up with the best design, the manufacturing engineer should have a good knowledge of the benefits and limitations of various composite manufacturing techniques. The team members should also be familiar with tools such as design for manufacturing (DFM), design for assembly (DFA), etc. for developing high-quality design. As compared to metals, composite materials offer the highest potential of utilizing DFM and part integration, and therefore can significantly reduce the cost of production.

Engineers utilizing isotropic materials such as aluminum and steel traditionally fabricate parts by first selecting raw materials from a design handbook based on performance requirements. Once the raw material is selected, the manufacturing process to fabricate the part is identified. This philosophy is not viable in the field of composite materials. With engineered composite materials, the material selection, design, and manufacturing processes all merge into a continuum philosophy embodying both design and manufacture in an integrated fashion. For example, a rod produced by filament winding, pultrusion, RTM, or braiding would impart distinct stiffness, damping, and mass characteristics due to different fiber and resin distributions and fiber volume fractions. Composites manufacturing processes create distinct microstructural properties in the product.

The best design example is Nature's design in which different artifacts are grown in the entire system as a single entity. In contrast, engineers fabricate various parts and assemble them together. At present, we do not have biological manufacturing processes but we have plenty of opportunities for innovation by learning and imitating the no-assembly designs of the natural world.[1] Designs in nature are strong but not necessarily stiff — they are compliant. Nature tries to make the design compliant, whereas engineers traditionally make the structure and mechanism stiff. Ananthasuresh and Kota[1,2] developed a one-component plastic stapler in which they replaced the conventional steel stapler with no-assembly design. Compliant mechanisms are single-piece, flexible structures that deliver the desired motion by undergoing elastic deformation as opposed to rigid body motion.

5.2 Design Problems

The defect or quality problem in the product is caused by three things: bad design, bad material, and wrong manufacturing process. For example, if the product is correctly designed, and if the manufacturing method is not correctly designed, then the product will be defective. Similarly, an incorrectly designed product will also result in quality problems despite having chosen the right materials and good manufacturing methods. The occurrence of these defects is caused by several factors inherent in the design.

A poor design can cause many problems in production plants. It not only increases the cost of the product, but also decreases product quality. Common design problems include loose parts, rattling, parts not aligned, tight parts, missing parts, labor-intensive assembly, too many machining operations, difficult to manufacture, difficult to assemble and bond, difficult to achieve quality, ergonomic problems, serviceability problems, etc. These design problems can be solved early in the design phase utilizing best practices. Poor quality results in a higher rejection rate and therefore a higher production cost. Product quality depends on how the product is designed.

5.3 What Is DFM?

DFM (design for manufacturing) can be defined as a practice for designing products, keeping manufacturing in mind. DFM starts by taking a plain sheet of paper and identifying a product's functional, performance, and other requirements. It utilizes rules of thumb, best practices, and heuristics to design the part. Best practices for a high-quality product design are to minimize the number of parts, create multifunctionality in the part, minimize part variations, and create ease of handling. DFM involves meeting the end-use requirements with the lowest-cost design, material, and process combinations.

In the past, several product problems arose because of poor design. The designers were not aware of the various manufacturing techniques available on the market, nor the capabilities of each manufacturing technique. As a result, products were heavy, had many parts and thus many assembly operations, and resulted in poor quality and increased cost. To effectively design the product, manufacturing knowledge needs to be incorporated into product design. The designer should know how the process and design interact. In general, the real challenge in designing composite products is to develop a good understanding not only of engineering design techniques, but also of processing and material information. The purpose of DFM is to:

- Narrow design choices to optimum design (Figure 5.1)
- Perform concept generation, concept selection, and concept improvement
- Minimize product development cycle time and cost
- Achieve high product quality and reliability
- Simplify production methods
- Increase the competitiveness of the company

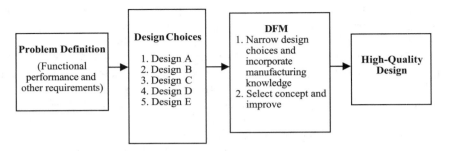

FIGURE 5.1
Design flow diagram in DFM.

- Have a quick and smooth transition from the design phase to the production phase
- Minimize the number of parts and assembly time
- Eliminate, simplify, and standardize whenever possible

5.4 DFM Implementation Guidelines

The main objective of DFM is to minimize the manufacturing information content in the product without sacrificing functional and performance requirements. DFM can also be applied for a product that is already in production or on the market. The main objective here will be to make the product more cost-competitive. The following DFM guidelines are applicable to products made of composites, metals, and plastics.

5.4.1 Minimize Part Counts

There is good potential for part integration by questioning the need for separate parts. At General Motors, Ford, Chrysler, GE, IBM, and other companies, DFM strategies have reduced the total number of part counts by 30 to 60% in many product lines. Composite materials offer good potential for part integration. Minimization of part counts can result in huge savings by eliminating the need for assembly, inventory control, storage, inspection, transportation, and servicing. According to Huthwaite,[3] "the ideal product has a part count of one." In general, more than one part is needed if there is a relative motion requirement, a different materials requirement, a different manufacturing requirement, or an adjustment requirement. An example of part integration is the steel identification badge clip that has four different parts but can be replaced by a single injection molded plastic part. Another example is the monocoque composite bicycle frame. Do not perform part intergration if design becomes overly complex, heavy, or difficult to manufacture.

Design for Manufacturing

A typical automobile, airplane, or luxury yacht consists of thousands of parts to meet various functional or performance needs. For example, a Heloval 43-meter luxury yacht from CMN Shipyards is comprised of about 9000 metallic parts for hull and superstructure and over 5000 different types of parts for outfitting.

To determine if a part is a potential candidate for elimination, the following questions should be asked:

1. Do the parts move relative to each other?
2. Is there any need to make parts using a different material?
3. Will the part require removal for servicing or repair?
4. Will there be a need for adjustment?

If the answers to the above questions are "no," then the part is a potential candidate for replacement. The following guidelines can be used to minimize the number of parts.

- Question and justify the need for a separate part. Ask the four questions above; and if the answer is "no," then redesign the product by eliminating the separate part.
- Create multifunctionality features in the part.
- Eliminate any product feature that does not add any value for the customer.
- Use a modular design.

5.4.2 Eliminate Threaded Fasteners

Avoid the use of screws, nuts, bolts, and other fasteners in the product. It is estimated that driving a screw into the product costs almost 6 to 10 times the cost of a screw. The use of fasteners increases inventory costs and add complexity in assembly. Fasteners are used to compensate for dimensional variation, to join two components, or for part disassembly. The use of fasteners creates the potential for a part to become loose during service. IBM has used this philosophy to redesign its printer, eliminating many screws and replacing them with snap-fit assembly. The resulting design had 60% less parts and 70% reduced assembly time.

Snap-fits are used with plastics or short fiber composite parts and provide ease of assembly due to the lack of any installation tool requirement. General concerns regarding the use of snap-fits include strength, size, servicing, clamp load, etc.

5.4.3 Minimize Variations

Part dimensional variation as well as property variation are the major sources of product defects and nonconformities. Try to use standard parts off-the-shelf and avoid the use of special parts. Eliminate part variations such as types of bushings or O-rings, seals, screws, or nuts used in one application. The same size would mean the same tool for assembly and disassembly. This guideline aims to reduce part categories and the number of variations in each category, thus providing better inventory control and part interchangeability.

5.4.4 Easy Serviceability and Maintainability

Design the product such that it is easy to access for assembly and disassembly. The part should be visible for inspection and have sufficient clearance between adjacent members for scheduled maintenance using wrench, spanner, etc.

5.4.5 Minimize Assembly Directions

For product assembly, minimize assembly direction. While designing the product, think about the assembly operations needed for various part attachments. It is preferable to use one direction; z-direction assembly operation allows gravity to aid in assembly. A one-direction assembly operation minimizes part movement as well as the need for a separate assembly station. It is better in terms of an ergonomics point of view as well.

5.4.6 Provide Easy Insertion and Alignment

When there are more than two parts in a product, the mating parts need to be brought close by performing insertion or alignment. Some guidelines for easy insertion and alignment are:

- Provide generous tapers, chamfers, and radii for easy insertion and assembly.
- Provide self-locating and self-aligning features where possible.
- Avoid hindrance and obstruction for accessing mating parts.
- Avoid excessive force for part alignment.
- Design parts to maintain location.
- Avoid restricted vision for part insertion or alignment.

5.4.7 Consider Ease for Handling

In an assembly plant, various parts are kept in separate boxes near the assembly station. Workers pick up those parts and assemble them using adhesive bonding or mechanical fastening or by slip-fit or interference-fit. Avoid using parts such as springs, clips, etc., which are easy to nest and become interlocked. It disrupts the assembly operation and creates irritation for the worker. For smooth assembly operation and ease of handling, parts should not be heavy and should not have many curves, thus reducing the potential

Design for Manufacturing

for entanglement. To avoid physical fatigue of the worker, part and assembly locations should be easy to access. Parts should be symmetric to minimize handling and aid in orienting. Add features that help guide the part to its desired location. The following suggestions can improve part handling. These suggestions are more applicable for a high-volume production environment.

- Minimize handling of parts that are sticky, slippery, fragile, or have sharp corners or edges.
- Keep parts within operator reach.
- Avoid situations in which the operator must bend, lift, or walk to get the part.
- Minimize operator movements to get the part. Avoid the need for two hands or additional help to get the part.
- Avoid using parts that are easy to nest or entangle.
- Use gravity as an aid for part handling.

5.4.8 Design for Multifunctionality

Once an overall idea of the product's functions is gleaned, one can design individual components such that they provide maximum functionality. It is preferable to use molding operations that provide net-shape or near-net-shape parts. For example, an injection molded composite housing part meets the structural requirement of the product and has built-in features for alignment, self-locating, mounting, and a bushing mechanism. This technique helps minimize the number of parts.

5.4.9 Design for Ease of Fabrication

In composite part fabrication, product design cannot be made effective without knowledge of the manufacturing operations. Each manufacturing process has its strengths and weaknesses. The product design should be tailored to reap the benefits of the selected manufacturing process. For example, if close tolerances are required on the inside diameter of a tube, then filament winding is preferred compared to a pultrusion process. The design should be simplified as much as possible because it helps in manufacturing and assembly and thus in cost savings. Workers and others who are dealing with the products can easily understand simplified design.

5.4.10 Prefer Modular Design

A module is a self-contained component that is built separately and has a standard interface for connection with other product components. For example, a product that has 100 parts can be designed to have four or five modules.

Each module can be independently designed and improved without affecting the design of the other modules. Modular design is preferred because it helps in the final assembly, as well as in servicing where a defective module can be easily replaced by a new module. Modular design can be found in aerospace, automotive, computer, and other products. For example, steering systems, bumper beams, and chassis systems are separate modules designed, produced, and improved upon by independent organizations and assembled in the vehicle. In each of these modules, there are many other modules, which are again designed by various groups of the organization.

5.5 Success Stories

There are many examples for which DFM techniques have minimized the number of parts and reduced assembly time. The monocoque composite bicycle frame is one example where several metal parts are replaced by one composite monocoque structure. In automotive applications, SMC parts have reduced total part counts by integration of various parts. Following are some of the success stories of DFM, where it helped in improving the product quality and lowering the product cost.

5.5.1 Composite Pickup Box

Ford, in partnership with The Budd Company's Plastics Division, has developed a composite pickup box for the Ford 2001 Explorer SportTrac. The SMC pickup box is close to 20% lighter than a typical steel box. The old process required 45 pieces of sheet metal to be assembled. With the new composite one-piece box, there are fewer pieces, fewer tools, and fewer assembly fixtures, and it takes up less floor space in the assembly plant, which results in cost savings.

Using a composite instead of steel yields an overall reduction in vehicle weight, resulting in increased fuel economy. A structural sheet molding composite (SMC) box inner does not trap water under a liner, thereby eliminating the risk of rust damage to the pickup bed.

The composite box exceeds the 150,000-mile durability requirements specified by Ford Truck for all pickup boxes. It is built "Ford Tough" and is far superior to steel for corrosion and dent resistance. Not only does the composite box reduce weight, improve fuel economy, increase durability, and cut down on production time and cost, but it is also recyclable.

5.5.2 Laser Printer

A laser printer has hundreds of parts and is assembled to meet the desired functionality. IBM, GE, and other companies have tried to reduce the number of parts in laser printers using DFM and DFA tools. In the new Optra S 1250 laser printer, the latest model manufactured by Lexmark International (Lexington, KY), GE Plastics (Pittsfield, MA) redesigned the chassis component using injection molded parts. GE Plastics reduced the part count from 189 for the sheet metal chassis to a 12-part fiberglass-reinforced polyphenylene oxide (PPO) design. The design guidelines used were part integration, minimization of number of parts, design for assembly, etc. The result: more than 50% reduction in assembly time and more than 20% savings in assembly cost.

5.5.3 Black & Decker Products

Black & Decker has used DFM to greatly reduce the number of hardware components purchased by the company.[4] For example, the list of plain washers was reduced from 448 to 7 (one material, one finish, one thickness) in future products. Similarly, the number of ball bearings was reduced from 266 to 12 (one seal, one lubricant, one clearance, metric only).

5.6 When to Apply DFM

DFM should be employed in the early stages of the product development process when decisions have the greatest impact on product cost and when there is the best chance for its implementation. It is too late to utilize DFM after the product design is released or when the part is in production. Any changes in the design at a later stage significantly increase the cost of product. DFM can also be used to improve the design of existing products to obtain cost benefits and increase market competitiveness. DFM helps in coming up with a final design in much less time than with traditional product design techniques.

5.7 Design Evaluation Method

An application has many requirements, as well as many design options to meet those requirements. Each design option must be evaluated for features and requirements that are important to customers. It is a challenging task to compare the various design options, but selection processes such as Pugh analysis and other techniques can simplify the task of selecting a best design. Here is a method that can be used for concept selection. According to this method, a matrix is created between criteria of selection and design options as shown in Table 5.1. Each design option is rated on a scale of 1 to 5 for

TABLE 5.1
Evaluation of Design Concepts

Factors	Weight (%)	Design "A"	Design "B"	Design "C"
Weight	15	3	4	3
Cost	20	4	5	3
Performance	10	3	4	3
Reliability	5	2	4	3
Noise	5	3	3	4
Assembly time	15	3	5	3
Robustness	7	3	4	3
Number of parts	10	2	4	3
Aesthetics	5	2	4	4
Ease of servicing	8	2	5	3
Total	**100**	**2.92**	**4.38**	**3.1**

various selection criteria. The weight assigned to each criterion depends on its importance for that application. Each rating is multiplied by weight and totaled for final selection. The design that obtains the highest point total is selected as the best design.

5.8 Design for Assembly (DFA)

It is found that the cost of assembly accounts for 40 to 50% of the total product costs in a wide variety of industrial products.[5] Therefore, there is a great need to lower the cost of assembly operations by reducing the number of parts and by using simple assembly operations for the remaining parts. The design for assembly (DFA) strategy is employed to design the assembly operation at minimum cost and maximum productivity. The difference between DFA and DFM is that DFA deals only with the assembly operation, whereas DFM deals with the entire manufacturing process. Therefore, DFA comes under the envelope of DFM. To obtain maximum benefit from DFA, the designer should have a good understanding of the various assembly operations and should justify the need for separate parts. Elimination of parts reduces assembly time and cost, and can sometimes eliminate an entire section of an automated assembly machine. The DFA method was developed by Boothroyd and Dewhurst and is described in detail in *Product Design for Assembly*.[5]

In the early stages of the design phase, the designer should decide which type of assembly method is going to be best for the product. The design then revolves around the capability and benefits of the selected assembly method.

Design for Manufacturing

The decision on the assembly method is based on costs, total number of parts in the product, production rate, etc. For example, if the annual volume of the product is only 1000, then manual assembly should be preferred. For an annual production volume of several million products, selection of automation equipment would be a better choice.

Boothroyd developed software for analyzing the assembly operation and presented a best way of assembling parts for minimum part movements and minimum actions. The software determines assembly time and cost for various design alternatives, and helps in selecting an optimum design and assembly sequence. Even for the same design, there are various ways and sequences that different parts can be assembled. Using the Boothroyd technique, each assembly sequence can be analyzed and a best sequence can then be selected.

DFA guidelines for reducing the part count include promoting part integration, unless there are relative motion and other requirements. In design for no assembly (DFNA), parts with relative motion between them can be consolidated into monolithic mechanical devices using jointless compliant mechanisms, provided the relative motion is small (e.g., in a stapler).

5.8.1 Benefits of DFA

The objective of DFA is to design the product for minimum assembly to avoid problems due to wear, lubrication, backlash, noise, and leakage. DFA results in the following benefits:

- Reduced numbers of parts
- Reduced assembly operations and part complexities
- Reduced assembly time and cost
- Ergonomically sound design
- Reduced product cost
- Reduced product development time
- Reduced capital investment
- Fewer design releases
- Weight saving
- Better inventory control
- Better quality

5.8.2 Assembly-Related Defects

There are many defects caused by assembly operations. After becoming aware of these defects, solutions to reduce these defects can be determined. Common defects include:

1. Part misaligned
2. Part damaged
3. Fastener-related defects
4. Missing parts
5. Part interchanged
6. Part interferences such as loose or tight part

In addition to the above defects, human errors are also a cause of assembly defects. Rook[6] observed that for 1 in 10,000 to 1 in 100,000 cases, manufacturing operations are typically omitted without detection. That is, omission errors themselves cause defect rates in the range of 10 to 100 ppm. Human error is primarily caused by:

- Forgetting to perform prescribed actions, resulting in missing parts
- Performing actions that are prohibited, such as incorrect lubrication, incorrect screw, incorrect material, or incorrect part selection
- Misinterpretation of manufacturing step

Other types of errors identified by Poka-yoke[7] are:

- Processing errors, such as overcooked or undercooked
- Errors caused during setting up of fixtures, tools, and workpiece
- Misoperation or adjustment mistakes
- Working on wrong workpiece
- Missing processing operations

The product-, process-, design-, and material-related defects are listed in Figure 5.2. The mistakes described above rarely happen but the collective effects of the above errors typically exceed one mistake per 1000 actions (1000 ppm). Automotive manufacturers and suppliers, sporting goods manufacturers, etc. make thousands of parts per day and go through millions of actions and assembly operations per day. The total number of part defects in such cases becomes significantly high. Many times, the defective part goes to the customer and causes a serious problem. Studies have shown that the majority of product defects are caused by mistakes.[8,9] To avoid such errors, error-proofing of product and process designs is performed. Error-proofing involves designing the product or process such that an operator or a machine *cannot* perform a mistake. Error-proofing operations include placing an alarm as a reminder for missing operation, keeping a checklist of operation and the addition of features for reminders, and redesigning the product and process for omission of the assembly operation.

In the aerospace industry, many prepreg layers are laid at various angles to make the composite part. Omission of a prepreg layer or incorrect placement

Design for Manufacturing

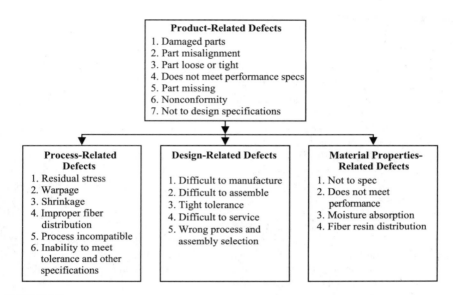

FIGURE 5.2
Typical product defects.

of a prepreg layer could significantly affect the performance of composite parts. To minimize such errors in a prepreg lay-up process, personnel from the quality control department check the lay-up sequence at certain intervals, as prescribed in the process sheet.

5.8.3 Guidelines for Minimizing Assembly Defects

To avoid mistakes during assembly operations, the parts should be designed keeping the worker in the mind. Workplace productivity can be enhanced by:

- Simplifying the product design
- Minimizing parts handling
- Reducing assembly operations and time
- Maximizing equipment uptime
- Eliminating ergo stressors
- Error-proofing the design and process

The workplace and the machines should be designed such that the materials, fasteners, and parts are as close as possible to a machine operator and that parts fall into their desired locations with little or no effort. To effectively design the workplace, factors such as material handling, ergonomics, and error-proofing should be given proper consideration. An action that causes little or no fatigue is given top priority.

References

1. Ananthasuresh, G.K. and Kota, S., Designing compliant mechanisms, *ASME Mech. Eng.*, p. 93, November 1995.
2. Kota, S., Synthesis of mechanically compliant artifacts: product design for no assembly, submitted to *ICED*.
3. Huthwaite, B., Design for competitiveness, *Bart Huthwaite Workshops*, Troy Engineering, Rochester, MI, 1988.
4. Bradyhouse, R., The rush for new products versus quality designs that are producible: Are these objectives compatible?, presented at the *SME Simultaneous Engineering Conference*, June 1, 1987, Society of Manufacturing Engineers, Dearborn, MI.
5. Boothroyd, G. and Dewhurst, P., *Product Design for Assembly*, Boothroyd and Dewhurst, Inc., Wakefield, RI, 1987.
6. Rook, L.W., Jr., Reduction of Human Error in Production, SCTM 93-62 (14), Sandia National Laboratories, Division 1443, June 1962.
7. Poka-yoke, Improving product quality by preventing defects, Nikkan Kogyo Shimbun, Ed., *Factory Magazine*, English translation copyright © 1988 by Productivity Press, Inc.
8. Hinckley, C.M. and Barkan, P., The role of variation mistakes and complexity in producing non-conformities, *J. Qual. Technol.*, 27(3), 242, 1995.
9. Hinckley, C.M., A Global Conformance Quality Model — A New Strategic Tool for Minimizing Defects Caused by Variation, Error, and Complexity, dissertation submitted to the Department of Mechanical Engineering, Stanford University, Stanford, CA, 1993.

Questions

1. Why is the knowledge of DFM important in product design?
2. Why is DFM more important in the area of composites manufacturing?
3. What are the common design problems?
4. In what ways does the minimization of part counts help a company?
5. What is the difference between DFM and DFA?
6. In an ideal product, how many parts should there be and why?
7. How would you determine whether a part is a potential candidate for elimination?
8. What are the common process-related defects?
9. What are the common assembly-related defects?

6
Manufacturing Techniques

6.1 Introduction

Every material possesses unique physical, mechanical, and processing characteristics and therefore a suitable manufacturing technique must be utilized to transform the material to the final shape. One transforming method may be best suited for one material and may not be an effective choice for another material. For example, wood is very easy to machine and therefore machining is quite heavily utilized for transforming a wooden block to its final shape. Ceramic parts are difficult to machine and therefore are usually made from powder using hot press techniques. In metals, machining of the blank or sheet to the desired shape using a lathe or CNC machine is very common. In metals, standard sizes of blanks, rods, and sheets are machined and then welded or fastened to obtain the final part. In composites, machining of standard-sized sheets or blanks is not common and is avoided because it cuts the fibers and creates discontinuity in the fibers. Exposed and discontinuous fibers decrease the performance of the composites. Moreover, the ease of composites processing facilitates obtaining near-net-shape parts. Composites do not have high pressure and temperature requirements for part processing as compared to the processing of metal parts using extrusion, roll forming, or casting. Because of this, composite parts are easily transformed to near-net-shape parts using simple and low-cost tooling. In certain applications such as making boat hulls, composite parts are made at room temperature with little pressure. This lower-energy requirement in the processing of composites as compared to metals offers various new opportunities for transforming the raw material to near-net-shape parts.

There are two major benefits in producing near-net- or net-shape parts. First, it minimizes the machining requirement and thus the cost of machining. Second, it minimizes the scrap and thus provides material savings. There are cases when machining of the composites is required to make holes or to create special features. The machining of composites requires a different approach than machining of metals; this is discussed in Chapter 10.

Composite production techniques utilize various types of composite raw materials, including fibers, resins, mats, fabrics, prepregs, and molding compounds, for the fabrication of composite parts. Each manufacturing technique requires different types of material systems, different processing conditions, and different tools for part fabrication. Figure 1.5 in Chapter 1 shows a list of the various types of most commonly used composites manufacturing techniques and Figure 2.1 in Chapter 2 shows the type of raw materials used in those manufacturing techniques. Each technique has its own advantages and disadvantages in terms of processing, part size, part shapes, part cost, etc. Part production success relies on the correct selection of a manufacturing technique as well as judicious selection of processing parameters. The main focus of this chapter is to describe emerging and commercially available manufacturing techniques in the field of thermoset- and thermoplastic-based composite materials. Various composites manufacturing techniques are discussed in terms of their limitations, advantages, methods of applying heat and pressure, type of raw materials used, and other important parameters. The basic knowledge of these processes will help in selecting the right process for an application. Section 6.2 briefly describes the manufacturing process selection criteria.

6.2 Manufacturing Process Selection Criteria

It is a monumental challenge for design and manufacturing engineers to select the right manufacturing process for the production of a part, the reason being that design and manufacturing engineers have so many choices in terms of raw materials and processing techniques to fabricate the part. This section briefly discusses the criteria for selecting a process. Selection of a process depends on the application need. The criteria for selecting a process depend on the production rate, cost, strength, and size and shape requirements of the part, as described below.

6.2.1 Production Rate/Speed

Depending on the application and market needs, the rate of production is different. For example, the automobile market requires a high rate of production, for example, 10,000 units per year (40 per day) to 5,000,000 per year (20,000 per day). In the aerospace market, production requirements are usually in the range of 10 to 100 per year. Similarly, there are composites manufacturing techniques that are suitable for low-volume and high-volume production environments. For example, hand lay-up and wet lay-up processes cannot be used for high-volume production, whereas compression molding (SMC) and injection molding are used to meet high-volume production needs.

6.2.2 Cost

Most consumer and automobile markets are cost sensitive and cannot afford higher production costs. Factors influencing cost are tooling, labor, raw materials, process cycle time, and assembly time. There are some composite processing techniques that are good at producing low-cost parts, while others are cost prohibitive. Determining the cost of a product is not an easy task and requires a thorough understanding of cost estimating techniques. The cost of a product is significantly affected by production volume needs as well. For example, compression molding (SMC) is selected over stamping of steel for the fabrication of automotive body panels when the production volume is less than 150,000 per year. For higher volume rates, steel stamping is preferred. Various cost-estimating techniques, as well as various parameters that affect the final cost of the products, are discussed in Chapter 11.

6.2.3 Performance

Each composite process utilizes different starting materials and therefore the final properties of the part are different. The strength of the composite part strongly depends on fiber type, fiber length, fiber orientation, and fiber content (60 to 70% is strongest, as a rule). For example, continuous fiber composites provide much higher stiffness and strength than shorter fiber composites. Depending on the application need, a suitable raw material and thus a suitable composite manufacturing technique are selected.

6.2.4 Size

The size of the structure is also a deciding factor in screening manufacturing processes. The automobile market typically requires smaller-sized components compared to the aerospace and marine industries. For small- to medium-sized components, closed moldings are preferred; whereas for large structures such as a boat hull, an open molding process is used. Table 6.1 reveals the suitability of composites manufacturing techniques in terms of product size.

6.2.5 Shape

The shape of a product also plays a deciding role in the selection of a production technique. For example, filament winding is most suitable for the manufacture of pressure vessels and cylindrical shapes. Pultrusion is very economical in producing long parts with uniform cross-section, such as circular and rectangular.

Table 6.1 characterizes each manufacturing method based on the above factors. The cost category of the part is shown when the manufacturing equipment is running at full capacity.

TABLE 6.1
Manufacturing Process Selection Criteria

Process	Production Speed	Cost	Strength	Size	Shape	Raw Material
Filament winding	Slow to fast	Low to high	High	Small to large	Cylindrical and axisymmetric	Continuous fibers with epoxy and polyester resins
Pultrusion	Fast	Low to medium	High (along longitudinal direction)	No restriction on length; small to medium size cross-section	Constant cross-section	Continuous fibers, usually with polyester and vinylester resins
Hand lay-up	Slow	High	High	Small to large	Simple to complex	Prepreg and fabric with epoxy resin
Wet lay-up	Slow	Medium	Medium to high	Medium to large	Simple to complex	Fabric/mat with polyester and epoxy resins
Spray-up	Medium to fast	Low	Low	Small to medium	Simple to complex	Short fiber with catalyzed resin
RTM	Medium	Low to medium	Medium	Small to medium	Simple to complex	Preform and fabric with vinylester and epoxy
SRIM	Fast	Low	Medium	Small to medium	Simple to complex	Fabric or preform with polyisocyanurate resin
Compression molding	Fast	Medium	Medium	Small to medium	Simple to complex	Molded compound (e.g., SMC, BMC)
Stamping	Fast	Low	Medium	Medium	Simple to contoured	Fabric impregnated with thermoplastic (tape)
Injection molding	Fast	Low to medium	Low to medium	Small	Complex	Pallets (short fiber with thermoplastic)
Roll wrapping	Medium to fast	Low to medium	High	Small to medium	Tubular	Prepregs

Manufacturing Techniques

The process selection criteria in Table 6.1 are useful in prescreening of the fabrication choices. For the final selection of a process, a detailed study in terms of the above variables (e.g., cost, speed, and size) is performed.

6.3 Product Fabrication Needs

To make a part, the four major items needed are:

1. Raw material
2. Tooling/mold
3. Heat
4. Pressure

Depending on the manufacturing process selected, a suitable raw material is chosen and laid on the tool/mold. Then, heat and pressure are applied to transform the raw material into the final shape. Heat and pressure requirements are different for different material systems. Solid materials such as metals or thermoplastics require a large amount of heat to melt the material for processing, whereas thermosets require less heat. In general, the higher the melting temperature of a material, the higher the temperature and pressure required for processing. For example, steel, which melts at 1200°C, requires higher temperatures and pressures to process the part. Aluminum, which melts at around 500°C, requires less heat and pressure for transforming the shape as compared to steel processing. Thermoplastics have melting temperatures in the range of 100 to 350°C and therefore require lesser amounts of heat and pressure as compared to steel and aluminum. Thermosets are in the liquid state at room temperature and therefore are easy to form and process. Thermosets require heat for rapid curing of the material. The temperature requirement for thermosets depends on resin formulation and cure kinetics. In composites, fibers are not melted and thus heat is required for proper consolidation of the matrix materials only.

The higher pressure and temperature requirements during a manufacturing process need strong and heavy tools, which increase the cost of tooling. In addition to higher tooling costs, the higher pressure and temperature requirements mandate special equipment, which is another source of increased processing cost. For example, the higher pressure requirement during SMC molding requires large and bulky equipment and usually costs more than $1 million. The ideal manufacturing process will be the one that requires extremely low amounts of heat and pressure and is quick to process in order to obtain significant processing cost savings.

Every process requires a set of tools to transform the raw material to the final shape. Therefore, the success of a production method relies on the quality of the tool. Section 6.4 discusses design parameters and fabrication methods for making various types of commonly used tools in the composites industry. The knowledge provided in the following section will be helpful in understanding the tooling needs for various composites manufacturing techniques.

6.4 Mold and Tool Making

The three most critical steps in developing a new product are product design, processing engineering, and mold engineering. Obviously, these are not independent, and product design engineers need to think about the constraints of the mold-makers and manufacturing engineers. Product designers can draw a wonderful design with very good aesthetics, but if the part cannot be manufactured economically, then there is no point to that great design.

Mold- and tool-making are a challenging segment of the composites manufacturing area. A tool transforms the raw material to a given shape. Without the tool or mold, the raw material cannot be shaped to the final dimension and size requirements of the part.

The type of tool requirement depends on the selection of a manufacturing technique. For example, filament winding uses mandrels for laying the raw material, RTM uses a closed mold, pultrusion uses a die, and the wet lay-up process uses an FRP mold for providing the desired shape in the final part. The quality and surface finish of the part heavily rely on the surface finish of the tool. Section 6.4.1 identifies the important criteria for mold design.

6.4.1 Mold Design Criteria

6.4.1.1 Shrinkage Allowance

In mold design, shrinkage of the composite material is taken into account to make sure that the end product is of the desired size and shape after the part is cured. Shrinkage is the reduction in volume or linear dimensions caused by curing the resin as well as by thermal contraction of the material. For a composite material as well as mold material, the shrinkage allowance is determined and factored into the design of the part and the mold.

6.4.1.2 Coefficient of Thermal Expansion of Tool Material and End Product

The coefficient of thermal expansion (CTE) is an important parameter for the mold design. Every material expands and contracts to a different extent when heated and cooled from a certain temperature. The CTE of the tool and the composite part should closely match to avoid residual stresses and

Manufacturing Techniques

dimensional inaccuracies in the end product. For a room-temperature cure system, the CTE consideration is not important.

6.4.1.3   Stiffness of the Mold

During part fabrication, the mold experiences significant pressure, especially in closed molding operations. Under such pressures, the mold should not deform; otherwise, it may cause distortion in the part. The mold should be stiff enough to take processing pressures.

6.4.1.4   Surface Finish Quality

The surface finish of the end product relies on the surface finish quality of the tool. To obtain Class A surface finish on the part, the tool surface should be of high quality. During part fabrication, the tool is waxed and dirt is removed to avoid the inclusion of any foreign material in the part. The die for the pultrusion process, and molds for making boat hulls and automotive parts have extremely high surface quality.

6.4.1.5   Draft and Corner Radii

On vertical surfaces, a 1° draft angle is recommended. A generous draft angle promotes better material flow, reduced warpage, and easier release from the mold. Sharp corners must be avoided during mold and part design. Minimum inside corner radii of 0.08 in. and minimum outside corner radii of 0.06 in. are recommended for better material flow along the corner as well as for ease in part removal.

6.4.2   Methods of Making Tools

6.4.2.1   Machining

Machining a block of material is more common in making molds and tools for small- to medium-sized parts. Mandrels for filament winding, dies for pultrusion, and molds for compression molding, RTM, and SRIM are made using this process. To make a mandrel, a steel rod is taken and then machined to obtain the desired diameter. The surface of the mandrel is then ground and chrome plated to get a smooth and glossy surface finish. To make molds for the various molding processes, metal blocks (mostly tool steels) are taken and machined using CNC and a grinding machine to get the desired shape. Data from finite element (FE) analysis or a CAD (computer-aided design) model is transferred into the machine and the desired surface is generated. For complicated shapes, electrical discharge machining (EDM) is used to generate the mold surface. The two main types of EDM are termed sinker or plunge, used for making mold or die cavities, and wire, used to cut shapes as needed for stamping dies. EDM is very common for making molds for injection molding operations. For closed molding operations such as RTM,

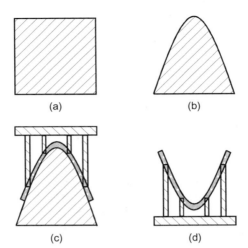

FIGURE 6.1
Illustration of steps in making a finished master model and laminated mold: (a) solid board; (b) master model after machining; (c) laminated molds with backup; and (d) finished mold.

SRIM, and injection molding, inlet and outlet ports or gates are provided for feeding and escaping of raw material. The outlet port (vent) has two functions: to allow air to escape from the mold and to ensure proper mold filling with the resin material. Heating and cooling devices are incorporated into the mold. Closed molding operations allow production of net-shape or near-net-shape parts.

For prototype building operations, wood, styrofoam, plastic, and other materials are machined and used as molds.

6.4.2.2 FRP Tooling for Open Molding Processes

Fiber reinforced plastic (FRP) toolings are primarily manufactured for open molding operations such as hand lay-up, wet lay-up, and spray-up processes. Open molds are made from a master pattern. Both the molds with which composite components are built and the master pattern (models or plug) from which the molds are created are critical to the quality of end product. The master pattern could be an existing part such as boat hull, automotive fender, or hatch cover, or could be made by machining a block of metal, wood, plastic, foam, or any other material as shown in the Figure 6.1. Creation of master model from a solid board by machine processing is shown in Figures 6.1a and b. The master model should be glossy and defect-free to reduce the amount of sanding and buffing on the mold.

Once the master model is ready, it is waxed with release agent for easy removal of the mold. The master is coated with release wax three or four times in alternate directions and allowed to harden after each layer is applied.

The next step is to apply a tooling gel coat on the surface of the master. The tooling gel coat provides a hard, glossy, and long-lasting surface on the

TABLE 6.2

CTE and Service Temperatures of Various Tooling Materials

Tooling Material	CTE (µin./in. -°F)	Maximum Service Temperature (°F)
Stainless steel	8–12	1000
Aluminum alloys	12–13.5	300–500
Room temp. cure carbon/epoxy prepreg	1.4	300–400
Intermediate temp. cure carbon/epoxy prepreg	1.4	300–400
Carbon/cyanate ester prepreg	1.5–2.0	450–700
Carbon/BMI prepreg	2.0–3.0	450–500
Room temp. cure glass/epoxy prepreg	7.0–8.0	300–400
Intermediate temp. cure glass/epoxy prepreg	7.0–8.0	300–400
Epoxy-based tooling board	30–40	150–400
Urethane-based tooling foam	35–50	250–300

mold. It is applied using a brush or spraying equipment. The gel coat is allowed to gel before applying any laminating material. To make sure the gel coat has properly gelled, the surface is lightly touched with a finger. If the finger does not stick or does not leave a slight fingerprint mark on the gel surface, it is ready for lamination. After the gel coat is ready, a spraygun is used to laminate short fiber composites. Prepreg material or wet fabric can also be used for lamination. For aerospace applications, prepregs are generally used for lamination. For making bathtubs or boat hulls, a spray-up process or combination of spray-up and wet fabric lamination is used for making the mold. For large and stiffness-critical structures, wood or foam core material is embedded into the lamination to achieve a sandwich structure. Various types of tooling materials, as listed in Table 6.2, are available as a backing material for the mold.

Once the mold is prepared, it is strengthened with a backup structure called a cradle to support the mold, as shown in Figure 6.1c. Egg-crate is also used to support the mold. The cradle can be constructed of wood or steel; however, it is important that the cradle be insulated from the mold at attachment. This can be accomplished using spacers between the mold and the cradle. Spacers can be made of cardboard, foam, or coremat and placed wherever the cradle comes into contact with the mold. The cradle is fixed into the mold using fabric and resin lamination. Care is taken so that resin does not seep through the spacer material.

Figure 6.2 is a photograph of a tool shop consisting of master models, laminated molds with backup structures, checking templates, etc. Figure 6.3 shows the prototype of a car shape generated by CNC machining of aero-mark 80 liquid board. Inexpensive and lightweight 8-lb density foam was used as a filler material and 2-in. thick liquid board was applied over the machined foam structure. The use of this lightweight foam over a solid modeling board eliminates the need for heavy and expensive base plates and produces substantial savings in material costs.

FIGURE 6.2
Master models and laminated molds. (Courtesy of Lucas Industries.)

FIGURE 6.3
Generation of a car shape from a liquid board. (Courtesy of Lucas Industries.)

6.4.3 Tooling Guidelines for Closed Molding Operations

For closed molding operations such as RTM, SRIM, and injection molding, the mold material should be strong enough to take the clamping force as well as injection pressure. Stresses induced by clamping forces on the mold can be high enough to cause appreciable distortion on the part and mold.

Manufacturing Techniques

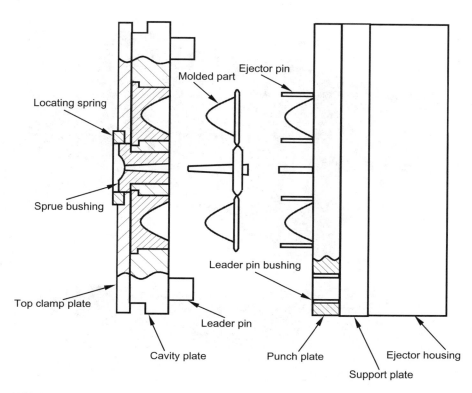

FIGURE 6.4
Two-plate mold for injection molding process.

They should be checked using conservative strength of materials analyses to ensure that mold part deflections do not cause out-of-tolerance moldings. For closed molding operations, creating a gate location or inlet and outlet port location is very critical in the mold design. For RTM processes, various computer models have been developed to numerically simulate resin flow inside the mold. These simulation models are used during the mold design phase to predict the optimal locations for inlet ports and vents, and optimal injection sequencing in the case of multiple ports, thus achieving the goals of minimal inlet pressure and fill time and the elimination of incomplete mold filling or dry spots. Simulation models identify potential trouble spots and allow the designer to evaluate various injection strategies on the computer — instead of the more costly empirical approach, where an expensive mold may have to be modified significantly or abandoned. Computer models are extensively used for thermoplastic injection molding processes. Molds for injection molding processes are very expensive and these computer models save significant mold design cost. Figure 6.4 shows a two-plate mold for injection molding process. The two-plate mold is a simple form and is most common in injection molding industries. For the two-plate mold in

110   Composites Manufacturing: Materials, Product, and Process Engineering

FIGURE 6.5
Tooling for fiberglass/epoxy snowboard. (Courtesy of Radius Engineering, Inc.)

Figure 6.4, edge gates are shown. The raw material is injected through a sprue bushing and reaches the cavities through runners. The purpose of the sprue, runner, and gate systems is to transfer liquid resin or melt uniformly and with minimum pressure and temperature drops in each cavity. Uniformity of flow refers to an equal flow rate through each gate and thus equal pressure at the cavity entrance. Multiple cavities are quite common in the injection molding industry. In RTM, the use of multiple cavities is less common and used when small parts are manufactured. For example, four hockey sticks can be manufactured simultaneously in an RTM process with a production cycle time of about 14 min from molding to deloading.[1] For making large parts in an RTM process, single-cavity molds are made with multiple ports. Multiple ports are used to speed cycle times as well as mitigate the need for high injection pressures. Figure 6.5 shows a mold for making snowboards. The molds were integrally heated and the lower mold half contains approximately 1400 vacuum vent holes for vacuforming a thermoplastic face sheet immediately prior to lamination of the fiberglass/epoxy urethane cored snowboard. Figure 6.6 shows match molds for making carbon fiber tennis rackets. Prepreg was used with an internal pressure bladder to make the part. Tooling for aerodynamic helicopter fairing components is shown in

Manufacturing Techniques

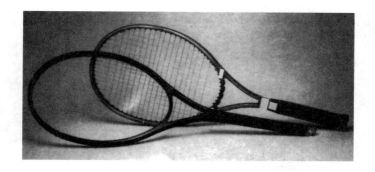

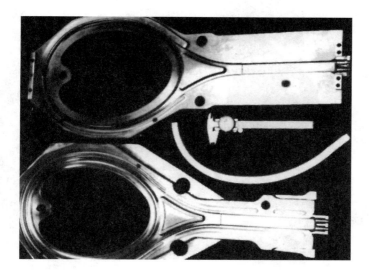

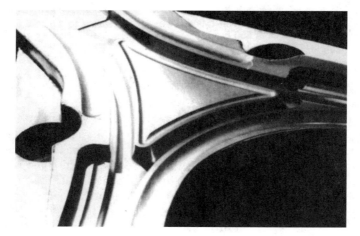

FIGURE 6.6
Tooling for carbon fiber tennis racquet. (Courtesy of Radius Engineering, Inc.)

FIGURE 6.7
Tooling for aerodynamic helicopter fairing components. (Courtesy of Radius Engineering, Inc.)

Figure 6.7. These cone and dome tools were used for the production of aerodynamic fairings for helicopter external fuel tanks. The fairings were constructed of prepregged fiberglass epoxy. High fiber volume and low void content parts were produced in the matched tool sets. Figure 6.8 shows aluminum tools for making helicopter tail rotor parts. The carbon/epoxy tail rotor part is shown at the bottom of the photograph. This part is made by the RTM process. The top two items in Figure 6.8 show the upper and lower halves of the mold and the third item from the top shows three aluminum mandrels placed inside the mold to create rib sections in the rotor part. Carbon fabrics are wrapped around these mandrels and placed inside the mold. The mold is then closed and epoxy resin is injected to consolidate the part. After curing, the part is removed from the mold as are the mandrels.

Manufacturing Techniques

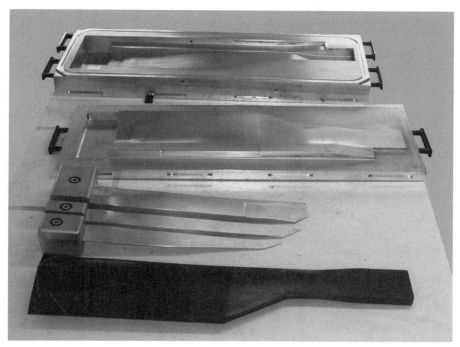

FIGURE 6.8
Tooling for making helicopter tail rotor. The carbon/epoxy tail rotor is shown at the bottom. (Courtesy of Radius Engineering, Inc.)

Guidelines for designing inlet ports or gate locations for closed molding operations include:

1. For thin-walled parts having large flow length-to-thickness ratios such as in hockey sticks, create two or more gates to obtain equal flow distances and to avoid flow distribution problems.

2. Locate gates where the flow of resin or melt proceeds uniformly along the greatest dimension as shown in Figure 6.9. For a mold with a single gate, locate the gate on the short side; and for multiple gates, locate gates on the long side, as shown in Figures 6.9, to divide the area into equal subsections. To make an 1-in. square, 52-in. long hockey sticks having 0.072-in. wall thickness, it might take about 30 min to complete the mold filling in an RTM process if resin is injected from one end.[1] Moreover, the rejection rate for production may be high in a single-gate mold because of dry spots. To speed the process, multiple ports along the length of the stick can be used. Multiple ports can be created using an injection runner parallel to the stick with a film gate that transmits resin from the runner to the preform.

3. Avoid placing gates on the exposed side of the part where the appearance of the part is crucial.

114   Composites Manufacturing: Materials, Product, and Process Engineering

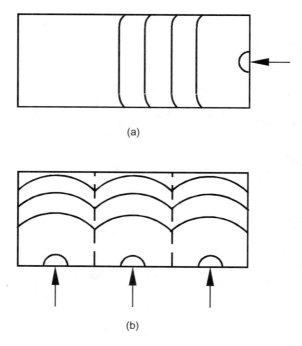

FIGURE 6.9
Illustration of flow front for single and multiple cavities: (a) flow front in a single gate at short edge; and (b) flow front in multiple gates.

4. Locate gates in such a way that weld lines are avoided. This is critical in injection molding as well as compression molding of SMC parts. Weld lines result when two flow fronts come together head-on or when two parallel streams merge (Figure 6.10). Weld lines are the weaker area of the part and affect the appearance. A high filling rate tends to reduce the adverse effects of weld lines.

5. Position gates so that the air displaced during resin flow is naturally impelled through the vent (outlet port) or through the parting plane.

6. Locate gates in such a way so as to avoid stagnant areas that are filled late and can have voids or dry spots (Figure 6.11). As shown in Figure 6.11b, incoming resin can jet across the mold, leaving unfilled zones of trapped air or voids. This scenario should be avoided, especially in the case of injection molding of short fiber composites. In Figure 6.11a, the resin front experiences flow resistance and avoids the problem of jetting.

7. Locate gates in the thickest section to avoid incomplete filling or sink marks. Then use a high injection speed to ensure that the furthermost thin section is filled.

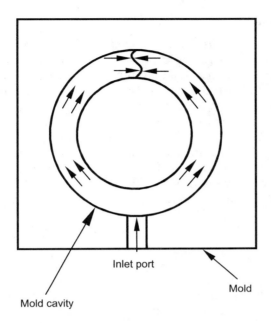

FIGURE 6.10
Illustration of weld lines during mold filling process.

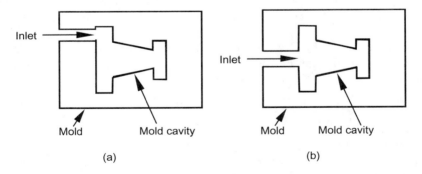

FIGURE 6.11
Recommended (a) and not recommended (b) gate locations.

6.5 Basic Steps in a Composites Manufacturing Process

There are four basic steps involved in composites part fabrication: wetting/impregnation, lay-up, consolidation, and solidification. All composites manufacturing processes involve the same four steps, although they are accomplished in different ways.

6.5.1 Impregnation

In this step, fibers and resins are mixed together to form a lamina. For example, in a filament winding process, fibers are passed through the resin bath for impregnation. In a hand lay-up process, prepregs that are already impregnated by the material supplier in a controlled environment are used. In a wet lay-up process, each fabric layer is wetted with resin using a squeezing roller for proper impregnation. The purpose of this step is to make sure that the resin flows entirely around all fibers. Viscosity, surface tension, and capillary action are the main parameters affecting the impregnation process. Thermosets, which have viscosities in the range of 10 e1 to 10e4 cp are easier to wet-out. Viscosities of thermoplastics fall in the range of 10e4 to 10e8 cp and require a greater amount of pressure for good impregnation.

6.5.2 Lay-up

In this step, composite laminates are formed by placing fiber resin mixtures or prepregs at desired angles and at places where they are needed. The desired composite thickness is built up by placing various layers of the fiber and resin mixture. In filament winding, the desired fiber distribution is obtained by the relative motions of the mandrel and carriage unit. In a prepreg lay-up process, prepregs are laid at a specific fiber orientation, either manually or by machine. In an RTM process, the preform has built-in fiber architecture, either from a braiding operation or from some other machine, and resin is injected to form the laminate.

The purpose of this step is to achieve the desired fiber architecture as dictated by the design. Performance of a composite structure relies heavily on fiber orientation and lay-up sequence.

6.5.3 Consolidation

This step involves creating intimate contact between each layer of prepreg or lamina. This step ensures that all the entrapped air is removed between layers during processing. Consolidation is a very important step in obtaining a good quality part. Poorly consolidated parts will have voids and dry spots. Consolidation of continuous fiber composites involves two important processes: resin flow through porous media and elastic fiber deformation.[2,3] During the consolidation process, applied pressure is shared by both resin and fiber structure. Initially, however, the applied pressure is carried solely by the resin (zero fiber elastic deformation). Fibers go through elastic deformation when the compressive pressure increases and resins flow out toward the boundary. There are various consolidation models[4,5] that ignore the fiber deformation and consider only resin flow.

6.5.4 Solidification

The final step is solidification, which may take less than a minute for thermoplastics or may take up to 120 min for thermosets. Vacuum or pressure is maintained during this period. The lower the solidification time, the higher the production rate achievable by the process. In thermoset composites, the rate of solidification depends on the resin formulation and cure kinetics. Heat is supplied during processing to expedite the cure rate of the resin. In thermoset resins, usually the higher the cure temperature, the faster the cross-linking process. In thermoplastics, there is no chemical change during solidification and therefore solidification requires the least amount of time. In thermoplastics processing, the rate of solidification depends on the cooling rate of the process. In thermoset composites, the temperature is raised to obtain faster solidification; whereas in thermoplastics processing, the temperature is lowered to obtain a rigid part.

The above four steps are common in thermoset as well as thermoplastic composites processing. The methods of applying heat and pressure, as well as creating a desired fiber distribution, are different for different manufacturing methods; this is discussed in Sections 6.8 and 6.9. Section 6.6 discusses the advantages and disadvantages of thermoset and thermoplastic composites processing techniques.

6.6 Advantages and Disadvantages of Thermoset and Thermoplastic Composites Processing

6.6.1 Advantages of Thermoset Composites Processing

The common thermoset resins are epoxy, polyester, and vinylester. These materials could be one-part or two-part systems and are generally in the liquid state at room temperature. These resin systems are then cured at elevated temperatures or sometimes at room temperature to get the final shape. Manufacturing methods for processing thermoset composites provide the following advantages.

1. Processing of thermoset composites is much easier because the initial resin system is in the liquid state.
2. Fibers are easy to wet with thermosets, thus voids and porosities are less.
3. Heat and pressure requirements are less in the processing of thermoset composites than thermoplastic composites, thus providing energy savings.
4. A simple low-cost tooling system can be used to process thermoset composites.

6.6.2 Disadvantages of Thermoset Composites Processing

1. Thermoset composite processing requires a lengthy cure time and thus results in lower production rates than thermoplastics.
2. Once cured and solidified, thermoset composite parts cannot be reformed to obtain other shapes.
3. Recycling of thermoset composites is an issue.

6.6.3 Advantages of Thermoplastic Composites Processing

The initial raw material in thermoplastic composites is in solid state and needs to be melted to obtain the final product. The advantages of processing thermoplastic composites include:

1. The process cycle time is usually very short because there is no chemical reaction during processing, and therefore can be used for high-volume production methods. For example, process cycle time for injection molding is less than 1 min and therefore very suitable for automotive-type markets where production rate requirements are usually high.
2. Thermoplastic composites can be reshaped and reformed with the application of heat and pressure.
3. Thermoplastic composites are easy to recycle.

6.6.4 Disadvantages of Thermoplastic Composites Processing

1. Thermoplastic composites require heavy and strong tooling for processing. Moreover, the cost of tooling is very high in thermoplastic composites manufacturing processes. For example, the tooling cost in the injection molding process is typically more than $50,000, whereas a mandrel for the filament winding process costs less than $500.
2. Thermoplastic composites are not easy to process and sometimes require sophisticated equipment to apply heat and pressure.

6.7 Composites Manufacturing Processes

Composites manufacturing processes can be broadly subdivided into two main manufacturing categories: manufacturing processes for thermoset composites and manufacturing processes for thermoplastic composites. In terms

Manufacturing Techniques

of commercial applications, thermoset composite parts dominate the composite market. About 75% of all composite products are made from thermoset resins. Thermoset composite processes are much more mature than their thermoplastic counterparts mainly because of the widespread use of thermoset composites as well as its advantages over thermoplastic composite processing techniques. The first use of thermoset composites (glass fiber with unsaturated polyester) occurred in the early 1940s, whereas the use of thermoplastic composites came much later.

The manufacturing processes described in this chapter are discussed under the following headings:

1. Major applications of the process
2. Basic raw materials used in the process
3. Tooling and mold requirements
4. Making of the part
5. Methods of applying heat and pressure
6. Basic processing steps
7. Advantages of the process
8. Limitations of the process

We first discuss the manufacturing processes for making thermoset composite parts under the above eight headings. Thermoplastic composite processes are then discussed.

6.8 Manufacturing Processes for Thermoset Composites

In terms of commercial applications, more than 75% of all composites are made of thermoset composites. Their uses predominate in the aerospace, automotive, marine, boat, sporting goods, and consumer markets. There are several dominant thermoset composite processing methods available on the market, each with its pros and cons. The advantages and limitations of each method are also included for each manufacturing process. The commercially available manufacturing techniques are described below. The order of description of a process below does not mean the order of importance of the process.

6.8.1 Prepreg Lay-Up Process

The hand lay-up process is mainly divided into two major methods: wet lay-up and prepreg lay-up. The wet lay-up process is discussed in Section 6.8.2.

Here, the prepreg lay-up process, which is very common in the aerospace industry, is discussed. It is also called the autoclave processing or vacuum bagging process. Complicated shapes with very high fiber volume fractions can be manufactured using this process. It is an open molding process with low-volume capability. In this process, prepregs are cut, laid down in the desired fiber orientation on a tool, and then vacuum bagged. After vacuum bagging, the composite with the mold is put inside an oven or autoclave and then heat and pressure are applied for curing and consolidation of the part.

The prepreg lay-up or autoclave process is very labor intensive. Labor costs are 50 to 100 times greater than filament winding, pultrusion, and other high-volume processes; however, for building prototype parts and small quantity runs, the prepreg lay-up process provides advantages over other processes.

6.8.1.1 Major Applications

The prepreg lay-up process is widely used in the aerospace industry as well as for making prototype parts. Wing structures, radomes, yacht parts, and sporting goods are made using this process. Figure 6.12 shows a variety of aircraft radomes such as sharknose, conical, varying lengths, solid laminates, and sandwich constructions with dielectrically loaded foam cores. Radomes are used at the nose and tail ends of aircraft. Figure 6.13 shows glass/epoxy/honeycomb sandwich fairings for the airbus A330/340 flap tracks. Figure 6.14 shows

FIGURE 6.12
Variety of aircraft radomes. (Courtesy of Marion Composites.)

Manufacturing Techniques 121

FIGURE 6.13
Large glass/epoxy/honeycomb sandwich fairings for the Airbus 330/340 flap tracks. (Courtesy of Marion Composites.)

FIGURE 6.14
Main landing gear door being prepared for second stage bond. (Courtesy of Marion Composites.)

landing gear doors for the C-17 airlifter. Each plane requires eight main gear doors and four nose gear doors. In Figure 6.14, the main landing gear door being prepared for second stage bond is shown.

6.8.1.2 Basic Raw Materials

Graphite/epoxy prepregs are the most commonly used materials for the prepreg lay-up process. Glass/epoxy and Kevlar/epoxy are also used but their use is much less than carbon/epoxy prepregs. The main reason is that carbon/epoxy is much lighter and stronger than other prepreg materials and provides greater mass savings in the component. Because this process is widely used in the aerospace industry, where weight is a critical design factor, carbon fiber prepreg is the material of choice. Moreover, in terms of cost, there is no significant price difference between carbon/epoxy prepregs and other prepregs.

Other than epoxy, high-temperature resins such as polyimides, polycyanate, and BMI are also used in prepreg systems.

6.8.1.3 Tooling Requirements

The tooling for the prepreg lay-up process is an open mold on which prepregs are laid in the desired fiber orientation and sequence. For prototype building purposes, tools are made by machining metals, woods, and plastics. For the manufacture of aerospace components, the tooling material is mostly the composite tooling material such as carbon/epoxy prepregs, carbon/cyanate ester prepregs, carbon/BMI prepregs, glass/epoxy prepregs, glass/cyanate ester prepregs, epoxy- and urethane-based tooling board, etc. Wide varieties of prepregs with room-temperature, intermediate-, and high-temperature cure, as listed in Table 6.1, are used for making the mold. Steel is also a common material for making tools for prepreg lay-up process.

6.8.1.4 Making of the Part

The raw material for this process is prepreg material, which is kept refrigerated. To make the composite part, prepreg is removed from the refrigerator and brought slowly to room temperature. In general, thawing is done in the original package to avoid condensation. Once the prepreg is brought to room temperature, it is cut to the desired length and shape. For cutting, the prepreg is placed on a cutting board and then, using a steel ruler and utility knife, the prepreg is cut. For aerospace applications, this operation takes place in a very neat and clean atmosphere under controlled humidity and temperature conditions. Dust is prohibited in the room. Workers are required to cover their heads, shoes, and body with clean clothing accessories. Figure 6.15 shows a cleanroom facility. For production parts of decent quantity, automated cutting machines are used for cutting prepregs. In this case, the prepreg is laid on the cutting table and using the reciprocating action of a knife, laser, or ultrasonic cutter, the prepreg is cut into the desired pattern. These machines are computer controlled and utilize software for ply cutting optimization. The software minimizes scrap and provides repeatability and consistency in the ply cutting operation. The machine can cut several layers of prepregs stacked together at one time and thus creates efficiency. Predominantly

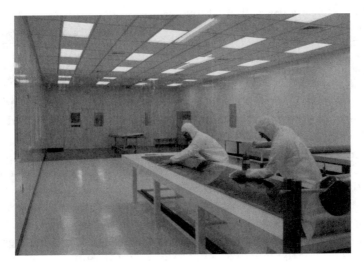

FIGURE 6.15
A cleanroom facility for composites part fabrication. (Courtesy of Lunn Industries.)

unidirectional fiber prepregs are used for part fabrication. Plies are cut in such a way as to provide the desired fiber orientation. In some cases, prepregs made of fabrics are used.

Part fabrication is done by laying the prepregs on top of an open mold. Release agent is applied to the mold for easy removal of the part. The backing film is first removed from the prepreg and then prepregs are laid in the sequence dictated in the manufacturing chart. For aerospace components and for parts of greater safety issues, quality control personnel check the ply sequence after every few layers are laid down. After applying each prepreg

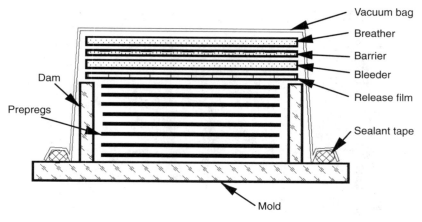

FIGURE 6.16
Vacuum bagging for prepreg lay-up process.

layer, it is necessary to ensure that there is no entrapped air. Squeezing rollers are used to remove entrapped air and to create intimate contact.

Once all the prepregs are laid in the desired sequence and fiber orientation, vacuum bagging preparations are made as shown in Figure 6.16 for curing and consolidation of the part. The steps required for vacuum bagging are:

1. Apply release film on top of all the prepreg. The release film is a perforated film that allows entrapped air, excess resins, and volatiles to escape.
2. Apply bleeder, a porous fabric, on top of the release film. The function of the bleeder is to absorb moisture and excess resin coming from the stack of prepregs.
3. Apply barrier film on top of the bleeder. The film is similar to release film except that it is not perforated or porous.
4. Apply breather layer, a porous fabric similar to the bleeder. The function of the breather is to create even pressure around the part and at the same time allowing air and volatiles to escape.
5. The final layer is a vacuum bag. It is an expendable polyamide (PA) film or reusable elastomer. This film is sealed on all sides of the stacked prepreg using seal tape. If the mold is porous, it is possible to enclose the entire mold inside the vacuum bag. Seal tape is a 0.5- to 1-in.-wide rubbery material that sticks to both the mold and the bagging material. A nozzle is inserted into the vacuum bag and connected to a vacuum hose for creating vacuum inside the bag.

Peel ply fabrics are applied on the top of prepreg layers if consolidated parts are to be adhesively bonded at a later stage. The peel ply creates a good bondable surface on the fabricated part. Sometimes, co-curing of the

Manufacturing Techniques 125

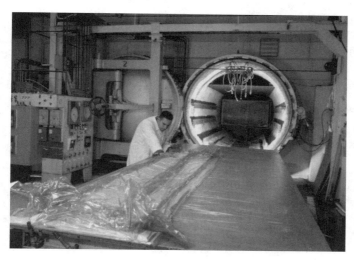

FIGURE 6.17
Vacuum bagged aerospace part ready to go inside an autoclave. (Courtesy of Lunn Industries.)

FIGURE 6.18
Manufacturing facility containing various autoclaves. (Courtesy of Marion Composites.)

various parts is done to eliminate any extra processing step and to reduce the number of parts. Figure 6.17 shows a vacuum bagged part ready to go inside an autoclave for curing process. Figure 6.18 shows a manufacturing facility containing various autoclaves.

See Section 6.8.1.5 on how heat and pressure are created during the prepreg lay-up process. Once the part is cured, the vacuum bag is removed and the part is taken out.

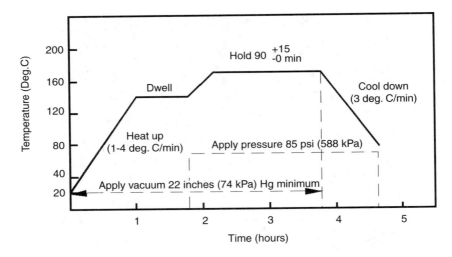

FIGURE 6.19
Typical cure cycle during the autoclave process.

6.8.1.5 Methods of Applying Heat and Pressure

After lamination and bagging, the mold is placed inside an autoclave for curing and consolidation. An autoclave, similar to a pressure vessel, can maintain the desired pressure and temperature inside the chamber for processing of the composite. A typical cure cycle is shown in Figure 6.19. The cure cycle depends on the type of resin material and the thickness and geometry of the part.

The pressure is created in two ways: using the vacuum bag as well as the external pressure inside the autoclave. The vacuum bag creates a vacuum inside the bagging material and thus helps in proper consolidation. To create vacuum inside the bag, the nozzle in the bagging system is connected to the vacuum pump using a hose. The vacuum pump generates the desired vacuum.

External pressure inside the autoclave is created by injecting pressurized air or nitrogen. Nitrogen is preferred for cases in which curing is done at high temperature to avoid burning or fire. Thus, the external pressure outside the bag and the vacuum inside the bag creates sufficient pressure to compact the laminate against the mold and create intimate contact between each layer.

The heat for curing comes from heated air or nitrogen. The pressurized gas supplied to the chamber comes heated to increase the temperature inside the autoclave. Cartridge heaters can also be placed inside the autoclave for increasing the temperature. The temperature and pressure inside the autoclave are controlled by computer-controlled equipment located outside the autoclave. The user sets up the cure profile as shown in Figure 6.19 and the computer controls both parameters using an on/off switch.

As shown in Figure 6.19, vacuum is applied in the bagging system first and then the temperature is raised to a level to increase the resin flow. The heating rate is usually 2°C/min to 4°C/min. After dwelling for some time at the dwell temperature, the temperature is further raised to another level for curing of the composites. During this stage, pressure is applied to the outside of the bagging system and maintained for about 2 hr, depending on the requirements.

6.8.1.6 Basic Processing Steps

The basic steps in making composite components by prepreg lay-up process are summarized as follows.

1. The prepreg is removed from the refrigerator and is kept at room temperature for thawing.
2. The prepreg is laid on the cutting table and cut to the desired size and orientation.
3. The mold is cleaned and then release agent is applied to the mold surface.
4. Backing paper from the prepreg is removed and the prepreg is laid on the mold surface in the sequence mentioned in the manufacturing chart.
5. Entrapped air between prepreg sheets is removed using a squeezing roller after applying each prepreg sheet.
6. After applying all the prepreg sheets, vacuum bagging arrangements are made by applying release film, bleeder, barrier film, breather, and bagging materials as mentioned in Section 6.8.1.5.
7. The entire assembly is then placed into the autoclave using a trolley if the structure is large.
8. Connections to thermocouples and vacuum hoses are made and the autoclave door is closed.
9. The cure cycle data are entered into a computer-controlled machine and followed.
10. After cooling, the vacuum bag is removed and the part is taken out.

6.8.1.7 Typical Manufacturing Challenges

Some of the challenges that manufacturing engineers face during the prepreg lay-up process are listed below.

1. Maintaining accurate fiber orientations in the part is difficult because prepregs are laid down by hand. Automated tape placement equipment can be used for precise fiber orientation control.
2. Obtaining void-free parts is a challenge during this process. Voids are caused by entrapped air between layers.
3. Achieving warpage- or distortion-free parts during the prepreg lay-up process is challenging. Warpage is caused by built-in residual stresses during processing.

6.8.1.8 Advantages of the Prepreg Lay-Up Process

The prepreg lay-up process is very common in the aerospace industry and offers the following advantages:

1. It allows production of high fiber volume fraction (more than 60%) composite parts because of the use of prepregs. Prepregs usually have more than 60% fiber volume fraction.
2. Simple to complex parts can be easily manufactured using this process.
3. This process is very suitable for making prototype parts. It has the advantage of low tooling cost but the process requires high capital investment for the autoclave.
4. Very strong and stiff parts can be fabricated using this process.

6.8.1.9 Limitations of the Prepreg Lay-Up Process

Although prepreg lay-up is a mature process, it has the following limitations:

1. It is very labor intensive and is not suitable for high-volume production applications.
2. The parts produced by the prepreg lay-up process are expensive.

6.8.2 Wet Lay-Up Process

In the early days, the wet lay-up process was the dominant fabrication method for the making of composite parts. It is still widely used in the marine industry as well as for making prototype parts. This process is labor intensive and has concerns for styrene emission because of its open mold nature. In this process, liquid resin is applied to the mold and then reinforcement is placed on top. A roller is used to impregnate the fiber with the resin. Another resin and reinforcement layer is applied until a suitable thickness builds up. It is a very flexible process that allows the user to optimize the part by placing different types of fabric and mat materials. Because the reinforcement is placed manually, it is also called the hand lay-up process. This process requires little capital investment and expertise and is therefore easy to use.

6.8.2.1 Major Applications

On a commercial scale, this process is widely used for making boats, windmill blades, storage tanks, and swimming pools. Because of its process simplicity and little capital investment, this process is widely used for making prototype parts. Test coupons for performing various tests for the evaluation of reinforcements as well as resins are made using this process. Simple to complex shapes can be made using this process. Figure 6.20 shows a 41-ft, 9-in. and 27-ft, 1-in. sport boat and Figure 6.21 shows a 41-ft cruiser boat

Manufacturing Techniques

FIGURE 6.20
Sports boats having a 41-ft, 9-in. centerline length (top) and 27-ft, 1-in. centerline length (bottom); approximate dry weights are 13,100 lb (5942 kg) and 5250 lb (2381 kg), respectively. (Courtesy of Thunderbird Products, Decatur, IN.)

FIGURE 6.21
A 41-ft cruiser boat; approximate dry weight is 18,520 lb (8401 kg). (Courtesy of Thunderbird Products, Decatur, IN.)

FIGURE 6.22
A 72-ft yacht. (Courtesy of Mikelson Yachts, San Diego, CA.)

FIGURE 6.23
Crew members applying cross-linked foam core to a 72-ft yacht. (Courtesy of Mikelson Yachts, San Diego, CA.)

made using this process. A 72-ft yacht is shown in Figure 6.22. Application of foam core to a 72-ft yacht is shown in Figure 6.23. The hulls are constructed by laminating fiberglass layers with core materials such as balsa or foam. Built-in wooden frames may be provided to strengthen the hull. In a single skin panel without any core material, the longitudinal and transverse framings are complex and heavier than sandwich panel. Sandwich panel can support larger spans than single skin panels.

Manufacturing Techniques

6.8.2.2 Basic Raw Materials

Woven fabrics of glass, Kevlar, and carbon fibers are used as reinforcing material, with E-glass predominating in the commercial sector. Epoxy, polyester, and vinylester resins are used during the wet lay-up process, depending on the requirements of the part. Polyester resin is the most common resin in building boats and other commercial items. Glass rovings are also used in the making of boat hulls. The roving is chopped using a spraygun and laid over the mold.

6.8.2.3 Tooling Requirements

The mold design for the wet lay-up process is very simple as compared to other manufacturing processes because the process requires mostly a room-temperature cure environment with low pressures. Steel, wood, GRP, and other materials are used as mold materials for prototyping purposes. The mold can be a male or female mold. To make shower bathtubs, a male mold is used. In the boating industry, a single-sided female mold made from FRP (fiber-reinforced plastic) is used to make yacht hulls. The outer shell of the mold is stiffened by a wood frame. The mold is made by taking the reversal of a male pattern. Several different hull sizes can be made using the same mold. The length of the mold is shortened or lengthened using inserts and mold secondaries such as windows, air vents, and propeller tunnels.

6.8.2.4 Making of the Part

A schematic of the wet lay-up process is shown in Figure 6.24, where the thickness of the composite part is built up by applying a series of reinforcing layers and liquid resin layers. A roller is used to squeeze out excess resin and create uniform distribution of the resin throughout the surface. By the squeezing action of the roller, homogeneous fiber wetting is obtained. The

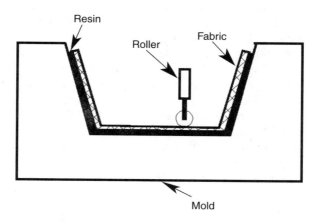

FIGURE 6.24
Schematic of the wet lay-up process.

part is then cured mostly at room temperature and, once solidified, it is removed from the mold.

The overall process cycle time is dictated by the size of the component as well as the resin formulation used. For large-sized structures such as boats, room-temperature curing is commonly used. If the laminate to be made is thick, then the wall thickness is built up in stages to allow the exotherm to take place without overheating. Under these circumstances, it is common to finish off the day's work with a peel ply, which is subsequently removed to expose a clean and better surface for bonding the next layer.

Quality control in the wet lay-up process is relatively difficult. The quality of the final part is highly dependent on operator skill. The process remains an important one for the boat-building process, although increasingly stringent emissions regulations are forcing several manufacturers to explore the use of closed mold alternatives such as RTM and VARTM.

To obtain further insight into the wet lay-up process, boat hull fabrication using this technique is discussed here.

To make a boat hull using this process, a release agent is applied to the mold surface to facilitate the demolding operation. A gel coat is then applied using a brush or spraygun. The gel coat improves the surface finish quality and provides coloring as needed. A polyester gel coat is commonly used and includes a thixotropic additive and the pigmentation for the desired color finish. The gel coat is then cured to avoid print-through of the laminate. The gel coat provides a Class A surface finish on the hull surface. Once the gel coat hardens, a skin coat is applied using a spray-up process to obtain improved corrosion and chemical resistance.[6] The skin coat is nothing more than a layer of chopped glass mat with vinylester resin. Although vinylester is almost double the cost of polyester resin, it is used because it has better corrosion resistance than polyester. After applying the skin coat, it is cured overnight. The cured skin coat acts as a barrier to the structural laminate.

Lamination begins the next day by cutting stitched bidirectional fabric and laying it on the mold surface in such a way that it covers the entire surface. Usually, two stitched bidirectional fabric layers, depending on the size of the hull, are placed on the mold as a first laminating skin. Resin is then worked into the reinforcement using a brush, roller, or flow coater. Extra fabric is placed into spots that require additional strength, such as the centerline and seacock penetrations. It depends on worker skill to make sure that there is no dry fiber or entrapped air. The process of applying the resin and uniformly wetting the fiber is very labor intensive. To make the process more efficient, the fabric is first wetted on a table and then placed on the mold, or an in-house impregnating machine is used to uniformly apply the resin on the fabric and then it is placed on the mold. In an impregnating machine, as shown in Figure 6.25, the fabric passes through two rollers, which wet the fabric. A fabric impregnator precisely controls the resin-to-glass ratio. With this machine, rapid lamination can be achieved. The machine can wet laminate about 1000 lb per hour.

Manufacturing Techniques

FIGURE 6.25
Fabric impregnator demonstrating impregnation of a glass fabric. (Courtesy of Merritt Boat and Engine Works, Pompano Beach, FL.)

FIGURE 6.26
Demonstration of vacuum bagging of core materials. Vacuum bagging provides a tight skin-to-core bond and low resin content. (Courtesy of Merritt Boat and Engine Works, Pompano Beach, FL.)

After fiber wet-out, the laminate is allowed to cure. Following the curing process, adhesive is spread over the laminate and then balsa core, or any suitable core, is applied to create a sandwich structure. The entire structure is then vacuum bagged as shown in Figure 6.26. The vacuum bagging is done at a pressure of about 15 in. Hg for 2 hr. Vacuum bagging ensures good contact between the core and the laminate. On top of the core material, the final skin layer is laminated by placing about two bidirectional fabric layers. The fabric is impregnated with the resin in the same manner as the first laminated skin was prepared. The laminate is then cured at room temperature. The total thickness of the hull thus obtained is about 1.5-in. Bulkheads, motor mounts, stringers, and decks are assembled while the hull is still in the mold.

The major limitation of wet lay-up process is that the molding has only one smooth surface. The lack of control over part thickness, void fraction, fiber content, and surface quality on the rear face means that applications are typically for very low stressed parts where dimensional accuracy is non-critical.

6.8.2.5 Methods of Applying Heat and Pressure

The wet lay-up process is normally done under room-temperature conditions. The resin is normally left at room temperature for a day or for overnight curing, depending on the resin chemistry. The cure time can be shortened by blowing warm air on the laminate. Pressure is applied using rollers during lamination. During the curing process, there is no pressure, or sometimes vacuum bagging is used to create good consolidation between the layers as well as to remove entrapped air. If the part size is small, it can be put into an autoclave and external pressure is applied. Post-curing of an entire assembly is sometimes done to improve part performance.

6.8.2.6 Basic Processing Steps

The major processing steps in the wet lay-up process include:

1. A release agent is applied to the mold.
2. The gel coat is applied to create a Class A surface finish on the outer surface. The gel coat is hardened before any reinforcing layer is placed.
3. The reinforcement layer is placed on the mold surface and then it is impregnated with resin. Sometimes, the wetted fabric is placed directly on the mold surface.
4. Using a roller, resin is uniformly distributed around the surface.
5. Subsequent reinforcing layers are placed until a suitable thickness is built up.
6. In the case of sandwich construction, a balsa, foam, or honeycomb core is placed on the laminated skin and then adhesively bonded. Rear-end laminated skin is built similar to how the first laminated skin was built up.
7. The part is allowed to cure at room temperature, or at elevated temperature.

6.8.2.7 Advantages of the Wet Lay-Up Process

The wet lay-up process is one of the oldest composite manufacturing techniques with the following advantages:

1. Very low capital investment is required for this process because there is negligible equipment cost as compared to other processes.
2. The process is very simple and versatile. Any fiber type material can be selected with any fiber orientation.
3. The cost of making a prototype part is low because a simple mold can be used to make the part. In addition, the raw material used for this process is liquid resin, mat, and fabric material, which are less expensive than prepreg materials.

6.8.2.8 Limitations of the Wet Lay-Up Process

The wet lay-up process has the following limitations:

1. The process is labor intensive.
2. The process is mostly suitable for prototyping as well as for making large structures.
3. Because of its open mold nature, styrene emission is a major concern.
4. The quality of the part produced is not consistent from part to part.
5. High fiber volume fraction parts cannot be manufactured using this process.
6. The process is not clean.

6.8.3 Spray-Up Process

The spray-up process is similar to the wet lay-up process, with the difference being in the method of applying fiber and resin materials onto the mold. The wet lay-up process is labor intensive because reinforcements and resin materials are applied manually. In the spray-up process, a spraygun is used to apply resin and reinforcements with a capacity of 1000 to 1800 lb material delivered per hour. In this process a spraygun is used to deposit chopped fiber glass and resin/catalyst onto the mold. The gun simultaneously chops continuous fiber rovings in a predetermined length (10 to 40 mm) and impels it through a resin/catalyst spray onto the mold. The spray-up process is much faster than the wet lay-up process and is a less expensive choice because it utilizes rovings, which is an inexpensive form of glass fiber.

6.8.3.1 Major Applications

The spray-up process is used to make small to large custom and semi-custom parts in low- to medium-volume quantities. Where the strength of the product is not as crucial, spray-up is the more suitable option. Bathtubs, swimming pools, boat hulls, storage tanks, duct and air handling equipment, and furniture components such as seatings are some of the commercial uses of this process.

6.8.3.2 Basic Raw Materials

The reinforcement material for this process is glass fiber rovings, which are chopped to a length of 10 to 40 mm and then applied on the mold. For improved mechanical properties, a combination of fabric layers and chopped fiber layers is used. The most common material type is E-glass, but carbon and Kevlar rovings can also be used. Continuous strand mat, fabric, and various types of core materials are embedded by hand whenever required. The weight fraction of reinforcement in this process is typically 20 to 40% of the total weight of the part.

The most common resin system used for the spray-up process is general-purpose or DCDP polyester. Isophthalic polyester and vinylesters are also used in this process. Fast-reacting resins with a pot life of 30 to 40 min are typically used. The resin often contains a significant amount of filler. The most common fillers are calcium carbonate and aluminum trihydrate materials. In filled resin systems, fillers replace some of the reinforcements; 5 to 25% filler is used by weight.

6.8.3.3 Tooling Requirements

The mold used in this process is identical to that used in the wet lay-up process. Male and female molds are used, depending on the application. Tubs and showers utilize male molds, whereas boat hulls and decks utilize female molds. To make bathtubs, FRP molds are used. The method used to make the mold is described in Section 6.4.3.

6.8.3.4 Making of the Part

The processing steps used in the spray-up process are very similar to those in the wet lay-up process. In this process, the release agent is first applied to the mold and then a layer of gel coat is applied. The gel coat is left for 2 hr, until it hardens. Once the gel coat hardens, a spraygun is used to deposit the fiber resin mixture onto the surface of the mold. The spraygun chops the incoming continuous rovings (one or more rovings) to a predetermined length and impels it through the resin/catalyst mixture as shown in Figure 6.27. Figure 6.28 shows the application of chopped fibers and resin by a robot. Resin/catalyst mixing can take place inside the gun (gun mixing) or just in front of the gun. Gun mixing provides thorough mixing of resin and catalyst inside the gun and is preferred to minimize the health hazard concerns of the operator. In the other type, the catalyst is sprayed through two side nozzles into the resin envelope. Airless sprayguns are becoming

Manufacturing Techniques

137

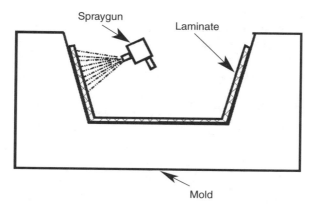

FIGURE 6.27
Schematic of the spray-up process.

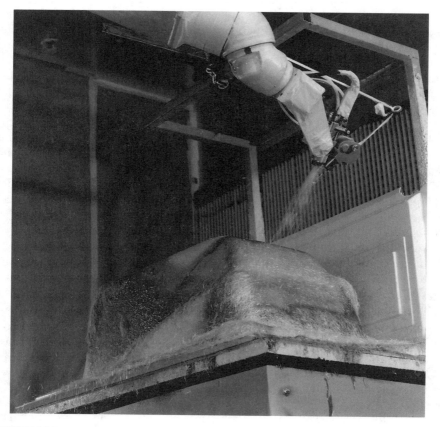

FIGURE 6.28
Robotic spray-up process for making a bathtub. The robot is applying chopped fiberglass with gel. (Courtesy of Fanuc Robotics.)

popular because they provide more controlled spray patterns and reduced emission of volatiles. In an airless system, hydraulic pressure is used to dispense the resin through special nozzles that break up the resin stream into small droplets which then become saturated with the reinforcements. In an air-atomized spraygun system, pressurized air is used to dispense the resin.

In the spray-up process, the thickness built up is proportional to the spraying pattern, and the quality of the laminate depends on operator skill. Once the material is sprayed on the mold, brushes or rollers are used to remove entrapped air as well as to ensure good fiber wetting. Fabric layers or continuous strand mats are added into the laminate, depending on performance requirements. The curing of the resin is done at room temperature. The curing of resin can take 2 to 4 hr, depending on the resin formulation. After curing, the part is demolded and tested for finishing and structural requirements.

To gain a better understanding of the spray-up process, the process is described here for the fabrication of bathtubs. There are two types of bathtubs available on the market: one with a gel coat surface finish and the other with an acrylic finish.[7] In both of these products, the gel coat or acrylic is first applied on the mold and then the composite material is applied as a backing laminate material. The thicker acrylic finish base is sturdier than a gel coat-finished tub but they are about $100 more expensive. For this reason, acrylic-based tubs are dominant in luxury items such as whirlpool baths, whereas gel coat-finished tubs dominate the broader commodity market.[7] In the manufacturing of gel coat-finished bath tubs, the first step is to apply the release agent to the male mold and then apply the isopolyester gel coat. This gel coat is highly pigmented with titanium white (titanium dioxide) and filled with minerals, talcs, and silica to reduce styrene content. The purpose of the gel coat is to get a very high polished surface finish on the part. Once the gel coat hardens, a black-pigmented barrier coat is applied to stop fiber print-through. Both the gel coat and the barrier coat are then cured in an oven.

In acrylic-finished tubs, instead of male mold, a female mold is used to make the tubs. The acrylic sheet is heated first and then vacuum formed onto the female mold. After the acrylic sheet solidifies, it is removed from the mold. The hardened acrylic sheet is now sufficiently rigid to work as a male mold. The acrylic sheet is supported on a matched form and then composite materials are applied. The acrylic sheet becomes an integral part of the tub. From here onward, all the manufacturing steps for gel coat- and acrylic-based tubs are the same.

For the spray-up process, dicyclopentadiene (DCPD) polyester resins are used. Calcium carbonate and aluminum trihydrate fillers are added into the resin and mixed using a high shear mixing unit. A wax-type additive is added into the resin to suppress styrene emission during lamination. The wax rises to the laminating surface during the cure cycle and creates a barrier film, which reduces styrene evaporation to less than 20%. The mixed resin is pumped to the holding tank, which is connected to the spraygun. A fiberglass chopper, which chops the glass rovings, is mounted on the spraygun. Then the mixture of resin, catalyst, and chopped fiber glass is sprayed onto the barrier coat in a fan pattern. The method of mixing the resin and catalyst depends on the type of spraygun, as previously discussed. After each layer of lamination, workers roll out the entire laminate. Increased attention is paid to radii and corners so that a smooth and even surface is obtained.

Once the first skin or laminate is built up, corrugated material, foam, or wood is applied to key parts as a core material to make it a sandwich structure. The core material is applied to flat areas, bend areas, and the bottom of the part. The part is then cured in an oven and brought to room temperature. After curing, a second skin or laminate is formed using the same procedure. The same material is used during spray-up process to form the second laminate. The part is again oven cured and brought to room temperature. The mold is removed and it is waxed and polished for the next manufacturing cycle.

Finishing work is done on the tub by trimming the edges and drilling holes for drains and grab bars. Other secondary operations are performed according to the requirements of the product. The part is then weighed for material control purposes and inspected for dimensional tolerances, structural soundness, and surface finish quality. Finally, the product is tagged for identification, crated, and shipped to the warehouse.

6.8.3.5 Methods of Applying Heat and Pressure

The spray-up process is very economical because it does not have a high need for heat or pressure. The part is room-temperature cured or sometimes oven cured for higher production volume. No pressure is applied during the curing process. After spraying the resin and reinforcement, rollers are used to remove the entrapped air as well as to create an even and smooth laminate surface.

6.8.3.6 Basic Processing Steps

The steps used in the spray-up process are almost the same as for the wet lay-up process, except for the method of creating the laminates. The basic steps are as follows:

1. The mold is waxed and polished for easy demolding.
2. The gel coat is applied to the mold surface and allowed to harden before building any other layer.
3. The barrier coat is applied to avoid fiber print through the gel coat surface.
4. The barrier coat is oven cured.
5. Virgin resin is mixed with fillers such as calcium carbonate or aluminum trihydrate and pumped to a holding tank.

6. Resin, catalyst, and chopped fibers are sprayed on the mold surface with the help of a hand-held spraygun. The spraygun is moved in a predetermined pattern to create uniform thickness of the laminate.
7. A roller is used for compaction of sprayed fiber and resin material as well as to create an even and smooth laminate surface. Entrapped air is removed.
8. Where desirable, wood, foam, or honeycomb cores are embedded into the laminate to create a sandwich structure.
9. The laminate is cured in an oven.
10. The part is demolded and sent for finishing work.
11. Quality control personnel inspect the part for dimensional tolerances, structural soundness, and good surface finish quality, and then approve or reject the part, depending on its passing criteria.

6.8.3.7 Advantages of the Spray-Up Process

The spray-up process offers the following advantages:

1. It is a very economical process for making small to large parts.
2. It utilizes low-cost tooling as well as low-cost material systems.
3. It is suitable for small- to medium-volume parts.

6.8.3.8 Limitations of the Spray-Up Process

The following are some of the limitations of the spray-up process:

1. It is not suitable for making parts that have high structural requirements.
2. It is difficult to control the fiber volume fraction as well as the thickness. These parameters highly depend on operator skill.
3. Because of its open mold nature, styrene emission is a concern.
4. The process offers a good surface finish on one side and a rough surface finish on the other side.
5. The process is not suitable for parts where dimensional accuracy and process repeatability are prime concerns. The spray-up process does not provide a good surface finish or dimensional control on both or all the sides of the product.

6.8.4 Filament Winding Process

Filament winding is a process in which resin-impregnated fibers are wound over a rotating mandrel at the desired angle. A typical filament winding process is shown in Figures 6.29 and 6.30, in which a carriage unit moves

Manufacturing Techniques 141

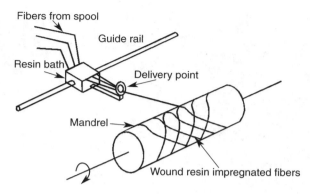

FIGURE 6.29
Schematic of the filament winding process.

 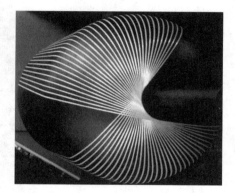

FIGURE 6.30
Demonstration of the filament winding operation. (Courtesy of Entec Composite Machines, Inc.)

back and forth and the mandrel rotates at a specified speed. By controlling the motion of the carriage unit and the mandrel, the desired fiber angle is generated. The process is very suitable for making tubular parts. The process can be automated for making high-volume parts in a cost-effective manner. Filament winding is the only manufacturing technique suitable for making certain specialized structures, such as pressure vessels.

6.8.4.1 Major Applications

The most common products produced by the filament winding process are tubular structures, pressure vessels, pipes, rocket motor casings, chemical storage tanks, and rocket launch tubes. Some filament wound parts are shown in Figure 6.31. The introduction of sophisticated filament winding machines and dedicated CAD systems has enabled more complex geometries to be produced and many of the original geometric limitations have now been overcome. Bent shapes, connecting rods, bottles, fishing rods, golf

FIGURE 6.31
Filament wound parts. (Courtesy of Advanced Composites, Inc.)

shafts, pressure rollers, bushings, bearings, driveshafts (industrial and automotive), oil field tubing, cryogenics, telescopic poles, tool handles, fuse tubes, hot sticks (non-conducting poles), conduits, fuse lage, bicycle frames and handle bars, baseball/softball bats, hockey sticks, fishing rods, ski poles, oars, tubes, etc. are currently produced using filament winding techniques.

Filament wound glass reinforced plastic (GRP) is used for water supply piping systems. It provides clean and lead-free piping system. Filament wound pipe reduces the pumping energy required to move water by 10 to 35%, due to its smooth interior surfaces compared to concrete or ductile iron pipe. The weight of GRP pipe is one fourth that of ductile iron, one tenth that of concrete, and thus provides added advantages in transportation and installation of these pipes. Figure 6.32 shows a composite production riser used for deep-water applications. It offers significant cost benefits to offshore platform and riser systems with no reduction in system reliability. The composite production riser's light weight and reduced stiffness relative to steel riser systems lead to platform size reduction and reduction of top tensioning requirements. High-pressure composite accumulator bottles are shown in

Manufacturing Techniques

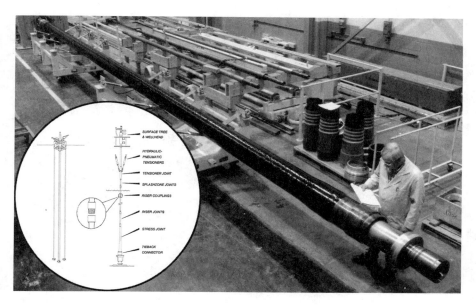

FIGURE 6.32
Composite production riser made by the filament winding operation. (Courtesy of Lincoln Composites.)

Figure 6.33. These bottles are used in the offshore oil industry and offer significant weight savings relative to steel bottles.

6.8.4.2 Basic Raw Materials

In general, starting materials for filament winding are continuous fibers (yarns) and liquid thermoset resins. Yarns are kept in spool form at the back rack and passed through a resin bath located in the carriage unit. Fibers get wet as they pass through the resin bath. Glass, carbon, and Kevlar fibers are used for the filament winding process but glass fibers are more common because of its low cost. Epoxy, polyester, and vinylester are used as resin materials. Glass fibers with polyester resins are widely used for low-cost applications. Glass with epoxy is used in spoolable filament-wound tubes for offshore applications. Sometimes, prepreg tows are used as starting materials. The use of prepreg tows provides uniform fiber distribution and resin content throughout the thickness of the part. The filament winding process is also used for making preforms for the RTM process.

6.8.4.3 Tooling

The most common tooling material for the filament winding process is a steel mandrel. Steel mandrels are chrome plated in certain applications to get a high-gloss finish on the inside surface of the composite structure as well as to aid in easy removal of the mandrel. Aluminum is also used for

FIGURE 6.33
Composite accumulator bottles. (Courtesy of Lincoln Composites.)

making mandrels. For some applications, such as pressure vessels, the mandrel is not removed and becomes an integral part of the composite structure. The non-removal mandrel provides an impermeable layer/barrier surface on the composite inner surface and thus avoids leakage of compressed gas or liquid inside the pressure vessel. Typically, metals and thermoplastic materials are used as barrier materials. Plaster of Paris and sand are also used to make destructible/collapsible mandrels. For protyping purposes, wood, plastics, and cardboard can be used.

Cylindrical mandrels are inexpensive compared to tooling costs for other manufacturing processes. A 48- to 60-in.-long steel mandrel with a 1- to 2-in. diameter costs between $50 and $100. A chrome-plated mandrel costs between $250 and $500 for the same size.

6.8.4.4 Making of the Part

To make filament wound structures, a mandrel is place on the filament winding machine as shown in Figures 6.29 and 6.30. A filament winding machine is similar to a lathe machine where the head and tail stocks are used to hold the mandrel and the cutting tool is replaced by a payout eye on the carriage unit. The mandrel rotates and the carriage unit moves relative to the mandrel to lay down the resin-impregnated fibers at a specific angle. Various types of computer-operated filament winding machines are available in the market, ranging from two-axes to six-axes filament winding machines.

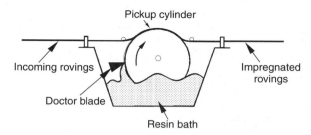

FIGURE 6.34
Doctor blade arrangement in the filament winding operation.

The carriage unit can move along the x, y, and z axes as well as rotate about these axes. In two-axes filament winding machines, the mandrel rotates and the carriage unit moves back and forth only in one direction.

Before winding begins, the mandrel is coated with release agent. Sometimes, a gel coat is applied on the top of the release agent to get high surface finish quality on the interior surface of the composite. Once the mandrel is prepared, it is placed between the head and tail stocks of the machine. During wet winding, fiber yarns, which are placed in spool form at the creels, are passed through the resin bath located in the carriage unit and then to the mandrel through the payout eye. To achieve good fiber wet-out, air held inside the fiber bundle must be removed and replaced with the resin. The impregnation system must facilitate the break-up of any film formers on the bundle for resin ingress. To achieve good impregnation, several things are done: rovings are kept at constant tension, rovings are passed through guided pins, a doctor blade at resin bath is used (Figure 6.34). If the tension on the rovings is too small, the laminate is not fully compacted and creates an excess resin region on the laminate. If the tension is too high, it can cause fiber breakage or resin-starved areas near inside layers. The use of a doctor blade arrangement scrapes off excess resin and creates a uniform resin layer. Laminate is formed after a series of relative motions between the mandrel and the carriage unit. To get the desired winding, the machine operator inputs various parameters such as pipe diameters, mandrel speed, pressure rating, band width, fiber angle, etc., depending on the software requirements. After creating the desired fiber angle distribution, the mandrel with the composite laminate is removed to a curing area where the laminate is cured at room temperature or at elevated temperature. For thick laminates, it may be necessary to wind in stages and allow the laminate to cure between winding operations. Once the part is cured, the mandrel is extracted using an extracting device. Sometimes, a small taper angle is provided in the mandrel for easy removal of the composite part.

To manufacture 28-in. diameter GRP pipes for water piping systems using the filament winding process, first chopped rovings (2-in. length) is applied on the mandrel to get equal properties in the axial and longitudinal directions.[8] Then, continuous fibers are wound along 90° to get hoop strength. After the hoop wound layers, a 2-in. wide nonwoven fiberglass surface mat having a density of 30 g/m² is applied. Orthophthalic polyester resin (30 to 40% by weight) is commonly used to make GRP pipes. The pipe is then moved to a curing station where it is cured at about 250 to 265°F using infrared heating on the outer layers of the laminate and induction heating of the steel band in the inside layers of the composites. After curing, pipes are cut to a desired length using a computer-controlled saw. For ease of installation, the edges of the pipe are chamfered to a 30° angle. The pipes are then taken out for finishing and pressure testing.

6.8.4.5 Methods of Applying Heat and Pressure

The pressure during filament winding is applied by creating fiber tension. In general, 1 lbf to 6 lbf fiber tension is created using some tensioning device or by passing the fibers through the carriage unit in such a way that it creates tension. Composites thus fabricated are cured at room temperature, or in an oven at a higher temperature. For large-volume production, the process of part fabrication is automated. In an automated line, the filament wound part with the mandrel is moved to a heated chamber using a robot. The part slowly moves in the heated chamber and comes out after partial or full cure of the composite part. The part is then sent to the mandrel extracting station where the mandrel is extracted and sent back to the filament winding machine for winding purposes. All of this can be done automatically.

Ultraviolet (UV) curing as well as electron beam curing are also performed during the filament winding process to cure the resin. Radiation energy is used for curing the resin in both UV and electron beam curing. UV radiation is at the low end of the radiation spectrum, releasing energy in the 1.7 to 6 electron volt (eV) range; whereas in electron beam radiation, it is in the 10,000 to 1 million eV range. To facilitate UV curing, an additive such as Accuset 303 is added to the resin formulation. The additive contains a photoinitiator that is sensitive to radiation energy. A light source such as an electrode-based mercury/vapor bulb is used to deliver the correct UV wavelength energy to initiate the curing action. Once exposed to the UV rays, the chemical structure of the photoinitiator breaks down into energized free radicals that actively seek new chemical bond sites within the resin mix. When the free radicals bond with other free radicals, the size of the polymer chain increases, causing the resin mix to polymerize into the solid state. UV curing has the drawback that it cannot be used for curing of laminates produced by other composite manufacturing processes because UV rays must "see" the material that needs to be cured. Moreover, it cannot penetrate beneath the top layer of pigmented or colored material, such as black carbon fiber. For this reason, UV curing is focused on continuous processing of materials such as filament winding.

The outer surface finish quality of filament wound parts is usually not good and requires extra machining and sanding of the outer surface. To create a good outer finish, the part is sometimes shrink taped after winding is complete or a teflon-coated air breather is applied on the outer surface to absorb excess resin.

6.8.4.6 Methods of Generating the Desired Winding Angle

The desired fiber architecture of the mandrel surface is generated by the relative motion of the mandrel and payout eye. There are several ways that winding motions can be determined to get the desired fiber angle distribution. Sometimes, the suppliers of the filament winding machine provide application programs for the winding of standard shapes such as rings, cylinders, bottles, and pressure vessels. In another approach, the winding motions are determined by the teach-in-programming technique, in which a desired trajectory of fiber path is first marked on the mandrel surface and then the mandrel and carriage units are incrementally moved by trial and error so that fiber filaments are laid on the marked trajectory.[9] Coordinates and thus the motion of the mandrel and carriage units are recorded. Sometimes, interactive graphics packages are used for smoothing and editing of the stored data.[10,11] Once the required data for one complete stroke are calculated, these data are generally repeated after indexing the mandrel. Teach-in-programming is still used in some places. Delivery point motion can also be determined by simulation of the filament winding process.[10]

The latest technology offers new-generation computer-controlled filament winding machines with floppy disk and hard disk drives and RS 232 input ports.[12,13] The equipment is configured such that it relies on real-time control of several servo axes, analog and digital outputs, and tension controllers. The winding motion is generated from data transferred from other sources using a floppy disk. Manufacturing engineers use computer graphics to simulate the winding pattern, record the data, and then feed the resulting data to the computer through a line or floppy disk.

CAD/CAM systems have also been developed for filament winding,[14,15] which integrate a three-dimensional surface modeler, specific filament winding software, and design software. A three-dimensional surface modeler is used to model component geometries and to determine the fiber orientation by performing stress analysis and using failure criteria. Once the winding path is known, the filament winding software simulates the winding process by defining successive straight lines tangential to the winding path at each point. The ends of these straight lines define the path of delivery point in the mandrel frame of reference. At this stage, an intersection calculation is applied to check for possible regions of collision. Once an acceptable winding circuit is generated, delivery point locations are converted from the mandrel frame of reference to a coordinate system that corresponds to the kinematics of the filament winding machine. The most difficult task here is to convert these data into the five or six simultaneous axes of the filament winding machine. Machine motion for each degree of freedom is then determined in such a way that the net effect of motion of each degree of freedom should match with the desired movement of the delivery point location. Machine motion by this technique is sometimes quite complex and requires a sophisticated filament winding machine with a large number of degrees of freedom.

Robots have also been introduced into filament winding of small, complex-shaped structures.[16,17] In many cases of robotic filament winding and CNC (computer numerical controlled) filament winding, the motion of the delivery point is obtained directly from the stable fiber path predictions on the mandrel surface.[17] In this case, it is assumed that the distance between the delivery point and the mandrel surface is zero and that the delivery point moves on the mandrel surface along the desired fiber trajectory. In fact, because the distance between the delivery point and the mandrel surface is not zero, the accuracy of fiber placement is less than that offered by regular filament winding machines.[17] Other possible restrictions that might prevent the robot arm from performing its prescribed task are collisions with the mandrel or other objects, singularity positions or joint angle limitations of the robot arm, and machine dynamics.[16,17]

Mazumdar and Hoa[18-23] have developed a series of kinematic models to determine the mandrel and carriage motions for generating desired fiber angle distributions on cylindrical, noncylindrical, axisymmetric, and nonaxisymmetric mandrel shapes. Their method relies on a geometric approach in which geometrical and trigonometrical relations are used to determine the winding motion. Using their model, a simplest form of filament winding machine having two degrees of freedom can generate the desired winding angle on cylindrical, noncylindrical, axisymmetric, and nonaxisymmetric shapes. For some cylindrical mandrels with polygonal cross-sections such as rectangular or hexagonal, the method requires some simple manual calculations without the use of a computer to determine the winding motion. For cylinders with polygonal cross-sections, the winding motion prediction is 100% accurate. For mandrels with curved surfaces, a small computer program is developed to determine the mandrel and carriage unit velocities for the desired winding angle.

6.8.4.7 Basic Processing Steps

To more easily understand the entire process, the major steps performed during the filament winding process are described here. These steps are common in all wet filament winding processes.

1. Spools of fiber yarns are kept on the creels.
2. Several yarns from spools are taken and passed through guided pins to the payout eye.
3. Hardener and resin systems are mixed in a container and then poured into the resin bath.
4. Release agent and gel coat (if applicable) are applied on the mandrel surface and the mandrel is placed between the head and tail stocks of the filament winding machine.
5. Resin-impregnated fibers are pulled from the payout eye and then placed at the starting point on the mandrel surface. Fiber tension is created using a tensioning device.

Manufacturing Techniques 149

FIGURE 6.35
Demonstration of fiber laydown on a mandrel. (Courtesy of Lincoln Composites.)

6. The mandrel and payout eye motions are started. The computer system in the machine creates winding motions to get the desired fiber architecture in the laminate system, as shown in Figure 6.35.
7. Fiber bands are laid down on the mandrel surface. The thickness builds up as the winding progresses.
8. To obtain a smooth surface finish on the outer surface, a teflon-coated bleeder or shrink tape is rolled on top of the outer layer after winding is completed.
9. The mandrel with the composite laminate is moved to a separate chamber where the composite is cured at room temperature or elevated temperature.
10. After curing, the mandrel is extracted from the composite part and then reused. For certain applications, the mandrel is not removed and it becomes an integral part of the composite structure.

6.8.4.8 Advantages of the Filament Winding Process

Filament winding has gained significant commercial importance due to its capability in laying down the fibers at a precise angle on the mandrel surface. Filament winding offers the following advantages.

1. For certain applications such as pressure vessels and fuel tanks, filament winding is the only method that can be used to make cost-effective and high-performance composite parts.
2. Filament winding utilizes low-cost raw material systems and low-cost tooling to make cost-effective composite parts.
3. Filament winding can be automated for the production of high-volume composite parts.

6.8.4.9 Limitations of the Filament Winding Process

Filament winding is highly suitable for making simple hollow shapes. However, the process has the following limitations.

1. It is limited to producing closed and convex structures. It is not suitable for making open structures such as bathtubs. In some applications, filament winding is used to make open structures such as leaf springs, where the filament wound laminate is cut into two halves and then compression molded.
2. Not all fiber angles are easily produced during the filament winding process. In general, a geodesic path is preferred for fiber stability. Low fiber angles (0 to 15°) are not easily produced.
3. The maximum fiber volume fraction attainable during this process is only 60%.
4. During the filament winding process, it is difficult to obtain uniform fiber distribution and resin content throughout the thickness of the laminate.

6.8.5 Pultrusion Process

The pultrusion process is a low-cost, high-volume manufacturing process in which resin-impregnated fibers are pulled through a die to make the part. The process is similar to the metal extrusion process, with the difference being that instead of material being pushed through the die in the extrusion process, it is pulled through the die in a pultrusion process. Pultrusion creates parts of constant cross-section and continuous length.

Pultrusion is a simple, low-cost, continuous, and automatic process. Figure 6.36 illustrates a typical pultrusion process in which resin-impregnated yarns are pulled through a heated die at constant speed. As the material passes through the heated die, it becomes partially or completely cured. Pultrusion yields smooth finished parts that usually do not require post-processing.

6.8.5.1 Major Applications

Pultrusion is used to fabricate a wide range of solid and hollow structures with constant cross-sections. It can also be used to make custom-tailored

Manufacturing Techniques

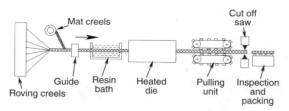

FIGURE 6.36
Illustration of a pultrusion process.

FIGURE 6.37
Typical pultruded shapes. (Courtesy of GDP, France.)

parts for specific applications. The most common applications are in making beams, channels, tubes, grating systems, flooring and equipment support, walkways and bridges, handrails, ladders, light poles, electrical enclosures, etc. Typical pultruded shapes are shown in Figure 6.37. Pultruded shapes are used in infrastructure, automotive, commercial, and other industrial sectors. Figure 6.38 shows a fiberglass grating system. The grating systems are lightweight, longlasting, and provide easy installation. Nonconductive fiberglass covers for high-voltage rails for a rapid transit system are shown in Figure 6.39, and roll-up fiberglass trailer doors and Z-bar separators in sidewall panels are shown in Figure 6.40. Figure 6.41 shows a complete fiberglass stair system having structural members, railings, treads, and a platform. Pultruded parts are used in the above applications.

FIGURE 6.38
Fiberglass grating and handrail systems using pultruded parts. (Courtesy of Creative Pultrusions, Inc.)

6.8.5.2 Basic Raw Materials

Pultrusion is typically used for making parts with unidirectional fibers. E-glass, S-glass, carbon, and aramid fibers are used as reinforcements, the most common type being E-glass rovings. Fabrics and mats are also used to add bidirectional and multidirectional strength properties. Unsaturated polyester is the most common resin material for the pultrusion process. Pultrusion offers an attractive performance-to-price ratio as well as easy processing. Vinylesters and epoxies can be used for improved properties but the processing of these resins becomes difficult. Moreover, the pulling speeds with these resins are lower because of lower resin reactivity.

FIGURE 6.39
Nonconductive fiberglass covers for high-voltage rails on a rapid transit system. (Courtesy of Creative Pultrusions, Inc.)

FIGURE 6.40
Lightweight roll-up fiberglass trailer doors and Z-bar separators in sidewall panels. (Courtesy of Creative Pultrusions, Inc.)

Various types of fillers are added to the polyester resin to improve the insulation characteristics, chemical resistance, and fire resistance, and to lower the overall cost. Calcium cabonates are added to lower the cost of the pultruded part. Calcium carbonate is a very inexpensive material and is a major filler in SMC compounds. It improves the whiteness (opacity) of the part. Alumina trihydrate and antimony trioxide are used for fire retardancy. Aluminum silicate (kaolin clay) provides enhanced insulation, opacity, surface finish, and chemical resistance.

FIGURE 6.41
All-fiberglass stair systems having structural members, railings, treads, and platform. (Courtesy of Creative Pultrusions, Inc.)

6.8.5.3 Tooling

For the pultrusion process, steel dies are used to transform resin-impregnated fibers to the desired shape. Dies have a constant cross-section along their length, except for some tapering at the raw material entrance. The dies are heated to a specific temperature for partial or complete cure of the resin. Tooling costs depend upon the complexity of the part as well as the volume requirement. The cost of the die ranges from $4000 to $25,000, depending on the size and cross-section of the part. Tooling life is generally in excess of 150,000 linear feet before major re-work is required. Tools are frequently re-chromed to minimize destructive wearing. The total life could go up to 750,000 linear feet until further re-work in the die becomes impractical.

Dies are segmented for easy assembly, disassembly, machining, and chrome plating. At the joining of these segments, surface marks (parting lines) are created on the pultruded part. A parting line appears as a slightly raised line on the surface.

6.8.5.4 Making of the Part

To make composite parts using the pultrusion process, spools of rovings are placed on the creel similar to the filament winding process and then reinforcements are passed through a resin bath where fibers are impregnated with the resin. There are two major impregnation options. In the first option, rovings are passed through an open resin bath, as shown in Figure 6.36. The reinforcement can pass horizontally inside the bath (Figure 6.36) or up and down through a guiding mechanism. Reinforcements pass horizontally when bending is avoided. Fabrics and mats are usually passed horizontally.

Open resin bath impregnation is the most common method because of its simplicity. In this method, impregnation takes place by capillary action. In the second option, reinforcement passes through a cavity where resin is injected under pressure. This system utilizes a different kind of die, which has a tapered cavity for impregnation. The advantages of this method are no or minimum styrene emission and low resin loss. However, this method requires expensive dies. Once the reinforcement becomes impregnated, it is less sensitive to friction and can be guided using sheet-metal guides. In general ceramic guides are used because of the abrasive nature of dry fibers. Reinforcements thus impregnated are passed through a heated die. The die has a slight taper at the entrance and a constant cross-section along its length. The resin cures and solidifies as it passes through the heated die. The length of the die depends on resin reactivity, part thickness, and production rate requirements. The higher the resin reactivity, the shorter the die length requirement.

The solidified material is pulled by caterpillar belt pullers or hydraulic clamp pullers. These pullers are mounted with rubber-coated pads that grip the composite material. The puller is distanced from the die in such a way that the composite material cools off enough to be gripped by the rubber pads.

Pultrusion provides lowest cost composite parts because of the highly automated nature of the process as well as lower fiber and resin costs as compared to prepregs and fabrics. Because the pultrusion process is a continuous process, literally any length can be produced. However, parts produced are cut to predetermined lengths using an automatic saw or manual hacksaw. The length of the pultrusion facility from creel to saw is typically on the order of 10 m. It can be longer, depending on part complexity and length requirements.

Following are some of the considerations while manufacturing and designing pultruded parts.

6.8.5.4.1 Wall Thickness

Wherever possible, select uniform thickness in the cross-section because it provides uniform cooling and curing, and thus avoids the potential of residual stress and distortions in the part. Moreover, uniform thickness will provide uniform shrinkage in the part and thus will limit the warpage in the product. Typically, 2 to 3% shrinkage occurs in the pultruded part. Also, maintain symmetry in the cross section for minimal distortion.

For high-volume production, the thickness of the part is critical because the curing time and therefore the rate of pull depend on the thickness of the part. For example, a 0.75-in. thick cross-section can be produced at a rate of approximately 9 in./min, whereas a 0.125-in. thick cross-section can provide a production rate of 3 to 4 ft/min. Therefore, if a design requires high rigidity in the part, then it can be achieved by creating deeper sections with thinner wall or by including ribs in the cross-section. Similarly, if there is a choice between selecting a thick rod or tube, select the tube because it offers a higher production rate, lower cost, and higher specific strength.

6.8.5.4.2 Corner Design

In a pultruded part, avoid sharp corners and provide generous radii at those corners. Generous radii offer better material flow at corners as well as improve the strength by distributing stress uniformly around the corner. A minimum of 0.0625-in. radius is recommended at corners.

Another important consideration in the design of corners is to maintain uniform thickness around the corner. This will avoid the build-up of resin-rich areas, which can crack or flake off during use. Moreover, uniform thickness will provide uniformity in fiber volume fraction and thus will help in obtaining consistent part properties.

6.8.5.4.3 Tolerances, Flatness, and Straightness

Dimensional tolerances, flatness, and straightness obtained in pultruded parts should be discussed with the supplier. Standard tolerances on fiberglass pultruded profiles have been established by industry and ASTM committees. Refer to ASTM 3647-78, ASTM D 3917-80, and ASTM D 3918-80 for standard specifications on dimensional tolerances and definitions of various terms relating to pultruded products.

Pultrusion is a low-pressure process and therefore does not offer tight tolerances in the part. Shrinkage is another contributing factor that affects tolerances, flatness, and straightness.

The cost of a product is significantly affected by tolerance requirements. Tight tolerance implies higher product cost. Therefore, whenever possible, provide generous tolerances on the part as long as the functionality of the product is not affected.

6.8.5.4.4 Surface Texture

Pultrusion is a low-pressure process and typically provides a fiber-rich surface. This can cause pattern-through of reinforcing materials or fibers getting easily exposed under wear or weathering conditions. Surfacing veils or finer fiber mats are used as an outer layer to minimize this problem. To create good UV and outdoor exposure resistance, a 0.001- to 0.0015-in. thick layer of polyurethane coating is applied as a secondary operation.

6.8.5.5   Methods of Applying Heat and Pressure

During pultrusion, there is no external source to apply presure for consolidation. Therefore, this process is known as a low-pressure process. The resin-impregnated rovings or mat, when passed through a restricted passage of the die, gets compacted and consolidated. The die is heated to a temperature and applies heat to incoming material for desired cure. The heat in the die cures the resin. The part coming out of the die is hot and is allowed to cool before it is gripped by the puller.

Manufacturing Techniques 157

6.8.5.6 Basic Processing Steps

The major steps performed during the pultrusion process are described here. These steps are common in most pultrusion processes:

1. Spools of fiber yarns are kept on creels.
2. Several fiber yarns from the spool are taken and passed through the resin bath.
3. Hardener and resin systems are mixed in a container and then poured in the resin bath.
4. The die is heated to a specified temperature for the cure of resin.
5. Resin-impregnated fibers are then pulled at constant speed from the die, where resin gets compacted and solidified.
6. The pultruded part is then cut to the desired length.
7. The surface is prepared for painting. Surface preparation is an important element to perform finishing operations because the pultrusion process utilizes internal mold releases. These mold releases are a form of wax that form a film on the outer surface of the part. This film can be removed by solvent wiping, sanding, or sandblasting. Solvent wiping is the simplest method of surface preparation. Several solvents (e.g., toluene, xylene, methylene chloride, or acetone) can be used for this purpose.

6.8.5.7 Advantages of the Pultrusion Process

Pultrusion is an automated process with the following advantages:

1. It is a continuous process and can be compeletely automated to get the finished part. It is suitable for making high-volume composite parts. Typical production speeds are 2 to 10 ft/min.
2. It utilizes low-cost fiber and resin systems and thus provides production of low-cost commercial products.

6.8.5.8 Limitations of the Pultrusion Process

Pultruded components are used on a large scale in infrastructure, building, and consumer products because of lower product cost. However, pultrusion has the following limitations.

1. It is suitable for parts that have constant cross-sections along their length. Tapered and complex shapes cannot be produced.
2. Very high-tolerance parts on the inside and outside dimensions cannot be produced using the pultrusion process.
3. Thin wall parts cannot be produced.

158   Composites Manufacturing: Materials, Product, and Process Engineering

4. Fiber angles on pultruded parts are limited to 0°. Fabrics are used to get bidirectional properties.
5. Structures requiring complex loading cannot be produced using this process because the properties are mostly limited to the axial direction.

6.8.6 Resin Transfer Molding Process

The resin transfer molding (RTM) process is also known as a liquid transfer molding process. Although injection molding and compression molding processes have gained popularity as high-volume production methods, their use is mostly limited to nonstructural applications because of the use of molding compounds (short fiber composites). In contrast to these molding processes, the RTM process offers production of cost-effective structural parts in medium-volume quantities using low-cost tooling. RTM offers the fabrication of near-net-shape complex parts with controlled fiber directions. Continuous fibers are usually used in the RTM process.

In the RTM process, a preform is placed into the mold cavity. A matching mold half is mated to the first half and the two are clamped together. Then, using dispensing equipment, a pressurized mixture of thermoset resin, a catalyst, color, filler, etc., is pumped into the mold using single or multiple ports in the mold. After curing for 6 to 30 min, depending on the cure kinetics of the mixture, the part is then removed from the mold. Thus, RTM results in the production of structural parts with good surface finish on both sides of the part.

The main issues in the RTM process are resin flow, curing, and heat transfer in porous media. The process involves injecting a precatalyzed thermosetting resin under pressure into a heated mold cavity that contains a porous fiber preform. During mold filling, the resin flows into the mold and experiences exothermic curing reactions, causing its viscosity to increase over time and finally solidification. After the fiber preform is completely saturated with resin, cure reactions continue past the gel-point to form a cross-linked polymer.

The RTM process is a closed mold operation in which a dry fiber preform is placed inside a mold and then the thermoset resin is injected through an inlet port until the mold is filled with resin. The resin is then cured and the part is removed from the mold.

6.8.6.1 Major Applications

The RTM process is suitable for making small- to large-sized structures in small- to medium-volume quantities. RTM is used in automotive, aerospace, sporting goods, and consumer product applications. The structures typically made are helmets, doors, hockey sticks, bicycle frames, windmill blades, sports car bodies, automotive panels, and aircraft parts. Some aircraft structures made by the RTM process include spars, bulkheads, control surface ribs and stiffeners, fairings, and spacer blocks.

Manufacturing Techniques

FIGURE 6.42
Braided and resin transfer molded carbon/epoxy toroidal hub for satellite application. (Courtesy of Fiber Innovation, Inc.)

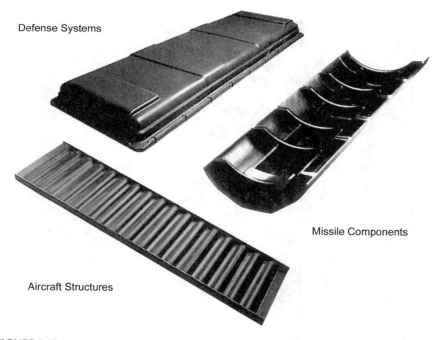

FIGURE 6.43
Resin transfer molded aerospace components. (Courtesy of Intellitec.)

Figure 6.42 shows a braided and resin transfer molded carbon/epoxy toroidal hub for satellite application. Some of the resin transfer molded aerospace components are shown in Figure 6.43. Figure 6.44 shows typical small- to medium-sized, precision reinforced resin transfer molded parts. Figure 6.45

FIGURE 6.44
Typical small- to medium-sized resin transfer molded parts. (Courtesy of Liquid Control Corp.)

FIGURE 6.45
One-piece monocoque bicycle frame. (Courtesy of Radius Engineering, Inc.)

shows a one-piece carbon fiber bicycle frame made by the RTM process. The monocoque frame does not use metallic joints to connect the various tubes. Figure 6.46 shows a composite fork for the bicycle. The fork molds use an electroformed nickel face. The structural backing material is formulated to

Manufacturing Techniques

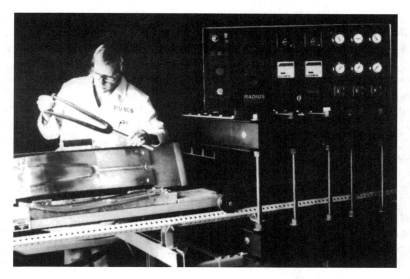

FIGURE 6.46
Composite fork for the bicycle. (Courtesy of Radius Engineering, Inc.)

match the coefficient of thermal expansion of the nickel face and provide high thermal conductivity. Integral electrical heating and water cooling provide extremely rapid thermal cycling. Air cooling is used to control the temperature rise at the end of the rapid heat-up to prevent temperature overshoot. Water cooling is used to obtain rapid cool-down at the end of the high-temperature hold. The molds operate in a shuttle press with an integral process control system.

6.8.6.2 Basic Raw Materials

For the RTM process, fiber preforms or fabrics are used as reinforcements. There are several types of preforms (e.g., thermoformable mat, conformal mats, and braided preforms) used in the RTM process. The thermoformable mats are produced by more or less randomly swirling continuous yarns onto a moving carrier film or belt and then applying a binder, which is typically a thermoplastic polymer, to loosely hold the mat together. The mat is cut to the desired size and formed under heat and pressure to the shape of the mold. In this process, scraps are produced to get the final shape preform because complicated three-dimensional shapes are produced by cutting the flat mat and then forming it under heat and pressure. Once a shape is formed, the preform maintains its shape. In conformal mats, fabrics or chopped strand mats are placed on the outer layer and spun-bonded core material is placed as core material. These layers are stitched together to form conformal mats. Spun material is placed as a core material to create a low-permeability region in the inside region. In braided preforms, fibers are woven over a mandrel to obtain a three-dimensional fiber architecture. For low-volume applications, weaves, braids, and mats are utilized; and for high-volume applications, random fiber preforms are used. Glass, carbon, and Kevlar are used as reinforcing fibers to make the preform, E-glass being the most common. Various methods of making preforms are discussed in Chapter 2.

A wide range of resin systems can be used, including polyester, vinylester, epoxy, phenolic, and methylmethacrylate, combined with pigments and fillers including alumina trihydrate and calcium carbonates. The most common resins used for the RTM process is unsaturated polyester and epoxies. Epoxy with carbon fiber is very common in the aerospace industry. The use of epoxies and other high-viscosity resins requires changes in equipment to meter and condition the resin prior to injection. New epoxy resins are being developed to provide fast cure, thus increasing the production rate.

Filler may be added to the resin during the RTM process. The main purpose of adding filler is to lower the cost of the part. The cost of calcium carbonate filler is only $0.05/lb, whereas epoxy resin costs about $2 to $10/lb. Microballoons can also be used as filler material but costs range between $3 and $4/lb. Microballoons are expensive on a weight basis but on a volume basis, 1 lb of microballoon can replace 100 lb of filler material and can turn out to be cheaper. When mixing filler material such as ground calcium carbonate with the resin, precautions are taken to ensure that filler size does not exceed 10 μm. A larger filler size creates a filtering problem with preforms. A filler size of 5 to 8 μm is recommended so that the filler can move with the resin without any problem inside the fiber architecture. Mixing the filler with the resin increases the viscosity of the resin and slows the production rate. It also significantly increases the weight of the part. The weight increase may be 30%, but the volume increase may be only 12% by adding the filler.

6.8.6.3 *Tooling*

The RTM process provides the advantage of utilizing a low-cost tooling system as compared to other molding processes such as injection and compression molding, the reason being that the pressure used during the RTM process is low compared to the pressure requirements of compression and injection molding processes. Because of this, tooling need not be strong and heavy. At the same time, low-cost tooling provides benefits of low initial investment for prototype building and for production run. Another benefit of RTM compared to filament winding, pultrusion, and other open molding processes is that the closed nature of the RTM process provides a better work environment, a factor of growing importance in light of increasingly stringent regulations concerning styrene emissions.

The mold for the RTM process is typically made of aluminum and steel but for prototype purposes plastic and wood are also used. Figure 6.46 shows the mold for making composite forks, whereas Figures 6.5 and 6.6 show molds for making snowboards and tennis racquets, respectively. Figure 6.47 shows a mold used to make complicated parts, such as shown in Figure 6.48 for the RTM process. The mold is generally in two halves containing single

Manufacturing Techniques 163

FIGURE 6.47
Match molds. (Courtesy of GKN Aerospace.)

FIGURE 6.48
Resin transfer molded part. (Courtesy of GKN Aerospace.)

or multiple inlet ports for resin injection and single or multiple vents for air and resin outlet. The design of the inlet port and vent locations is discussed in Section 6.4.3. The mold design is stiffness critical. The wall thickness of

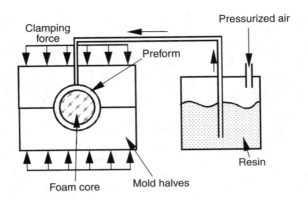

FIGURE 6.49
Schematics of the RTM process.

the mold should be sufficiently rigid to take all the pressure exerted during processing. The mold design should also take into account handling and thermal considerations. Thermal properties of the mold and composite materials affect the dimensional tolerances of the finished part and therefore the coefficients of thermal expansion for the mold and part need to be considered while designing the mold. The tool should be designed to ensure that it can be clamped and sealed properly.

The cost of molds for production runs falls in the range of $2,000 to $50,000, depending on the complexity and size of the part. To make prototype parts or to study resin flow behavior, the cost of a mold can fall in the range of $200 to $5,000. Transparent materials such as acrylic are used for studying resin flow behavior and other characteristics during processing.

6.8.6.4 Making of the Part

In the RTM process, a preform of dry fiberglass mat or fabric is positioned in the cavity of a matched mold as shown in Figure 6.49. Cores and inserts are inserted into the preform as required. Typically, balsa and foam cores are used as core materials. Honeycomb cores are not used in the RTM process because the open surface of the honeycomb material does not restrict the resin flow inside the core. Moreover, cutting the honeycomb core into complex shapes is challenging. There are some commercially available polypropylene cores that have surfacing veils, which restrict the flow of resin inside the core and can be used in the RTM process. Insertion of the core material makes the structure lightweight and strong by creating a sandwich construction. The purpose of putting inserts is to create a fastening mechanism in the structure. Once the reinforcing and core materials are placed into the cavity, then the mold is closed. The mold can be closed using hydraulic or pneumatic presses or by using clamps along the edges. A RTM molding press is shown in Figure 6.50. A typical production flow diagram during the RTM process is shown in Figure 6.51. It shows the sequence of material flow

Manufacturing Techniques

FIGURE 6.50
RTM molding press. (Courtesy of Radius Engineering, Inc.)

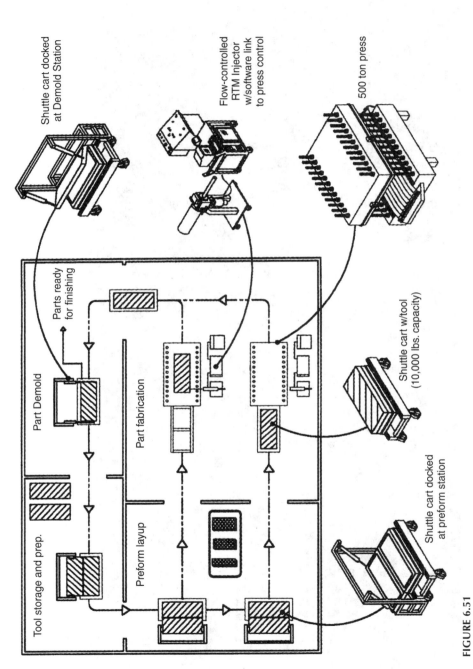

FIGURE 6.51
A typical production flow diagram during the RTM process. (Courtesy of Radius Engineering, Inc.)

Manufacturing Techniques 167

FIGURE 6.52
Molds and mandrels for making aircraft flaps. (Courtesy of Radius Engineering, Inc.)

from preform making to final part fabrication. Figure 6.52 shows an anodized aluminum mold for making aircraft flaps. The bottom of Figure 6.52 shows aluminum mandrels, which are used to create internal ribs in the flap. Carbon fabric is wrapped around these mandrels and placed inside the mold. The ends of the mandrels are placed inside the end blocks shown on the right and left sides of the mandrels to make sure that these mandrels do not move during the injection process. Mandrels, end blocks, and molds are made of aluminum and become black because of anodization. The reinforcements with end blocks and mandrels are then placed inside the mold and the mold is closed. The flap made by this mold is shown in Figure 6.53. In Figure 6.53, the inboard flap (right-hand side) and outboard flap (left-hand side) are shown with spoilers at the front of these flaps. The trim tab is shown at the extreme left of the figure. The press, which made these flaps, is shown at the back of the flaps.

After closing the mold, liquid resin is pumped under low to moderate pressure into the mold cavity using dispensing equipment. Custom-built dispensing equipment is available on the market for RTM purposes. RTM dispensing equipment is shown in Figure 6.54. The RTM machine can efficiently and accurately meter, mix, and inject materials, from a few ounces to hundreds of pounds, into low-pressure closed tools. The machines are designed to handle polyesters, methacrylates, epoxies, urethanes, and other two-component resin systems. In a typical dispensing equipment, resin and catalyst are stored in tanks A and B and mixed through a static mixer just before injection. In the simplest form of dispensing equipment, preformulated resin is stored in a pressure pot and then pressurized air is used to inject the resin into the mold as shown in Figure 6.49. Single or multiple ports are used for resin injection. For small components, a single port is typically used. For long and large structures, multiple ports are used for uniform resin distribution as well as for faster process cycle time. In general,

168   Composites Manufacturing: Materials, Product, and Process Engineering

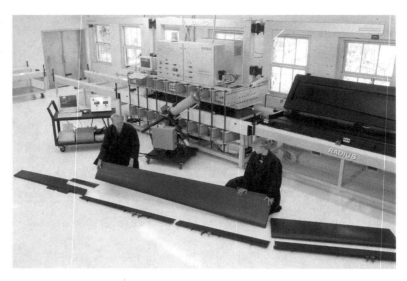

FIGURE 6.53
Outboard and inboard aircraft flaps, spoilers, and trim tabs. (Courtesy of Radius Engineering, Inc.)

FIGURE 6.54
RTM dispensing equipment. (Courtesy of Liquid Control Corp.)

the resin is injected at the lowest point of the mold and flows upward against gravity to minimize entrapped air. Vents are located at highest point of the mold.

In an RTM process, resin flow and fiber wet-out are critical. Resin flow within the RTM mold is determined by various parameters, including injection pressure, vacuum in the mold, resin temperature, viscosity, and preform permeabilities. Preform permeability depends on fiber material type, fiber architecture, fiber volume fraction, through-ply vs. in-plane flow, and several other factors. During mold filling, the resin follows the path of least resistance and experiences difficulty when impregnating tightly packed yarns and reinforcing fibers.

Various computer models have been developed to numerically simulate and visualize resin flow through preforms. These models are used during the mold design phase to predict the optimal locations for inlet ports and vents and optimal injection sequencing, thus achieving the goals of minimal inlet pressure and fill time and elimination of incomplete mold filling or dry spots. These models typically rely on a two-dimensional version of Darcy's law and finite element analysis to describe resin flow through the preforms. Because of the two-dimensional nature of these models, they are limited to simple geometries with uniform thickness.

Dry spots (or improper wet-out) are the biggest challenge for the RTM process. Dry spots in a composite structure lead to part rejection and therefore are directly related to production yield. To aid good resin flow and to avoid dry spots, vacuum at the vents is sometimes applied to displace air within the reinforcements. Vacuum also helps in rapid mold filling. Mold filling is achieved rapidly before the onset of cross-linking. Once the mold is completely filled with resin, it is allowed to cure rapidly for faster demolding. For unsaturated polyesters and vinylesters, cross-linking starts at room temperature, whereas other resins require heat for rapid curing. During curing, the vent is closed and a certain back-pressure is maintained at inlet ports until the resin gels. Once the resin is cured, the mold is opened and the part is removed.

6.8.6.5 Methods of Applying Heat and Pressure

The pressure during the RTM process is applied in the mold using resin injection pressure. This injection pressure helps the resin to flow inside the mold through porous media and allows the resin to fill the cavity. The RTM equipment has a compressor that injects resin at a certain pressure. Typically, the injection pressure is low and in the range of 10 to 100 psi (69 to 690 kPa). The injection pressure determines the flow rate of the resin, and thus the mold filling time. After the mold is completely filled with the resin, the pressure in the mold during curing is kept at around 2 to 10 psi. Resin injection pressure depends on the resin viscosity, mold size, permeability of porous media, mold fill time needed, and cure kinetics of the resin. The resin viscosity for RTM processing is kept low, typically between 100 and 500 cP

FIGURE 6.55
Preform fabrication using vacuum debulking under controlled-temperature parameters. (Courtesy of Intellitec.)

(centipoise) so that it does not go beyond the pumping capabilities of the dispensing equipment.

Temperature selection during RTM processing depends on the type of resin. The resin supplier recommends specific processing conditions, including the preheat temperature, mold temperature, and curing temperature.

6.8.6.6 Basic Processing Steps

For simplicity in understanding the complete RTM process, the basic steps for the fabrication of a composite component are shown in Figures 6.55 through 6.58. Figure 6.55 shows preform fabrication using vacuum debulking under controlled-temperature parameters. The preform is loaded into a mold as shown in Figure 6.56. Figure 6.57 illustrates injection of the resin into mold under exacting processing parameters. A typical RTM mold demonstrating standard design features, such as oil connections, injection runner system, and self-clamping/loading devices, is shown in Figure 6.58. After processing is complete, the part is removed from the mold, as shown in Figure 6.59, and then machining and finishing operations are performed. The amount of machining and finishing depends on the complexity of the part. Figure 6.60 represents a manufacturing cell showing multicavity semi-automated tooling for the production of engine vanes. This manufacturing cell is used to make high-volume RTM parts and utilizes the "drop and shoot" molding technique. Using this technique, Intellitec annually produces more than 35,000 engine vanes for AlliedSignal, and this number can be increased by adding more tools or by adding more shifts.

The steps during the RTM process are summarized below:

1. A thermoset resin and catalyst are placed in tanks A and B of the dispensing equipment.
2. A release agent is applied to the mold for easy removal of the part. Sometimes, a gel coat is applied for good surface finish.

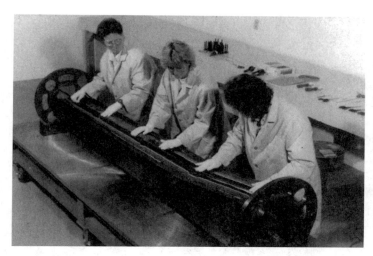

FIGURE 6.56
Preform loading into a mold using innovative mold rotating fixture. (Courtesy of Intellitec.)

FIGURE 6.57
Resin injection into a mold by a technician. (Courtesy of Intellitec.)

3. The preform is placed inside the mold and the mold is clamped.
4. The mold is heated to a specified temperature.
5. Mixed resin is injected through inlet ports at selected temperature and pressure. Sometimes, a vacuum is created inside the mold to assist in resin flow as well as to remove air bubbles.
6. Resin is injected until the mold is completely filled. The vacuum is turned off and the outlet port is closed. The pressure inside the mold is increased to ensure that the remaining porosity is collapsed.
7. After curing for a certain time (6 to 20 min, depending on resin chemistry), the composite part is removed from the mold.

172   Composites Manufacturing: Materials, Product, and Process Engineering

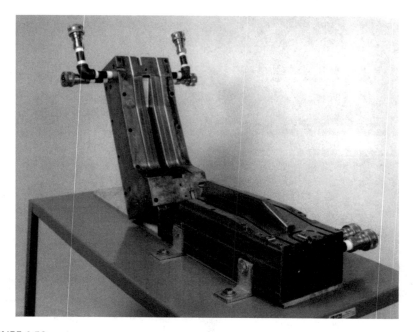

FIGURE 6.58
A typical RTM mold demonstrating standard design features such as oil connections, injection runner system, and self-clamping/loading devices. (Courtesy of Intellitec.)

FIGURE 6.59
Demolding component in a dedicated tool breakout cell. (Courtesy of Intellitec.)

Manufacturing Techniques 173

FIGURE 6.60
Manufacturing cell depicting multicavity semi-automated tooling for the manufacture of engine vanes. (Courtesy of Intellitec.)

6.8.6.7 *Advantages of the Resin Transfer Molding Process*

Recently, RTM has gained importance in the composites industry because of its potential to make small to large complex structures in a cost-effective manner. RTM provides opportunities to use continuous fibers for the manufacture of structural components in low- to medium-volume environments. Some of its major advantages over other composites manufacturing techniques include:

1. Initial investment cost is low because of reduced tooling costs and operating expenses as compared to compression molding and injection molding. For this reason, prototypes are easily made for market evaluation. For example, the dish antenna was first made using an RTM process to validate the design features before capital investment was made for compression molding of SMC parts.
2. Moldings can be manufactured close to dimensional tolerances.
3. RTM processing can make complex parts at intermediate volume rates. This feature allows limited production runs in a cost-effective manner. This lends benefits to the automotive market, in which there is a growing need toward lower production volumes per car model and quicker changes to appeal to more niche markets.

4. RTM provides for the manufacture of parts that have a good surface finish on both sides. Sides can have similar or dissimilar surface finishes.
5. RTM allows for production of structural parts with selective reinforcement and accurate fiber management.
6. Higher fiber volume fractions, up to 65%, can be achieved.
7. Inserts can be easily incorporated into moldings and thus allows good joining and assembly features.
8. A wide variety of reinforcement materials can be used.
9. RTM offers low volatile emission during processing because of the closed molding process.
10. RTM offers production of near-net-shape parts, hence low material wastage and reduced machining cost.
11. The process can be automated, resulting in higher production rates with less scrap.

6.8.6.8 Limitations of the Resin Transfer Molding Process

Although RTM has many advantages compared to other fabrication processes, it also has the following limitations.

1. The manufacture of complex parts requires a good amount of trial-and-error experimentation or flow simulation modeling to make sure that porosity- and dry fiber-free parts are manufactured.
2. Tooling and equipment costs for the RTM process are higher than for hand lay-up and spray-up processes.
3. The tooling design is complex.

A comparison of RTM with other molding processes is presented in Table 6.3.

TABLE 6.3

Comparison of RTM with Other Molding Processes

Molding Process	Production Rate/Year	Cycle Time (min)	Emission Concerns	Two-Sided Part Possible?
RTM	200–10,000	6–30	No	Yes
Open molding (hand lay-up, spray-up)	100–500	60–180	Yes	No
SRIM	More than 10,000	3–15	No	Yes
Compression molding (SMC, BMC)	More than 10,000	1–10	No	Yes
Injection molding	More than 20,000	0.5–2	No (very safe)	Yes

6.8.6.9 Variations of the RTM Process

There are several variations of the RTM process that are used in the commercial sector. Some of them are described below.

6.8.6.9.1 VARTM

VARTM is an adaptation of the RTM process and is very cost-effective in making large structures such as boat hulls. In this process, tooling costs are cut in half because one-sided tools such as open molds are used to make the part. In this infusion process, fibers are placed in a one-sided mold and a cover, either rigid or flexible, is placed over the top to form a vacuum-tight seal. A vacuum procedure is used to draw the resin into the structure through various types of ports. This process has several advantages compared to the wet lay-up process used in manufacturing boat hulls. Because VARTM is a closed mold process, styrene emissions are close to zero. Moreover, a high fiber volume fraction (70%) is achieved by this process and therefore high structural performance is obtained in the part.

6.8.6.9.2 SCRIMP

SCRIMP stands for Seemann Composite Resin Infusion Molding Process, a patented technology of the Seemann, TPI, and HardCore Composites companies. SCRIMP works especially well for medium- to large-sized parts. The process is similar to the VARTM process. In this process, a steady vacuum of 27 to 29 in. Hg is drawn to first compact the layers and then to draw resin into the layers. In this way, fiber compaction occurs before the filaments and core are infused with resin, thus eliminating voids and ensuring accurate placement. In addition, SCRIMP uses a patented resin distribution system consisting of a special resin flow medium combined with simple mechanical devices as shown in Figure 6.61. The major use of the SCRIMP process is in the marine industry.

Applications for the SCRIMP process include marine docking fenders for seaports, windmill blades, satellite dishes, railroad cars, buses, customized car bodies, amusement park rides, physical therapy pools, and aerospace components. The biggest advantages of the SCRIMP process compared to the wet lay-up process lie in its weight control and absence of styrene emissions. Moreover, dry lay-up of fabric and core materials saves labor and time during processing.

6.8.7 Structural Reaction Injection Molding (SRIM) Process

The SRIM process is similar to the RTM process, with the difference being in the resin used and the method of mixing resins before injection. In the SRIM process, two resins A and B are mixed in a mixing chamber at a very high velocity just before injecting into the mold (Figure 6.62). The resin flows at a speed of 100 to 200 m/s and collides in the mixing chamber. The pressure

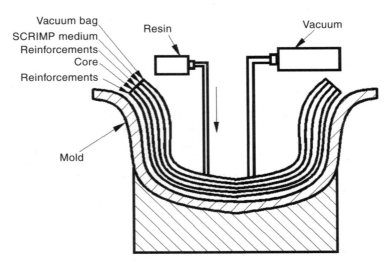

FIGURE 6.61
Schematic of the SCRIMP process.

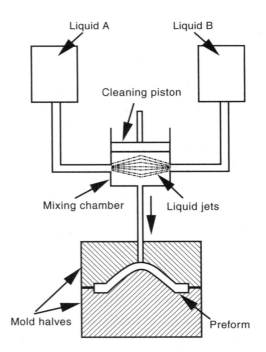

FIGURE 6.62
Schematic of the SRIM process.

generated during collision is in the range of 10 to 40 MPa although the resin is injected into the mold at a pressure of less than 1 MPa. Low pressure is used to prevent the wash-out of fibers at the injection port. The resin used for the SRIM process is of very low viscosity and the most common resin is polyisocyanurate (based on polyurethane chemistry). This mixed resin is injected into the mold, which contains fiber preforms. The preform used for the SRIM process can be made of short or long fibers.

The SRIM process was developed from the RIM (reaction injection molding) process, in which two resins are mixed at high velocity in a mixing chamber and then the resin is injected into a closed mold where there is no preform. The components made by the RIM process is weak because of absence of fibers. The RIM process is used for making automotive components because it provides a high-volume production capacity. In RIM and SRIM processes, cross-linking of the polymers is initiated by the rapid mixing of two resins; therefore, this process does not require heat to start the curing (cross-linking) process. The cross-linking process is very rapid and the resin begins to gel in few seconds after injection into the mold. The mold is heated to a certain temperature to aid the rapid cross-linking of the resin. To ensure complete cross-linking, the two resins must be mixed in the correct ratio. This requirement (the correct amount of mixing) as well as fast impingement increase the cost of RIM equipment.

To add strength to RIM parts, the RRIM (reinforced reaction injection molding) process was developed. In RRIM, short or milled glass fibers are added to one of the resin components before being mixed into the chamber. The fiber lengths are kept less than 0.02 in. (0.5 mm) in order to keep the resin viscosity low. RRIM parts are stiffer and more damage tolerant than RIM parts; however, from a structural point of view, RRIM parts are weaker. The RRIM process is used in automotive applications to make body panels and fascia. To utilize the short processing cycle time of the RIM process, SRIM was developed. SRIM is basically the combination of RIM and RTM, and is used for making structural parts of reasonable strength.

6.8.7.1 Major Applications

The SRIM process is used in applications (e.g., the automotive industry) where high-volume production at low cost is required. SRIM parts can be made in a process cycle time of 1 to 5 min, depending on part size. Recently, the Automotive Composites Consortium (general partnership of Daimler Chrysler, Ford, and General Motors) utilized the SRIM process to make pickup truck boxes. Other applications include bumper beams, fascia, and body panels.

6.8.7.2 Basic Raw Materials

The reinforcements for this process are preforms made of short or long fibers. Glass is the most common fiber type used in this process. With the drop in carbon fiber price, carbon fiber is also being considered for the SRIM process. The main resin materials for this process are polyurethane-based resins (polyisocyanurate resin). The resin viscosity for the SRIM process is quite low (10 to 100 cP) compared to the RTM process (100 to 1000 cP). The reaction rate for the resins used in the SRIM process is much faster than the polyester and vinylester resins that are mostly used in the RTM process.

6.8.7.3 Tooling

SRIM utilizes a closed mold similar to the RTM process but tools for the SRIM process are heavier and more expensive than RTM tools. In general, the tools are made of steel. Refer to Section 6.8.6.3 for more information on tooling requirements.

6.8.7.4 Making of the Part

The procedure to make SRIM parts is very similar to the RTM process, the main difference being that in SRIM, the resin dispensing equipment and resin mixing procedure are different. Moreover, the resin used in SRIM is chemically more reactive than in RTM. In the SRIM process, the preform is typically made of short fibers; whereas in RTM, the preform is usually made of continuous fibers (braided preforms) or short fibers.

Figure 2.18 in Chapter 2 depicts a schematic for making preforms. Single or multiple guns direct the glass fibers and thermoplastic-fiber binder onto a preform screen. Vacuum keeps the fibers in place. The amount of thermoplastic binder is minimal, just enough to hold the preform together. Once the preform is prepared, it is placed in the mold for resin injection. The mold is clamped and resin is rapidly injected. The clamping force needed for the SRIM process is less than that for compression and injection molding processes and greater than the RTM process. Care is taken to avoid fiber washaway. Because the chemical reactivity of the resin is very high, the resin must completely fill the mold before it starts gelling. The process cycle time is usually 1 to 5 min, depending on the size and geometry of the part. Once the part has solidified, the mold is unclamped and the part is removed from the mold.

6.8.7.5 Methods of Applying Heat and Pressure

The resins used in the SRIM process have a high reaction rate and therefore the mold cavity needs to be filled rapidly. This also means that the rate of resin injection is very high in the SRIM process. Because of the high injection rate, care should be taken to avoid the fiber wash-out due to advancing resin front. Fiber wash-out can be minimized by having a low fiber volume fraction and low resin viscosity. The resin is injected into the mold using dispensing equipment. The inlet pressure for the SRIM process falls in the same range as for the RTM process. For curing of the resin, no heat is required.

6.8.7.6 Basic Processing Steps

The major manufacturing steps used during the SRIM process can be summarized as follows:

1. The preform is prepared on-site or bought from a supplier.
2. The release agent is applied to the mold and the preform is placed on the mold.
3. The mold is clamped.
4. The resin inlet hose is connected to the inlet ports of the mold.
5. The mold is preheated according to requirements.
6. Resin mixing is initiated by operating the dispensing equipment.
7. Resin is injected into the mold.
8. After mold filling and curing of the resin, the mold is declamped.
9. The composite part is removed from the mold.

6.8.7.7 Advantages of the SRIM Process

SRIM is becoming the process of choice for applications where high-volume production is required. SRIM provides the following advantages:

1. It is very suitable for making high-volume structural parts at low cost, in particular for making automotive parts.
2. Small- to large-sized parts with complex configurations can be made with this technique.

6.8.7.8 Limitations of the SRIM Process

SRIM is highly suitable for high-volume applications. However, SRIM has the following limitations:

1. It requires a large capital investment in equipment.
2. The tooling cost for the SRIM process is high.
3. A high fiber volume fraction cannot be attained by this process; the maximum fiber volume fraction achievable is about 40%.

6.8.8 Compression Molding Process

Compression molding is very popular in the automotive industry because of its high volume capabilities. This process is used for molding large automotive panels. Sheet molding compounds (SMCs) and bulk molding compounds (BMCs) are the more common raw materials for compression molding. Compression molding is also used for making structural panels using prepregs and core materials; but because of the popularity of compression molding of SMC, the molding of SMC is discussed here.

Compression molding is popular in the auto industry because of its similarity to the stamping process. The auto industry has been using the stamping process for a long time and has built good know-how for this process. Similarly today, compression molding has become quite a mature process for the auto industry. In compression molding of SMC, the final component is produced in one molding operation, whereas in the stamping operation, the steel sheet metal goes through a series of stamping processes to get the final shape. Therefore, compression molding provides several advantages over the stamping process and saves significant cost in terms of molds and equipment. One of the advantages of SMC over steel lies in its ability to include ribs and bosses in a third dimension. Holes, flanges, shoulders, and nonuniform thicknesses can be created during the molding process, thus avoiding secondary operations such as welding, drilling, and machining.

Compression molding is used for making Class A surfaces. For Class A surfaces, the overall percentage of fiber content is limited to 30% to optimize smoothness of the surface as well as to reduce fiber read-through. Therefore, there is a trade-off in getting mechanical property enhancement or Class A surface quality. Short ribs are used on body panels and other one- and two-piece components are used to increase the stiffness of these panels. Ribs on Class A panels should be used cautiously because of their potential to cause sink marks. However, with proper design, sink marks formed on the top side can be eliminated or minimized.

6.8.8.1 Major Applications

Compression molding of SMC is used for making one- and two-piece panels for automotive applications. One-piece panels are used where the panel is fixed and can be supported around the majority of its periphery. Roof panels, quarter panels, fenders, and add-ons such as spoilers, ground-effects packages, and limited-access panels are examples of this category. Two-piece panels are used for closure panels such as doors, hoods, and decklids. These panels are usually supported by hinges with body structures and therefore should have enough rigidity of their own. The stiffness of two-piece panels is obtained by joining inner and outer panels as shown in Figure 6.63.

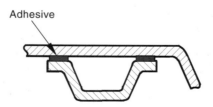

FIGURE 6.63
Compression molded two-piece panel. Two-piece panels rely on closed sections for stiffness.

Manufacturing Techniques

Other applications of the compression molding process include skid plates, military drop-boxes, pickup box components, radiator supports, heavy-truck sleeper cabs, engine components such as rocker covers/oil pans, personal water crafts (PWCs), and home applications such as showers/tubs. Electrical applications of compression molding processes are enclosures, fuses, outdoor lamps, lamp housings, switches, street light canopying, and more.

6.8.8.2 Basic Raw Materials

SMC, BMC, and TMC are used as initial raw materials for compression molding operations. These molding compounds are described in detail in Chapter 2. The SMC is obtained by mixing liquid resin, fillers, and fibers into a sheet product that is usually about 4 mm (0.16 in.) thick and 1.2 m (47.25 in.) wide. SMC is stored in rolled form or in a stack of rectangular or square pieces. SMC has a limited shelf life and the part should be usually manufactured within 2 weeks of manufacturing the molding compound. The material is stored at B-stage.

These molding compounds are fairly inexpensive material and typically cost $0.70 to $1.50/lb. The main ingredients in these materials are glass fibers, polyester resins and calcium carbonate; up to 50% fillers can be used in SMC.

6.8.8.3 Making of the Part

In compression molding operation, the SMC is cut into rectangular sizes and placed on the bottom half of the preheated mold as shown in Figure 6.64. These rectangular plies are called charge. The charge usually covers 30 to 90% of the total area, and the remaining area is filled by forced flow of the charge. The amount of charge is determined by calculating the final volume or weight of the part. The mold is closed by bringing the upper half of the mold to a certain velocity. Typically, the working speed of the mold is 40 mm/s with SMC and 80 mm/s with GMT. GMT is made using glass fiber mat and thermoplastic resins such as polypropylene (PP). Compression molding of GMT is discussed in the thermoplastic manufacturing processes section (Section 6.9).

In compression molding, the molds are usually preheated to about 140°C. With the movement of mold, the charge starts flowing inside the mold and fills the cavity. The flow of the molding compound causes removal of entrapped air from the mold as well as from the charge. After a reasonable amount of cure under heat and pressure, the mold is opened and the part is removed from the mold. Ejector pins are often used to facilitate easy removal of the part. Typical mold cycle times are about 1 to 4 min for a one-piece panel, and 1 to 2 min for a two-piece panel. Two-piece panels require an assembly time of 1 to 2 min. Figure 6.65 shows a molding press with ejector pins ejecting an automotive part. An SMC molded automotive part is shown in Figure 6.66.

The parts made by compression molding are usually thin as compared to RTM, injection molding, and other manufacturing processes. For thin sections, the temperature across the thickness remains uniform and is in the

182 Composites Manufacturing: Materials, Product, and Process Engineering

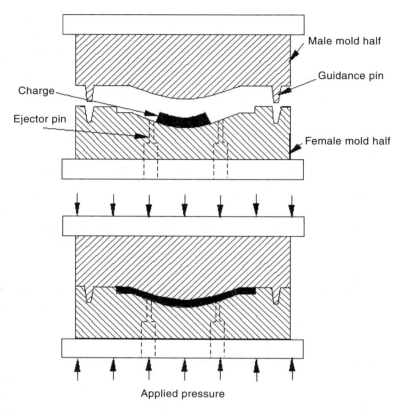

FIGURE 6.64
Schematic of the compression molding process.

neighborhood of the mold temperature. Uniform temperature across the thickness allows uniform curing in the part and thus avoids residual stress in the part caused by curing. For thick cross sections, the temperature distribution is not uniform across the thickness. In thick sections, the layer adjacent to the mold reaches the mold temperature quickly and remains uniform. It takes some time for the centerline layer to reach the mold temperature because of the low thermal conductivity of the charge material. However, because of the exothermic curing reaction of resin material, the temperature of the centerline layer goes above the mold temperature. As curing completes, the centerline temperature flattens to mold temperature.

Compression molding of SMC has the potential for several defects. The defects could be on the outer surface (such as an unacceptable surface finish or fiber read-throughs) or internal defects (such as porosities, blisters, weld lines, and warpage). Porosities are caused by air entrapment, whereas blisters are interlaminar cracks formed at the end of molding due to excessive gas pressure inside the molded part. This internal gas pressure may result from unreacted styrene monomer in undercured parts or from large pockets

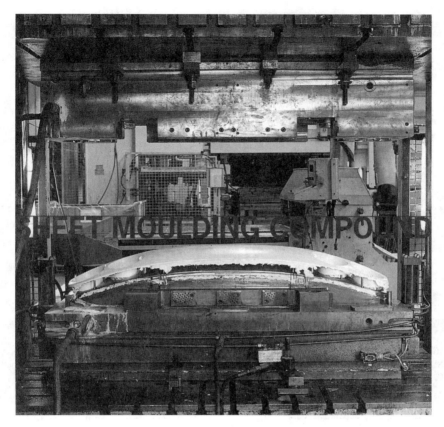

FIGURE 6.65
Compression molding press. (Courtesy of Ranger Group, Italy.)

of entrapped air between the stacked layers. Weld lines are formed when two flow fronts meet inside the mold. Weld lines have poor mechanical properties normal to the weld line because the fibers tend to align themselves along the weld line. Warpage in the composite results from residual stress. Typically, parts with nonuniform cross-section have variations in cooling rate between sections of different thickness. This differential cooling rate is a potential source of residual stress and thus warpage in the component.

SMC panels usually replace steel panels in the auto industry. The tensile modulus of SMC R25 is 1.2 to 1.8 Msi, SMC R50 is 1.8 to 2.8 Msi, and steel is 30 Msi. In one-for-one replacement, the SMC cross section needs to be larger than the comparable steel section to provide the same stiffness. In many cases, closed sections, as shown in Figure 6.67 for two-piece panels, require section enlargement only in one direction (thickness direction) if loads are applied in one plane. If section enlargement is not possible, then equal stiffness is achieved through the use of ribs, doublers, or added higher modulus fibers. Figure 6.67 shows the rib geometry for Class A surface and

FIGURE 6.66
SMC molded automotive component. (Courtesy of Ranger Group, Italy.)

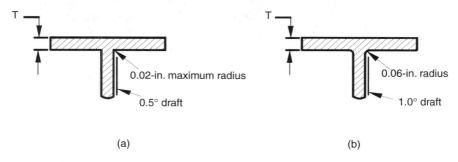

FIGURE 6.67
Rib geometry for various types of surfaces: (a) Class A surface and (b) nonappearance surface.

nonappearance surface. The base of ribs opposite the Class A surface should be 75% of the nominal panel thickness.[24] The minimum allowable thickness for ribs is 0.06 in. A minimum of 0.5° of draft per side is required to allow for part removal from the mold. In low visibility areas, the design criteria for ribs are relaxed, as shown in Figure 6.67b, to enhance moldability and structural performance.

Dimensional control and part repeatability of a process are very important requirements for the automotive industry. Some sources of variation in compression molded parts that affect process repeatability include:

1. *Press parallelism.* The thickness variation in a molded part depends on press parallelism tolerances as well as mold tolerances. Each equipment supplier has its own tolerances on parallelism. Advance-control presses provide low tolerance variation, but they are expensive. For two-piece panels, thickness variation can be compensated by varying bond line thickness.
2. *Mold tolerances.* Every mold and tool has tolerances that can affect part tolerances. The use of computer math data in cutting the part surface of the mold increases the accuracy of the mold over large surface areas.
3. *Molded datum features.* In dimensioning, geometric characteristics of a part are established from a datum. A datum feature is an actual feature of a part that is used to establish a datum. The use of molded datum features will increase the repeatability in bonded assemblies. The molded datums reduce the need to locate and maintain datums on the secondary fixtures and thus reduce additional sets of tolerances.
4. *Material shrinkage.* Shrinkage is the reduction in volume or linear dimension caused by curing of the resin as well as by thermal contraction of the material. Materials expand when heated and shrink when cooled. If the molded panel contracts more than the mold material when it is cooled from mold temperature to room temperature, then it is called shrinkage in the part. If the molded part shrinks less than the mold material, it is called expansion. Some formulations can exhibit an apparent expansion of up to 0.15%. Curing shrinkage occurs because of the rearrangement of polymer molecules into a more compact mass as the curing reaction proceeds. The volumetric shrinkage of cast polyester and vinylester resins is in the range of 5 to 12%, whereas for epoxy, it is 1 to 5%. The addition of fibers and fillers reduces the amount of shrinkage. A typical non-Class A formulation will have an apparent shrinkage of more than 0.05%.

Molds for compression molding are built by keeping in mind the type of material formulations to be used in making parts. Therefore, it is important to verify part fit if material formulation is changed after the mold is machined.

It is very important to know how to create a Class A surface during the molding operation. The following are some guidelines for making a Class A surface in molded parts[16]:

1. *Avoid flatness.* On flat panels, reflected highlights often make the panel look distorted and therefore a contoured surface is recommended to eliminate objectionable highlights on high-gloss surfaces.
2. *Maintain uniform thickness.* While designing the panel, maintain a uniform thickness across the length for uniform flow of material

as well as for uniform cooling rate. This will avoid the risk of part distortion and telegraphing of thickness changes through the surface. For one-piece Class A panels, the nominal thickness is typically 0.14 in. For two-piece panels, it is 0.08 to 0.12 in. for outer panels and 0.07 to 0.10 in. for inner panels.

3. *Inner and outer radii.* Sharp corners must be avoided during design. A minimum inside corner radius of 0.08 in. and minimum outside corner radius of 0.06 in. are recommended for better material flow along the corner as well as for ease in part removal. The general rule for paintability is to make inner and outer radii as large as possible without sacrificing part function.

4. *Create mash-offs for hole location.* A mash-off is a localized reduction in panel thickness for creating holes during secondary operation. It is a membrane-like thin area that can be easily removed. Charge can be placed directly over the mash-off area. This promotes a more homogeneous material flow through the hole location, which improves mechanical properties around the hole.

6.8.8.4 Mold Design

Compression molding is a match mold operation in which male and female molds are prepared. Because the charge is placed on the bottom half of the mold prior to molding, there is no inlet port as in the RTM and SRIM processes. Entrapped air and excess resin escapes from the parting line as shown in Figure 6.68. The shear edge design shown in the figure reduces the need for edge trimming and thus aids in near-net-shape part molding. The shear edge closes before the flow front reaches the parting line and therefore does not allow reinforcements to escape. A stop block is designed to prevent damage to the mold during the mold closing operation.

The mold is usually made of steel and is nickel or chrome plated to improve the surface finish of the molded part and the wear resistance of the mold.

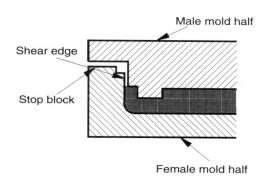

FIGURE 6.68
Shear edge in compression mold.

Mold halves are guided as shown in Figure 6.68 to maintain good alignment of mold halves and thus to minimize thickness tolerances on the part. Mechanical or pneumatic ejector pins are installed for easy removal of the molded part. The mold design is stiffness critical and therefore should be able to take the pressure exerted during molding without any appreciable deflection. The mold pressure is about 2 to 30 MPa (290 to 4350 psi). High pressure is required for parts that contain deep ribs and bosses.

6.8.8.5 Methods of Applying Heat and Pressure

The heat necessary for curing the resin during the compression molding process is supplied by the mold. The mold is usually preheated to about 120 to 170°C. Mold heating is performed either by circulating steam, oil, or by electrical cartridge. The pressure for uniform flow of charge material in the cavity is supplied by the relative movement of the mold halves. The upper half of the mold is moved at a speed of 40 mm/s during the molding operation. This relative movement of the mold halves causes the charge material to flow rapidly inside the cavity.

6.8.8.6 Basic Processing Steps

For a clear understanding of the process, the major steps performed during compression molding operations are descibed below.

1. The total volume or weight of the final part is calculated and, based on that, the amount of charge material is determined.
2. The charge material is brought from the storage area and cut to a specific size, for example, a 1 × 4-in. rectangular strip or 6 × 12-in. rectangular strip. The carrier film from charge material is removed.
3. The mold is preheated to about 140°C, or as required by the resin formulation.
4. The charge material is placed on the preheated lower mold half at locations determined by the manufacturing engineer. The charge locations are judiciously selected to get uniform flow across the surface as well as to achieve better mechanical performance. When high surface quality is required, the charge covers a small portion of the mold because flow promotes a good surface finish. When high mechanical performance is desired, the charge covers a larger area to minimize flow-induced reinforcement disorientation.
5. The upper half of the mold is closed rapidly at a speed about 40 mm/s. This rapid movement causes the charge to flow rapidly inside the cavity.
6. After about 1 to 4 min of cure cycle, the upper mold half is moved back and thus releases the pressure from the mold.
7. The part is demolded using ejector pins.

6.8.8.7 Advantages of the Compression Molding Process

Compression molding of SMC has become quite a popular process and has established a good name for itself in the automotive industry. Compression molding of SMC provides the following advantages.

1. It offers high-volume production and thus is very suitable for automotive applications. The mold cycle time is only 60 to 240 s (1 to 4 min).
2. It offers production of low-cost components at high volume because it utilizes SMC, which is fairly inexpensive.
3. The process offers high surface quality and good styling possibilities.
4. Multiple parts can be consolidated into one single molded part and thus is very advantageous compared to the metal stamping process.
5. Today, automotive companies are looking for ways to economically differentiate car and truck models with shorter production runs and more rapid design-to-production schedules. Compressive molding of SMC offers this benefit.

6.8.8.8 Limitations of the Compression Molding Process

Despite the many advantages of compression molding, this process has the following limitations.

1. The initial investment for the process is high because of high equipment and mold costs. However, this initial investment is low compared to sheet metal stamping processes.
2. The process is not suitable for making a small number of parts or for prototyping applications.
3. Compression molding of SMC provides nonstructural parts; but by utilizing ribs and stiffeners, structural parts can be manufactured.

6.8.9 Roll Wrapping Process

Roll wrapping is similar to prepreg lay-up with the exception that the tool is always a cylindrical or round tapered mandrel. The process requires a low initial investment and is suitable for high-volume production of tubular components. In this process, the prepreg is rolled over a removable mandrel and then shrink tape is wrapped for consolidation. The entire assembly is then batch cured for solidification.

6.8.9.1 Major Applications

The roll wrapping process is widely used for making golf shafts, fishing rods, bicycle frames, and other tubular shapes. Tens of millions of golf shafts as well as fishing rods are manufactured every year using this method. Some golf shaft manufacturing companies have the capacity to manufacture over 5 million golf shafts per year. Golf shafts are manufactured on a tapered mandrel with a wide variety of diameters and lengths to fit various user groups such as men, women, and children. The weight of a shaft falls in the range of 40 to 60 g. Fly rods used for fishing are also manufactured in a wide range, from lightweight rods suitable for trout fishing to rods capable of reeling in saltwater fish such as a 200-lb marlin. For a typical trout rod, wall thicknesses range from 0.12 in. at the butt to 0.011 in. at the tip.[25] At the tip of the rod, only 2 or 3 prepreg layers are used, whereas the butt carries about 8 to 20 prepreg layers. Similarly, there are casual anglers and serious anglers, and their needs are different.

6.8.9.2   Basic Raw Materials

The raw materials used for the roll wrapping process are, in general, carbon/epoxy unidirectional prepregs. Standard-modulus (33 to 35 Msi) carbon fibers to high-modulus (54 Msi) carbon fibers are used with various grades of epoxy ranging from standard to toughened. For manufacturing golf shafts, 0°, 90°, and 45° fiber orientations are used. For fishing rods, only the 0° fiber orientation is used. However, to obtain hoop strength, a lightweight glass scrim (about 35 g/m^2) is used. The weight of the prepreg is in the range of 100 to 175 g/m^2 excluding the weight of the glass scrim. The use of high-modulus fibers in the fly rod increases the thrust with which the rod propels the fly line forward and thus increases the casting distance. High-modulus fibers provide one major disadvantage due to lower tensile elongation. Lower elongation means the rod can easily break during bending while casting or reeling in a fish. This problem is alleviated by using the proper fiber orientation and taper rate from butt to tip.

6.8.9.3   Tooling

The roll wrapping process utilizes low-cost tooling systems. Steel mandrels in large quantities are used in this process. The cost of a steel mandrel for a golf shaft application ranges from $50 to $200. The mandrel for a golf shaft is a tapered mandrel with a maximum diameter of 1 in. and a taper angle of 10 to 15°. Standard dimensions are between 0.58 and 0.60 in. for the outside diameter at the butt end of the shaft, 0.335 in. at the tip of the driver, and 0.370 in. at the tip of an iron.

For making cylindrical parts using this process, a slight taper angle of 1 to 2° is given to aid in easy removal of the mandrel. For fly rods, the mandrel is about 8 to 10 ft long, 0.3 to 0.45 in. in diameter at the butt, and about 0.025 in. in diameter at the tip. Standard lengths for U.S. golf club shafts are 45 in. for drivers and 40 in. for irons.[25]

6.8.9.4   Making of the Part

Roll wrapping is used predominantly in the golf shaft and fishing industries and therefore this section emphasizes the production of golf shafts and

FIGURE 6.69
Prepreg cutting using a template. (From Benjamin, B. and Bassett, S., High Performance Composites, Ray Publishing, July/August 1996.)

fishing rods with this technology. Golf shaft and fishing rod manufacturers make a wide variety of product lines based on various design parameters; however, the two industries share similar manufacturing steps.

In the roll wrapping process, the prepreg sheet is first rolled out on a cutting table and then cut to rectangular or trapezoidal shapes, either manually or with the help of an automatic prepreg cutter. In manual cutting, a template and knife are used, as shown in Figure 6.69, to cut the prepreg. Any fiber orientations, including 0°, 90°, 45°, can be obtained with the help of this process. Templates provide a pattern for cutting the prepreg. For fishing rod applications, fiberglass scrim is placed on top of the prepreg with the fiberglass oriented at 90° relative to the caron fiber direction, which is at 0° fiber orientation.[25] The fiberglass provides hoop strength on the rod. Both the prepreg and scrim are cut simultaneously using the template. Trapezoidal shapes are cut in such a way as to provide more prepreg layer on the butt section than on the tip section. Placing a greater number of prepreg layers on the butt section is also achieved by graduating the length of the plies. One ply covers the entire length, whereas other plies cover shorter lengths. For aesthetic reasons, as well as to create fiber continuity on the exterior surface, the longest ply is placed on the outermost layer.

Wax or a release agent is applied to the mandrel surface and then placed on the rolling table. Figure 6.70 shows a worker ironing the prepreg for wrapping. Prepreg plies are stacked together and brought to the rolling table. The mandrel is placed over the prepreg layers at a specific angle and then rolled carefully. Figure 6.71 shows a rolling table and press for rolling the prepreg onto the mandrel. The press is shown at the back of rolling table in Figure 6.71. A clear picture of rolling table can be seen in Figure 6.72. The

Manufacturing Techniques 191

FIGURE 6.70
Worker irons leading edge of prepreg onto mandrel. (From Benjamin, B. and Bassett, S., High Performance Composites, Ray Publishing, July/August 1996.)

FIGURE 6.71
Mandrel and prepreg stack are placed on the rolling table. (From Benjamin, B. and Bassett, S., High Performance Composites, Ray Publishing, July/August 1996.)

prepreg is also rolled by hand instead of utilizing a press. While rolling, pressure is applied at the rolling table to ensure good prepreg and mandrel contact as well as good interlayer contact. The tip of the mandrel travels a shorter distance than the butt ends because of tapering of the mandrel. In the golf shaft industry, rolling is primarily done manually. There are

FIGURE 6.72
Rolling table. (Courtesy of Century Design, Inc.)

hydraulic press platens available for the rolling operation; these control the movement of butt and tip ends and apply pressure to create good compaction.

After rolling is completed, the mandrel is moved to a tape wrapping machine where shrink tape is wound over the prepreg layer. A shrink tape wrapper machine is shown in Figure 6.73. Shrink tape is made of polypropylene and is usually 0.5 in. wide and 0.005 in. thick. Shrink tape has the characteristic of shrinking at elevated temperature and thus applying pressure during curing. Once the shrink tape is applied, the entire assembly is placed in an air-circulating oven for batch curing. Approximately 100 to 400 shafts can be batch cured simultaneously. This is achieved by placing individual shafts on a rack and then inserting the entire rack inside the oven, as shown in Figure 6.74. The curing is done at 200 to 300°F for 2 to 4 hr, depending on the requirements of the resin system. As the outer tape shrinks tightly on the rolled prepreg, the air entrapped between the layers is squeezed out at the ends. After curing, the shrink tape is cut with a razor blade and removed manually. The mandrel is pulled out of the newly formed

Manufacturing Techniques 193

FIGURE 6.73
Tape wrapping machine. (Courtesy of Century Design, Inc.)

hollow tube. Figure 6.75 shows a hydraulic mandrel extractor and chain-driven mandrel puller. The exterior of the tube is lightly sanded to remove ridges of excess resin as well as to create good paintability.

The golf shafts or fishing rods thus prepared are trimmed to length and then visually inspected for defects such as fiber alignment or inclusion of foreign materials. Then the tube undergoes a bend test to ensure that it meets the flexibility standard for its model.

After the bend test, several finishing operations are performed on the tube, first by dipping the tubes in a sink containing an enamel coating. The tubes are taken from the sink and excess enamel is removed. The coating is then baked in an oven for about an hour at around 220°F. The enamel coating provides the tube with good aesthetics as well as impact resistance. Then the tube hardware is installed. For golf shafts, the grip is placed on the larger diameter end and the club is adhesively bonded at the narrower (bottom) end. For fishing rods, the grip is placed at the butt section and guides are installed along the length of the rod to provide a path for the fly line from the reel to the tip of the rod. For attaching the guides over rods, nylon thread is used to wrap the guide over the rod. The wrapped threads are then coated

FIGURE 6.74
Wrapped mandrels entering the oven for curing. (From Benjamin, B. and Bassett, S., High Performance Composites, Ray Publishing, July/August 1996.)

with two-part epoxy finish. This finish secures the placement of the guides and protects them from moisture.

6.8.9.5 Methods of Applying Heat and Pressure

In roll wrapping, pressure is applied by the shrink tape during the curing process. Shrink tape is wrapped over the prepreg using a tape wrapping machine (Figure 6.73) at a pressure of 10 psi. After wrapping, the entire assembly is placed inside an air-circulating oven at 200 to 300°F for 2 to 4 hr, depending on the process requirements. At high temperature, polypropylene tape shrinks and applies an additional 20 psi to the prepreg, thus giving good consolidation pressure. The pressure also helps in removing any entrapped air.

6.8.9.6 Basic Processing Steps

Roll wrapping has become popular in the sporting goods and recreation industries because of its capability to produce high-end structural parts — at high volume. The basic steps in roll wrapping are listed below for easy understanding:

1. The prepreg is removed from the refrigerator and placed at room temperature for thawing.
2. Thawed prepregs are rolled out on a cutting table. Prepregs are then cut into rectangular or trapezoidal shapes using a template or an automated cutting machine.
3. Wax or release agent is applied to the mandrel.

Manufacturing Techniques 195

FIGURE 6.75
Hydraulic mandrel extractor and chain-driven mandrel puller. (Courtesy of Century Design, Inc.)

4. The cut prepregs are brought to the rolling table and then stacked according to design requirements.
5. The mandrel is brought to the rolling table and placed on top of the stacked prepregs.
6. The mandrel is pressed against the table and then rolled. Because of the tackiness of the prepreg, it adheres to the mandrel and incoming prepreg layers. The rolling operation is performed either manually or using a rolling press.
7. The rolled prepreg is then taken to the tape wrapping machine for applying the shrink tape. The tape machine wraps the 0.5-in.-wide shrink tape along the length of the shaft under tension.
8. The shafts thus produced are placed on a rack.
9. The rack, which contains between 100 and 400 tubes, is then moved to an air-circulating oven for curing.
10. Curing is done for about 2 to 4 hr at 250°F.

11. The cured tubes are removed from the oven and the shrink tape is removed using a razor blade.
12. The mandrel is then pulled out from the composite tube thus fabricated.
13. The shaft is then lightly sanded to remove tape marks as well as ridges of excess resin.
14. The tube is then visually inspected for imperfections such as fiber alignment, insertion of foreign materials, etc. Standard tests are performed to meet product specifications.
15. Finishing operations such as painting and hardware attachment are done and then the product is shipped to the OEM (original equipment manufacturer).

6.8.9.7 Advantages of the Roll Wrapping Process

In the composites industry, the roll wrapping process competes with filament winding and pultrusion processes in tube manufacturing applications. Roll wrapping provides the following advantages over filament winding and pultrusion processes.

1. The initial capital investment in terms of equipment and tooling costs is very small.
2. The process provides a high fiber volume fraction compared to the filament winding and pultrusion processes.
3. The desired fiber orientation can easily be obtained in this process compared to the filament winding and pultrusion processes.
4. The process is very suitable for high-volume production methods.
5. The cost of prototyping is very low in the roll wrapping process.
6. Thin and tapered cross sections can easily be produced using this process.

6.8.9.8 Limitations of the Roll Wrapping Process

Although roll wrapping has several advantages, it suffers from the following limitations.

1. Complex parts cannot be manufactured by roll wrapping because it is limited to the production of tubular structures only.
2. Thick composite parts are not easily produced because there are limitations on applying pressure on the rolled part.

6.8.9.9 Common Problems with the Roll Wrapping Process

The manufacturing engineer faces the following challenges during fabrication of composite parts using the roll wrapping process.

1. It is difficult to obtain consistent fiber orientation because the process of applying the prepregs is done manually. This becomes an issue when various ply orientations are selected.
2. Proper consolidation beomes an issue for producing thick parts. In general, porosities and voids are formed between layers.

6.8.10 Injection Molding of Thermoset Composites

Injection molding is more common in the thermoplastics industry but has also been successfully used in the thermosets industry. It is a high-volume manufacturing process and is suitable for consumer, automotive, and recreation applications. In injection molding, a fixed amount of material is injected into the heated mold cavities. After the completion of cross-linking, the mold opens and the part is dropped into a receiving bin. Typically, the complete process takes about 30 to 60 s. Injection molding has the shortest process cycle time compared to any other molding operation and thus has the highest production rate. The production rate can be further increased by having a multiple-cavity mold. Injection molding is widely used for making small-sized parts, but it can also be used to make large structures. The details of injection molding are described in Section 6.9 on thermoplastic manufacturing processes because of its wide use in the thermoplastics industry.

6.8.10.1 Major Applications

Injection molding is used for high-volume applications. The typical parts produced are sink disposals, sewing machine parts, small power tools, small appliance motors, electrical plug fuses, and more. Glastic Corporation (Cleveland, Ohio) makes 1920 plug fuses per hour with this process.[26] They use 16-cavity molds, 0.5-lb shot size and have a process cycle time of 30 s (120 shots per hour). Glastic Corporation makes up to 30 million composite fuses per year in 14 different fuse configurations. Fiberglass-reinforced polyester (FRP) parts have replaced the brittle ceramic fuses and this saves about 20% in overall cost. Although the initial part cost is more with FRP fuses, the savings derive from less breakage during shipment and assembly. Figure 6.76 shows multi-pin connector insulators, high-voltage transformer bushings, and small engine components produced by the injection molding process.

6.8.10.2 Basic Raw Materials

The raw material for thermoset injection molding is bulk molding compound (BMC). BMC is usually sold in the form of a thick rope or log. It contains 15 to 20 wt% E-glass fibers (6 to 12 mm long), polyester resin, filler, additives, and pigment. For injection molding, the fiber length is kept short, typically between 1 and 5 mm. BMC is described in detail in Chapter 2. There are other injection molding compounds available based on epoxy, vinylester, and phenolic resins.

FIGURE 6.76
Injection molded components. (Courtesy of Cytec Fiberite.)

6.8.10.3 Tooling

Tool steel is used to make the molds for injection molding. Mold flow analysis is performed to come up with the right tool design. Tools for injection molding are very expensive and complex compared to other molding processes (e.g., RTM). Tools for injection molding are discussed in detail in Section 6.4.3.

6.8.10.4 Making of the Part

To make molded parts, bulk molding compound (BMC) is loaded into a hopper and forced by a piston stuffer into the barrel of the molding machine. The material is then conveyed through the barrel by a rotating screw or ram plunger as shown in Figure 6.77. The ram plunger type is preferred when higher mechanical properties in the part are desirable. In a screw-type plunger the screw may chop up the fiberglass and may affect the mechanical properties of the part.

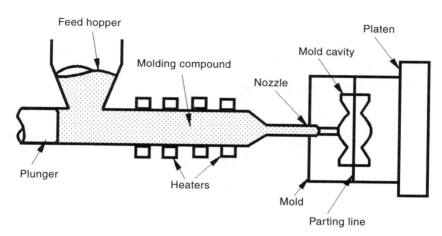

FIGURE 6.77
Schematic of the injection molding process.

The molding compound becomes less viscous as it passes through the heated barrel. When the fiber-filled resin reaches the end of the barrel, the plunger is stopped and then driven forward at high speed to inject the material into the heated mold cavities through sprues and runners. After material injection, a holding pressure in the cavity is maintained so that injected material does not back-flow. The duration to maintain heat and pressure depends on part size, resin chemistry, and wall thickness. After the part has cured, the mold is opened and the part is ejected. A typical mold cycle is 30 to 60 s with BMC.

The complete injection molding process is automatic once BMC is fed into the hopper. The operator enters process parameters into the machines's computer control panel and then the machine is started. The operator closely monitors the material flow, part molding, demolding, deflashing, etc. to make sure the process is continuous and runs without any problem.

For a better understanding of this process, the making of an electrical plug fuse using injection molding is described here. To fabricate a plug fuse, BMC is fed into the hopper and then the process parameters as shown in Table 6.4 are entered into the computer.[26] These process parameters are typical of an injection molding process but do not represent actual set-up conditions. The machine is turned on and left running until all parts have reached the proper operating temperature. All safety gates and lockouts are checked and closed. As the BMC passes through the heated barrel, its viscosity reduces and the material is ready to flow easily. The temperature of the barrel is closely controlled because the temperature difference between low viscosity and onset of cross-linking is less than 20°F. The temperature in the barrel is maintained close to the cross-linking temperature so that the cure time in the mold is reduced.[26] When the BMC reaches the mold and sees a temperature of 310°F, it starts polymerizing. The temperature of the runner as well

TABLE 6.4
Typical Injection Molding Process Parameters for the Manufacture of Plug Fuse

Process Parameters	Setting Value
Barrel temperature	125°F
Mold temperature	310°F
Shot size	0.5 lb
Injection pressure	10,000 psi
Holding pressure	5000 psi
Injection speed	3 s
Mold time	25 s
Clamp force	5 tons

as the mold is kept at 310°F so that the material in the runner also cures with the material in the mold. After a mold time of 25 s, the parts and runners become rigid and are ejected. It is to be noted here that the temperature of the mold in thermoset injection molding is higher than the barrel temperature, as opposed to thermoplastic injection molding where the mold temperature is lower than the barrel temperature. In thermoplastics, the material becomes rigid after it cools down from its melt temperature, whereas in thermosets the material becomes rigid after it starts curing.

The ejected parts are dropped into a receiving bin and from there are moved to a finisher — to deflash or remove material build-up at mold parting lines. The finisher uses plastic beads to deflash excess material; it takes approximately 2 to 4 min to complete the operation.

The method of applying heat and pressure, and the advantages and limitations of injection molding are described in Section 6.9.7 for injection molding of thermoplastic composites.

6.9 Manufacturing Processes for Thermoplastic Composites

In the composites industry, it is the thermoset composites that dominate the market. In 1998, thermoplastic composites represented about 25% of the total market for polymer-based structural composites. The use of thermoplastic composites is becoming popular in the aerospace and automotive industries because of their higher toughness, higher production rate, and minimal environmental concerns. In the commercial sector, the predominant thermoplastic manufacturing techniques include injection molding, compression molding, and, to some degree, the autoclave/prepreg lay-up process. However, most of the manufacturing processes (e.g., filament winding and pultrusion) available for thermoset composites are also used for the production of thermoplastic composite parts. There is a major difference in the processing

Manufacturing Techniques

of thermoplastic composites as compared to thermoset composites. In the case of thermoplastics, processing can take place in matter of seconds. The processing of thermoplastics is entirely a physical operation because there is no chemical reaction as there is in thermoset composites.

In this section, major thermoplastic processes, which are gaining importance in industry and academia, are the subject of discussion. Because most thermoplastic manufacturing processes are similar to their thermoset counterparts, the thermoplastic processes are not described in detail here. The reader is referred to Section 6.8 on thermoset manufacturing processes to gain a better understanding of thermoplastic manufacturing processes. Again, the order in which these manufacturing processes are described below does not represent the order of importance of the process.

6.9.1 Thermoplastic Tape Winding

Thermoplastic tape winding is also called thermoplastic filament winding. In this process, a thermoplastic prepreg tape is wound over the mandrel as shown in Figure 6.78. Instead of tape, commingled fibers can also be used. In thermoplastic tape winding, heat and pressure are applied at the contact point of the roller and the mandrel for melting and consolidation of thermoplastics. In this process, laydown, melting, and consolidation are obtained in a single step, which thereby avoids the curing stage that is necessary with thermosets. Thermoset-based filament winding and thermoplastic-based tape winding have their own advantages and disadvantages. Thermoset filament winding requires less sophistication, and the quality of consolidation achieved in the thermoset-wound component is very high compared to a thermoplastic-wound component. Advantages of tape winding over filament winding are described in Section 6.9.1.6.

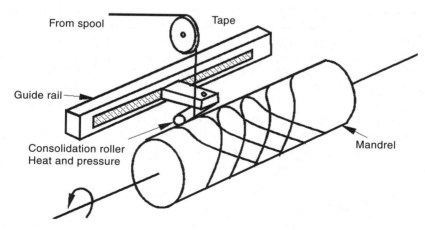

FIGURE 6.78
Schematic of thermoplastic tape winding.

FIGURE 6.79
Filament-wound thermoplastic composite component and XC-2 wear-resistance materials. (Courtesy of Cytec Fiberite.)

6.9.1.1 Major Applications

Several industries (e.g., Cytec Fiberite and Automated Dynamics Corp.) and many researchers have demonstrated the feasibility of the tape winding process by making prototype parts. However, the process has not been widely used for commercial applications. Tape winding can be used for making tubular structures such as bicycle frames and satellite launch tubes. The process has potential for making thick structures without building large residual stresses. Figure 6.79 shows a tape-wound part using APC-2 (carbon/PEEK) thermoplastics.

6.9.1.2 Basic Raw Materials

Thermoplastic prepreg tapes made of carbon, glass, and aramid as reinforcing fibers and various resins such as polyetheretherketone (PEEK), polyphenylene sulfide (PPS), polyamide (nylon 6), polyetherimide (PEI), polypropylene (PP), and polymethylmethacrylate (PMMA) are used for making tape-wound structures. The most common prepreg tapes are carbon/PEEK (APC-2), carbon/nylon, and carbon/PPS. The prepreg tape comes in a thickness of about 0.005 in. and in a width of 0.25 to 2.0 in., depending on the requirement. Some important properties of the above resins are shown in Table 6.5.

Commingled fibers are also used to demonstrate the feasibility of the thermoplastic tape winding process. In commingled fibers, reinforcement fibers are mixed with matrix fibers.

6.9.1.3 Tooling

The same type of mandrels as used in thermoset filament winding is used for tape winding. Because the properties of tape wound products depend on the cooling rate, the mandrel might have a heating mechanism to control

TABLE 6.5

Properties and Processing Characteristics of Key Thermoplastic Resins

Properties	PEEK	PPS	Nylon	PEI	PP	PMMA
Service temperature	250°C (480°F)	220°C (430°F)	70°C (160°F)	170°C (340°F)	55°C (130°F)	65°C (150°F)
Density (g/cc)	1.32	1.35	1.15	1.27	0.91	1.19
Processing temperature	385°C (725°F)	330°C (625°F)	275°C (525°F)	315°C (600°F)	175°C (350°F)	205°C (400°F)
Moisture absorption	Very low	Very low	High	Average	Low	Very low
Bonding characteristics	Poor	Poor	Poor	Good	Poor	Good

the degree of crystallinity in the semi-crystalline thermoplastic materials such as PEEK. The tool surface could be concave or convex, depending on the application need.

6.9.1.4 Making of the Part

Thermoplastic tape winding is similar to thermoset filament winding, in which the thermoplastic tape is wound over the mandrel. The difference between thermoplastic tape winding and thermoset filament winding is that in tape winding, a localized heat source and consolidation roller are required. Localized heating is used to melt the resin of incoming tape at the consolidation point, as shown in the Figure 6.78. The process is also called online consolidation because the incoming tape consolidates at the point where it is laid down. The consolidation roller is used to apply pressure at the consolidation point. Various types of heat sources are used to create localized heating of the material, as discussed in Section 6.9.1.5. The process is much cleaner because no liquid resin is used in this process.

During the thermoplastic tape winding process, localized melting causes a high rate of heating and cooling, which requires different processing conditions than compression molding and autoclave processing. For example, APC-2 (graphite/PEEK) is processed in the range of 380 to 400°C for autoclave, hot press, and diaphragm molding processes. Mazumdar and Hoa[27] found that a temperature greater than 600°C gave better interply consolidation during hot nitrogen gas-aided processing. No degradation in the material was found for processing at that high temperature. The temperature history during the winding process was found to be complex for online consolidation. For hot nitrogen gas aided processing, heating rates higher than 20,000°C/min, cooling rates more than 4000°C/min, and shorter melt times in the range of 1 to 4 s were obtained.[27] Shorter melt times and higher heating rates during thermoplastic tape winding limit the amount of diffusion; therefore, a higher processing temperature is needed for sufficient molecular interdiffusion. Higher processing temperatures would result in lower viscosity and higher diffusivity, which will cause a greater degree of resin flow and interdiffusion. A higher degree of resin flow and interdiffusion would result in better interply bond properties. Saint-Royre and co-workers[28] found that the minimum temperature required for good welding of polyethylene layers increases from 145 to 214°C with an increase in the heating rate from 20 to 400°C/min.

Custom machines are being built for the tape winding process. The machine has a built-in heat supply (hot nitrogen gas), consolidation roller, and tape cutter. These units are mounted on a robot. The robot moves back and forth on the mandrel surface while the mandrel is rotated. In some instances, the filament winding machine is modified to work as a tape winding machine.

Three important parameters — heat intensity, tape speed or winding speed, and consolidation force — have a significant effect on the quality of tape wound part. Optimum process parameters for each heat supply and material type need to be determined to obtain a good consolidated part. For laser processing of graphite/PEEK composites (APC-2), 60-W laser power and a tape speed between 13.00 and 20.00 mm/s gave good bond strength, as determined by Mazumdar and Hoa.[29] The experiment was done for making a ring with a 146.05-mm (5.75-in.) inside diameter. A consolidation force above 50.4 kN/m resulted in good interply bond strength. Here, the consolidation force was determined assuming linear contact between the roller and mandrel. For hot nitrogen gas aided processing, the optimum process conditions were found to be close to 264 scfh (specific cubic feet per hour), a nitrogen flow rate at 905°C, 13.3 mm/s tape speed, and 100 kN/m consolidation force per unit width of the tape.[27]

Fabrication of nonaxisymmetric shapes (e.g., an elliptical cross-section) is a challenge during the tape winding process for two reasons.[30] The first is that the contact point between the mandrel and the consolidation roller changes with the change in mandrel position, as shown in Figures 6.80 and 6.81. Moreover, the tape speed varies for a constant rotational speed of a

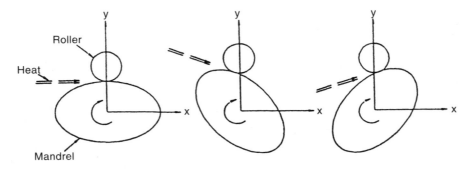

FIGURE 6.80
Contact point movement during tape winding for an elliptical mandrel. Mandrel and roller positions at different times of winding for the case when the roller is free to move only along the y-axis. (From Mazumdar and Hoa.[30])

Manufacturing Techniques

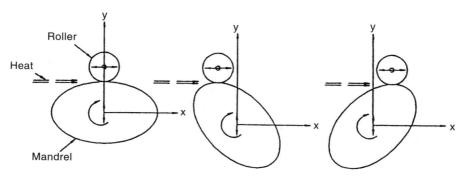

FIGURE 6.81
Mandrel and roller positions at different times of winding for the case when the laser heat source remains stationary and feeds horizontally. (From Mazumdar and Hoa.[30])

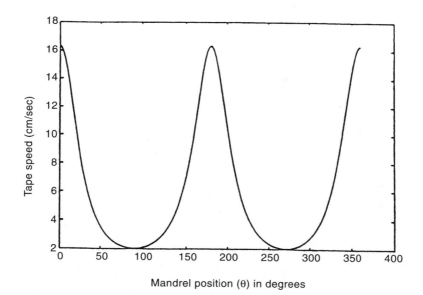

FIGURE 6.82
Effect of constant mandrel speed (10 rpm) on tape speed for an elliptical mandrel having a semi-major axis of 7.8 cm and a semi-minor axis of 3.9 cm. (From Mazumdar and Hoa.[30])

nonaxisymmetric mandrel, as shown in Figure 6.82. With the change in tape speed, the amount of heat at the contact point must change for uniform heating of the laminate; otherwise, component quality may differ along the circumference. Mazumdar and Hoa[30] performed kinematic analyses and determined that the mandrel velocity should change according to Figure 6.83 for constant tape speed for an elliptical mandrel. They also suggested that the amount of heat supply should remain constant during winding to obtain uniform bond quality around the circumference.

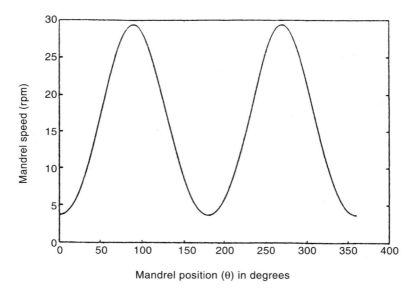

FIGURE 6.83
Mandrel speed for a constant tape speed of 6 cm/s for an elliptical mandrel having a semi-major axis of 7.8 cm and a semi-minor axis of 3.9 cm. (From Mazumdar and Hoa.[30])

6.9.1.5 Methods of Applying Heat and Pressure

Several heat sources can be used for online consolidation of incoming tape with preconsolidated laminate. Some examples are given below.

The heat necessary for melting and consolidation can be supplied by hot rollers, which could be induction or resistance heated. This heating needs to be provided from the top of incoming tape and cannot be applied between incoming tape and preconsolidated laminate. The problem with this kind of heating is that the thermoplastic material sticks to the surface of the roller and causes problems during winding.

Resistance heating can be used by passing an electric current through the carbon fibers. This is theoretically possible but is very complicated for online consolidation of continuous winding. Moreover, it does not work for glass and aramid fibers.

High-frequency waves can heat the material by causing the molecules in the thermoplastic to oscillate. However, this method only works with thermoplastics containing polar molecules. In addition, these waves are difficult to generate and to concentrate at a localized point, and can be hazardous to electronic equipment and to human beings.[31]

An open flame or an acetylene gas torch can be used as heating sources but they are usually so hot that they may degrade the polymer. Moreover, chances of spontaneous combustion are high.

Hot air or hot nitrogen gas can be used for heating purposes.[27] This is a cost-effective alternative for consolidation of laminate but has poor heat efficiency because the amount of energy that must be applied at the consolidation point is much smaller for localized heating. Hot air can be used for bonding low-temperature thermoplastics but has the potential of degrading the material because of the presence of oxidative atmosphere.

A laser has been used by several investigators to demonstrate the feasibility of online consolidation.[32,33] A laser provides a much cleaner atmosphere without any need for ventilation, and can localize the heat in a very small area. A laser has the additional advantage of real-time control during processing. The major problem with lasers is that the equipment is expensive. For space applications, the laser is very suitable for performing various operations (e.g., drilling, welding, trimming, joining, etc.) that are not possible by any other heat sources. The cost of the equipment can be sacrificed for such applications.

6.9.1.6 Advantages of the Thermoplastic Tape Winding Process

The advantages of tape winding are similar to those of filament winding with some limitations and added advantages. Tape winding provides the following advantages over and above those of filament winding.

1. Tape winding is a cleaner production method compared to thermoset filament winding.
2. Concave surfaces as well as nongeodesic winding are attainable in tape winding because the tape consolidates where it is laid down.
3. Thick and large composite structures can be formed without interrupting the process. It may not be convenient to wind them all at one time with thermoset filament winding because of exothermic reaction and residual stress generation.
4. Tape winding offers the ability to post-form the structure.
5. There is no styrene emission concern during the manufacturing process.
6. No secondary processing (e.g., oven curing) is necessary because the incoming tape consolidates where it meets the preconsolidated laminate.

6.9.1.7 Limitations of the Thermoplastic Tape Winding Process

Because of the significant limitations of this process, tape winding has not gained much commercial success. The process has the following limitations as compared to wet filament winding:

1. The process is complicated because it requires a localized heat source and a consolidation roller.
2. The process requires a high capital investment.
3. Getting a good consolidated part is a major challenge during the tape winding process.

4. The quality of products obtained by tape winding is inferior to that obtained by wet filament winding. During helical winding, voids and porosities are formed at the intersection of consolidated tapes.
5. The raw materials cost for tape winding is very high compared to wet filament winding.

6.9.2 Thermoplastic Pultrusion Process

This process is similar to the pultrusion process described in Section 6.8.5 for thermoset composites. In this process, commingled fibers or thermoplastic prepregs are pulled through a die to get the final product. Because of the high viscosity of thermoplastic resins, processing becomes difficult and requires a higher pulling force. This process provides a surface quality inferior to that provided by thermoset pultrusion. Thermoplastic pultrusion has gained little attention from industry and academia.

6.9.2.1 Major Applications

This process has not gained much attention in commercial applications because of inferior surface quality, poor impregnation, and difficulties in processing as compared to its thermoset counterpart. Thermoplastic composites are suitable for those applications that require reformability, higher toughness, recyclability, repairability, and high performance. On a commercial basis, rods, square and circular tubes, angles, strips, channels, rectangular bars, and other simple shapes have been produced using this process. Some of the pultruded thermoplastic composites components are shown in Figure 6.84 using FULCRUM, a trademark of the Dow Chemical Company.

6.9.2.2 Basic Raw Materials

The majority of thermoplastic resins can be used as a matrix material but most commonly used are nylons, polypropylene, polyurethane, PEEK, PPS, and PEI. Glass and carbon fibers have been used as reinforcements in most cases. Prepregs, commingled fibers, and powder-impregnated fibers made with the above matrix and reinforcing materials have been used.

6.9.2.3 Tooling

A steel die similar to that used in thermoset pultrusion is used for making thermoplastic pultruded parts. The die for thermoplastic pultrusion has significant tapers at the entrance for compaction of incoming material and has higher temperature requirements because of the higher processing temperature for thermoplastics. The length of the die is much less than its thermoset counterpart because of its shorter process cycle time. Thermoset

Manufacturing Techniques

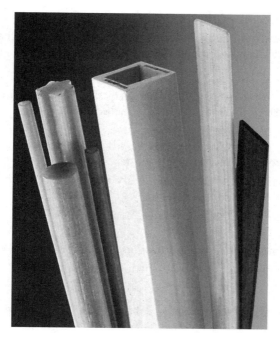

FIGURE 6.84
Pultruded thermoplastic composites components. (Courtesy of the Dow Chemical Company.)

resins usually require a longer curing time and therefore require a lengthier die. The die is usually made in two sections: one for heating and the other for cooling. The cooling section has constant cross-section.

6.9.2.4 Making of the Part

The method used in making the thermoplastic pultruded part is same as its thermoset counterpart. Parts are made by pulling the composite from a heated die. Compaction starts as the material reaches the tapered section of the die and continues until it leaves the die. The composite solidifies as it passes through the cooled portion of the die. The part is then cut to a specified length using a cutter.

6.9.2.5 Methods of Applying Heat and Pressure

The pressure for consolidation is applied by the die as the material passes through it. The die is heated to a processing temperature, usually 10°F lower than the melting temperature. Because the temperature requirements for processing thermolastics are high, preheaters are usually installed before the die to increase the processing speed.

6.9.2.6 Advantages of the Thermoplastic Pultrusion Process

Thermoplastic pultrusion provides advantages similar to its thermoset counterpart. However, it has some additional advantages:

1. The thermoset pultrusion process is usually limited to polyester and vinylester resins, whereas thermoplastic pultrusion can use a wide variety of resin materials, including PP, nylon, PPS, PEEK, polyurethane, PEI, etc.
2. Thermoplastic pultrusion is more advantageous where reformability and repairability are important.
3. The process is environment friendly and does not have any styrene emission concerns.
4. The part can be easily recycled.

6.9.2.7 Limitations of the Thermoplastic Pultrusion Process

Thermoplastic pultrusion provides the following limitations compared to its thermoset counterpart.

1. Processing of thermoplastic composites in a pultrusion environment is a big challenge because it requires high heat and pressure for consolidation.
2. The quality of surface finish is inferior compared to its thermoset counterpart.
3. Because of the high viscosity of resin material, the material does not flow easily. For this reason, complex shapes are difficult to produce.
4. The process requires a high capital investment.
5. The cost of initial raw materials for thermoplastic pultrusion is higher than for thermoset pultrusion.

6.9.3 Compression Molding of GMT

Compression molding of GMT (glass mat thermoplastic) is very similar to compression molding of SMC, with the only major difference being the type of raw material used in the process. In thermoplastic compression molding, GMT is used for making high-volume parts. This is the only thermoplastic manufacturing technique used in widespread commercial applications for making thermoplastic structural parts. The process is primarily used in the automotive industry. The process is two to three times faster than compression molding of SMC, as shown in Table 6.6.

6.9.3.1 Major Applications

With the ability to produce large parts in cycle times of less than 60 s, the GMT molding process is recognized as one of the most productive processes.

TABLE 6.6
Part Cycle Time for Various Manufacturing Processes

Manufacturing Process	Part Cycle Time (s)
Compression molding of GMT	30–60
Injection molding	20–60
Compression molding of SMC	80–150
Blow molding	60–140
SRIM	120–240

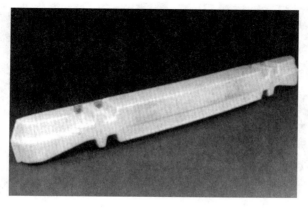

FIGURE 6.85
Automotive rear bumper beam. (Courtesy of Azdel Inc.)

This is the only thermoplastic manufacturing process used in industry for making structural thermoplastic composite parts. The process is used for making bumper beams, dashboards, kneebolsters, and other automotive structural parts. Figure 6.85 shows an automotive rear bumper beam having 5-mph capability. A painted kneebolster with molded-in texture to provide a Class A interior surface and heavy-truck dashboard are shown in Figure 6.86. Figure 6.87 shows an automotive seat back with part thickness less than 2 mm. Cut-outs are recycled for reuse. Consumer items such as office chairs (Figure 6.88) and helmets are also made using this process.

6.9.3.2 Basic Raw Materials

The raw material for thermoplastic compression molding process is GMT (glass mat thermoplastics). GMT is primarily made from polypropylene resin and continuous but randomly oriented glass fibers. Melt-impregnated GMT is the most common material form. Powder-impregnated, discontinuous fiber reinforced GMT is also used. A sheet of fiber glass impregnated with polypropylene is used in this process. These materials are available in pre-cut, pre-weighed pieces of sheet composite materials, called blanks, for processing.

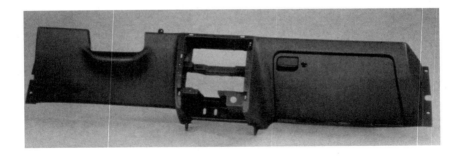

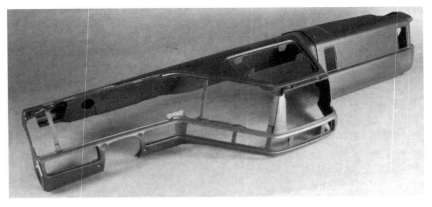

FIGURE 6.86
Compression molded automotive parts. Painted kneebolster with molded-in texture to provide Class A interior surface (top) and heavy-truck dashboard providing low-weight structural strength and high energy absorption (bottom). (Courtesy of Azdel Inc.)

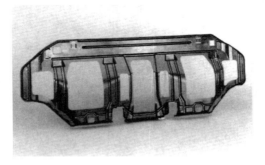

FIGURE 6.87
Automotive seat back with part thickness less than 2 mm. (Courtesy of Azdel Inc.)

The majority of parts are molded from a stack of multiple blanks of different sizes. The blanks are laid in the mold in a predetermined pattern called a blank layout pattern or blank design.

Manufacturing Techniques

FIGURE 6.88
Fully upholstered, single-piece office chair shell, molded with 40% composite. (Courtesy of Azdel Inc.)

6.9.3.3 Tooling

The mold used for compression molding of GMT is similar to that for SMC molding, with the difference being that the SMC molds are heated for processing, whereas in GMT processing the mold is water cooled. The molds have a heavy-duty guidance system and are designed not to close on stop blocks at the final part thickness. The mold is made of tool steel with shear edges hardened to prevent wear. Ejector systems are incorporated for part removal.

6.9.3.4 Part Fabrication

Compression molding of thermoplastic composites is a flow-forming process in which the heated composite sheet is squeezed between the mold halves to force resin and glass fibers to fill the cavity. Molding cycle times typically range from 30 to 60 s. Unlike SMC molding, GMT is heated in a conveyor-equipped oven above the melt temperature of the resin before it is laid on the mold cavity, as shown in Figure 6.89. A robot is used to place the heated blank in the mold. The mold is rapidly closed under pressure to form the

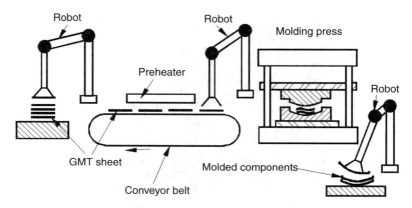

FIGURE 6.89
Schematic of the GMT process.

GMT and held closed until the part has solidified. In this process, the mold is not heated.

Blank design is critical to good part performance and molding efficiency. A well-designed blank pattern is necessary for producing parts with uniform glass distribution and optimum properties. The blank usually covers approximately half of the mold surface.

Because the entire process takes less than 1 min, the process is automized as shown in Figure 6.89. A robot picks up the blank and places it on a conveyor belt. The conveyor belt passes through a heated chamber to raise the temperature from room temperature to the polymer's melt temperature within 1 min. Because heating needs to be done quickly, the length of the heating chamber is significantly greater than the molding press and other equipment parts.

6.9.3.5 Methods of Applying Heat and Pressure

In compression molding of GMT, heat is supplied to the material before it is placed on the mold. The GMT blank is heated by infrared radiation or hot air while it is moved by a conveyor belt. For PP-based GMT, it is heated to about 200 to 230°C and then placed on the mold. The mold is water cooled and the temperature is kept between 30 and 60°C. The pressure for the flow of blank is applied by moving the upper mold half against the lower mold half. The working speed is typically 80 mm/s, whereas in SMC, it is 40 mm/s.

6.9.3.6 Advantages of Compression Molding of GMT

Compression molding of GMT has gained popularity in the automotive industry because of the following advantages:

1. This is one of the fastest techniques for making composite structural parts.
2. Because of higher productivity of the process, fewer tools and less labor are required.

6.9.3.7 Limitations of Compression Molding of GMT

Compression molding of GMT has the following limitations:

1. A high capital investment is required for the process.
2. The process is limited to high production volume environments.
3. The typical fiber volume fraction for this process is 20 to 30% because of the high viscosity of the resin.
4. The surface finish on the part is of an intermediate nature.

6.9.4 Hot Press Technique

This process is also called compression molding of thermoplastic prepregs, or the matched die technique. This process is similar to the sheet metal forming process. In this process, thermoplastic prepregs are stacked together and then placed between heated molds. Unlike GMT, the prepregs in this case are made with unidirectional continuous fibers. The fiber volume fraction is greater than 60%. This process is widely used in R&D environments to make flat test coupons.

6.9.4.1 Major Applications

This process is primarily used for making simple shapes such as flat laminates. The process has not gained much commercial importance. This process is used for making parts with constant thickness.

6.9.4.2 Basic Raw Materials

The raw materials used in this process are thermoplastic prepregs made with unidirectional fibers. Carbon fiber with PEEK (APC-2) and carbon with PPS are mostly used for this application. Instead of carbon, glass and Kevlar can also be used with polymers such as PP, nylon, and some other types of plastics.

6.9.4.3 Tooling

Molds for this process are made of stainless steel or aluminum. Because mostly flat laminates are made using this technique, the molds are very simple. In mold design, both part shrinkage and mold expansion and contraction are taken into consideration. It is desirable to have the same coefficient of thermal expansion between the composite and tooling materials when contoured surfaces are made. The mold design for this process is much simpler than for injection molding or RTM. Here, there is no need for a sprue, runner, or gate because there is no need for raw material injection into the mold. The mold is placed between two heated platens to form the part.

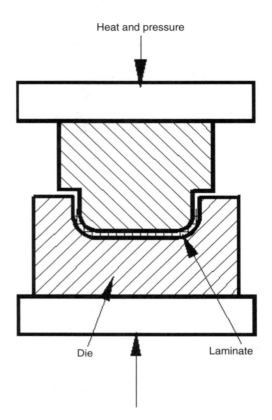

FIGURE 6.90
Schematic of the hot press technique.

6.9.4.4 Making of the Part

In the hot press method, a composite part is made by placing the prepregs in a mold as shown in Figure 6.90. In this case, the prepreg covers the entire mold surface. Thermoplastic prepregs do not have any tack and suffer from poor drapability. Because of these characteristics, only simple parts are made by this process. Prepregs are cut according to shape, size, and fiber angle requirements. They are stacked in the desired fiber orientation and welded together using a solder iron. Spot welding is done to make sure the prepregs do not move relative to each other during handling. Spot welding is done around edges as well as in the middle portion at 1 to 6 in. apart. After the prepregs are laid in the desired orientations, the laminate is placed inside the mold. Before placing the laminate, a release agent is applied to the mold surface for easy removal of the molded part. The mold is then closed and placed between two heated platens of the compression molding equipment. A slight pressure is applied until the material temperature reaches the melt temperature of the resin. Once the temperature reaches equilibrium, molding pressure is applied for a specified duration and then the mold is removed to a cold press. Pressure is maintained during the cooling process as well as until the part has completely solidified. Some hot press equipment has a built-in cooling system and, in that case, the mold is cooled right in the hot press without moving it. After cooling is performed, the mold is opened and the part is removed.

There are two parameters that result in a good consolidated part. First, there should be intimate contact between adjacent layers; and second, there should be sufficient heat and time for autohesion to take place at the interface. To create intimate contact, the air from the interface needs to be removed. Without two adjacent layers being physically in contact, the autohesion (molecular diffusion) process will not start. Because thermoplastic prepregs have poor drapability, the removal of air between two mating surfaces becomes a challenge. To create intimate contact, a suitable pressure is applied. Intimate contact and autohesion are described in detail in Chapter 7.

Several researchers have systematically studied the consolidation of APC-2 (carbon/PEEK) prepregs in hot presses and autoclaves.[34–43] In terms of temperature, most researchers agree that a temperature of 360 to 400°C provides a good consolidation, as suggested in manufacturer's data sheets. There was no effect of consolidation pressure on the quality of the bond in the pressure range from 2 to 40 atm, as found by some researchers. Regarding the time required for consolidation, most researchers preferred time in the range of 5 to 30 min. This time includes the time for the mold to reach the equilibrium temperature. Lee and Springer[39] found a time of less than 5 s for full autohesion.

6.9.4.5 Methods of Applying Heat and Pressure

In this process, the heat for processing is applied by placing the mold between two heated platens. The platens are electrically heated by turning the switch on and setting the desired temperature on the temperature gauge. Some presses are computer controlled, wherein the heating rate, cooling rate, and dwell time are all controlled by a computer. The heated platens first heat the mold and then heat the prepreg material. Pressure is applied by moving the bottom platen against the top platen. There is a pressure gauge that measures the applied pressure. The bottom platen is automatically moved by a switch, whereas the top platen usually remains stationary. Heat and pressure are maintained according to the manufacturer's recommended process cycle or based on experience and as dictated by process models.

6.9.4.6 Basic Processing Steps

The following manufacturing steps are commonly followed during the hot press technique:

1. The temperature is set on the temperature gauge and the heater is switched on.
2. The mold is cleaned and release agent is applied.
3. The prepreg is cut to the desired shape, size, and orientation.
4. The prepregs are laid on top of each other using spot welding.
5. The laminate is then placed into the mold and the mold is closed.
6. The mold is placed between two heated platens.
7. The bottom platen is raised against the top platen.
8. The desired pressure is applied in the mold by pressing the mold between the platens.
9. The temperature is raised close to the melt temperature of the thermoplastic.
10. The laminate is processed for approximately 4 to 10 min, depending on the process cycle requirements of the thermoplastic.
11. The mold is cooled at the desired cooling rate, typically at 2 to 10°C/min.
12. The pressure is released by moving the bottom platen downward.
13. The mold is removed from the press.
14. The mold is opened and the part is removed.

6.9.4.7 Advantages of the Hot Press Technique

This process is very suitable for performing research and development work and for making flat test coupons. The following are some of the advantages of the hot press technique:

1. A high fiber volume fraction is achieved by the hot press technique.
2. Small to big sized parts can be compression molded.
3. The parts are recyclable.

6.9.4.8 Limitations of the Hot Press Technique

The following are the limitations of the hot press technique:

1. The process is limited to making simple parts such as flat plates. The process has not gained much commercial importance.
2. Thick structures are not easily produced by this technique.
3. It is a challenge to create distortion- and warpage-free parts by this process.

6.9.5 Autoclave Processing

Autoclave processing of thermoplastic composites is similar to autoclave processing of thermoset composites. In this process, thermoplastic prepregs are laid down on a tool in the desired sequence and spot welded to make sure that the stacked plies do not move relative to each other. The entire assembly is then vacuum bagged and placed inside an autoclave. Following the process cycle, the part is removed from the tool. This process is similar to the hot press technique, with the only difference being the method of applying pressure and heat.

6.9.5.1 Major Applications

Autoclave processing is primarily used in the aerospace industry to make tougher composite parts. FRE Composites (St. Andre, Quebec, Canada) has made robot arms for the international space station using this technique. Robot arms were about 18 in. in diameter and 72 in. long. Carbon/PEEK composites were laid on a mandrel at the desired orientation. Prepregs were spot welded to lay the prepregs on top of each other until a suitable thickness was developed and then vacuum bagged. The entire assembly was then processed inside an autoclave.

6.9.5.2 Basic Raw Materials

The raw materials used in this process are thermoplastic prepregs made with unidirectional fibers. Carbon fiber with PEEK (APC-2) and carbon with PPS are mostly used for this application. Glass and Kevlar fibers are also used with polymers such as PP, nylon, and other types of plastics.

6.9.5.3 Tooling

Molds for this process are made of stainless steel or aluminum. The mold could be a mandrel or any open mold. In designing the mold, part shrinkage and mold expansion and contraction are taken into consideration.

6.9.5.4 Making the Part

In an autoclave process, a thermoplastic composite part is made by placing the prepregs on the surface of the open mold, similar to the thermoset autoclave process. Here, the difference is that the prepreg layers are spot welded to avoid relative motion between plies. Thermoset prepregs have good tack and drapability so prepreg lay-up is much easier in thermoset autoclave processing. Moreover, prepregs can easily be laid on a complex shape with thermoset prepregs, whereas the laying of thermoplastic prepregs is difficult on a complex shape. Before placing the thermoplastic tape, a release agent is applied to the mold surface for easy removal of the molded part. Once the prepreg layers are laid in the desired sequence, the entire assembly is vacuum bagged similar to the thermoset autoclave process. Vacuum bagging helps in removing the air from interfaces. To make thick cross sections such as a thick hollow cylinder, the air between prepreg layers is dispelled after every few layers of prepreg lay-up by applying vacuum. To improve the quality of consolidation, proper consideration is made to dispel most of the air during the prepreg lay-up procedure. This avoids wrinkling of fibers and also helps in creating intimate contact. Because thermoplastic prepregs do not have good tack and drapability, significant numbers of air pockets remain at the interface. Once all the tapes are laid down, entire assembly is vacuum bagged for consolidation. The vacuum bagging materials for thermoplastics processing have high temperature resistance because of a higher processing temperature. After vacuum bagging, the entire assembly is placed inside the autoclave.

Heat and pressure are applied similar to thermoset autoclave processing. Here, the duration of processing is much shorter because thermoplastic prepregs do not undergo any chemical reaction. Pressure is applied in two ways: (1) by creating a vacuum inside the vacuum bag, or (2) external pressure by the autoclave. The pressure serves two functions. It removes air from the interface and creates intimate contact. It also helps in the flow of resin material during consolidation. The resin flow fills up irregular spaces at the interface. Resin flow occurs once the temperature of the thermoplastic nears its melt temperature. Once two adjacent layers come into intimate contact, a bonding process between the interfaces begins. It has been suggested that the bonding is primarily caused by the autohesion process for two similar thermoplastic interfaces.[39-42] During autohesion, segments of the chain-like molecules diffuse across the interface. The extent of the molecular diffusion and hence the bond strength increase with time. This process is described in detail in Chapter 7. The amount of molecular diffusion depends on the temperature and duration, as is evident from Equations (6.1) and (6.2)[42-44]:

$$D_{au} = \chi \cdot t_a^{1/4} \quad (6.1)$$

where t_a is the time elapsed from the start of the autohesion process and χ is a constant. The parameter χ depends on the temperature T as shown in the following Arrhenius relation.

$$\chi = \chi_0 \exp\left(-\frac{E}{RT}\right) \quad (6.2)$$

where χ_0 is a constant, E is the activation energy, and R is the universal gas constant.

Depending on the temperature level, the autohesion process is completed in less than 1 min. After bond formation, the assembly is taken out and the vacuum bag is removed. The composite part is then removed from the tool.

Manufacturing Techniques 221

6.9.5.5 Methods of Applying Heat and Pressure

The procedure for applying heat and pressure is same as discussed in Section 6.8.1.5 for autoclave processing of thermoset composites. The heat is supplied for melting and for diffusion to occur at the interfaces, and pressure is applied for developing intimate contact and good consolidation. Heat and pressure are supplied by the autoclave. Usually, the temperature is raised close to the melt temperature of the thermoplastic. Vacuum is created inside the vacuum bagging system and external pressure is applied for consolidation. Lack of drapability and the higher viscosity of thermoplastic resins require higher consolidation pressures than for thermoset resins.

6.9.5.6 Basic Processing Steps

For easy understanding of the process, the complete process is broken into the following basic steps:

1. The tool is cleaned and release agent is applied.
2. The prepregs are cut to the size, shape, and fiber orientation as required for the component. The handling of thermoplastic prepregs is much easier than thermoset prepregs. Thermoplastic prepregs have an unlimited shelf life and there is no need for refrigeration or condensation prior to cutting the prepregs.
3. Prepregs are laid down on the tool and spot welded together. Thermoplastic prepregs pose difficulties in laying down the prepregs because of lack of tack and drapability. It is very important to remove air from interfaces during the lay-up process. With thermoplastic tapes, it is difficult to remove all the air from the interfaces because of the irregularity of the prepreg surface.
4. A vacuum bag is applied around the tool.
5. The entire assembly is placed inside the autoclave.
6. A vacuum hose is connected to the nozzle fitted into the assembly and then vacuum is created inside the bag.
7. The temperature and pressure inside the autoclave are raised to a processing level desired for the prepreg material.
8. After bond formation is complete, the entire assembly is cooled at the rate needed to attain the desired crystallinity. Typically, the cooling rate is 2 to 10°C/min for carbon/PEEK composites.
9. The vacuum bag is removed and then the composite part is removed from the tool.

6.9.5.7 Advantages of Autoclave Processing

Autoclave processing is probably the major manufacturing technique for the fabrication of thermoplastic composite parts in the aerospace industry. It has the following advantages:

1. It provides fabrication of structural composite components with a high fiber volume fraction.
2. It allows production of any fiber orientation.
3. It is simple. It is basically a replicate of autoclave processing of thermoset composites. The same autoclave equipment as in thermoset processing is used for thermoplastic autoclave processing. Equipment for thermoplastic filament winding and pultrusion is different than for its thermoset counterpart.
4. It is suitable for making prototype parts.
5. The tool design is simple for autoclave processing.

6.9.5.8 *Limitations of Autoclave Processing*

Autoclave processing of thermoplastic composites has the following limitations:

1. Due to lack of tack and drapability, prepreg lay-up during autoclave processing of thermoplastic composites is labor intensive as compared to its thermoset counterpart.
2. A high capital investment is required if the company must buy an additional autoclave.
3. Processing of thermoplastic composites is difficult as compared to thermoset composites. Higher temperatures and pressures are required due to the high melt temperature and higher viscosity of thermoplastics.

6.9.6 Diaphragm Forming Process

Thermoplastic forming techniques have gained a lot of interest from investigators because of their potential for forming complex parts in a high production volume environment. There are several types of forming processes available. Herein, we discuss the diaphragm forming process in detail. Unlike thermoplastic filament winding, pultrusion, and autoclave processes, the diaphragm forming process is a unique process in that it is not adapted from thermoset technology. This process was specifically developed to work with thermoplastic prepregs. In the diaphragm forming process, prepreg layers in the form of a composite sheet are placed between two flexible diaphragms and then formed under heat and pressure against a female mold. The prepreg layers float freely between the two constrained diaphragms. Diaphragm forming has been the subject of extensive study by researchers.[45–50]

6.9.6.1 *Major Applications*

This method has not yet gained much commercial importance. Several investigators have worked on this process for making complex parts such as helmets, trays, corrugated shapes, etc.

Manufacturing Techniques

6.9.6.2 Basic Raw Materials

The raw material for this process is same as for the hot press technique. Here, the composite sheet is formed by stacking unidirectional prepreg materials in a desired sequence and orientation. The plies are spot welded, usually around perimeters to adhere one layer to another. Carbon/PEEK (APC-2), carbon/PPS, carbon/nylon, and glass/nylon prepregs are commonly used for making composite sheets.

6.9.6.3 Tooling

For this process, female molds are created using tool steel. Vacuum ports are fabricated in the mold for creating vacuum between the lower diaphragm and the female mold surface. The diaphragm materials are usually superplastic aluminum alloys, polyimide films, Upilex, or sheet rubber. The diaphragm controls the forming process. The stiffness of the diaphragm is a critical factor in getting a good quality part. For simple shapes, compliant diaphragms are preferred; whereas for complex shapes, stiffer diaphragms are selected.

6.9.6.4 Making of the Part

Diaphragm forming is similar to the thermoforming process for plastics. In thermoforming, the polymeric sheet is preheated above T_g of the polymer and then placed inside the mold. The sheet is clamped around the edges and formed into the mold cavity by the application of either vacuum, pressure, or both (Figure 6.91). After forming, the part is cooled under pressure below the T_g of the resin and then removed from the mold. Thermoforming is used for making plastic cups, trays, packages, toys, etc. In diaphragm forming, the sheet is formed by laying unidirectional prepregs on top of each other and then spot welding them together to form the laminate. The thermoplastic composite sheet is formed under heat and pressure in a mold, as

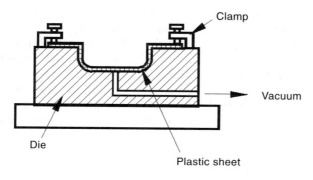

FIGURE 6.91
Schematic of the thermoforming process.

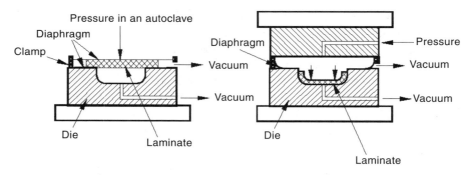

FIGURE 6.92
Schematic of the diaphragm forming process.

shown in Figure 6.92. There are two major differences between thermoforming of plastics and diaphragm forming of composites. In diaphragm forming, the composite sheet is heated close to the melt temperature, whereas in thermoforming, the plastic sheet is heated above the T_g of plastics. In diaphragm forming, the thermoplastic composite sheet cannot be clamped around the perimeter as in thermoforming. The fibers in individual plies do not allow for stretching without breaking the fibers. In diaphragm forming, the composite sheet is placed between two highly flexible diaphragms, which are clamped around the perimeter. Vacuum is used to evacuate the air between the diaphragms. There are several ways heat and pressure are applied during the diaphragm-forming process. In one version, the diaphragm with stacked composite sheet is placed inside an oven and heated to the melt temperature of the resin material. Then the assembly is quickly placed in a female mold where a vacuum is created between the mold and the diaphragm, and pressure is applied above the upper diaphragm for forming the composite sheet. As the forming pressure is applied, the deformation of diaphragms creates a biaxial tension in the composite sheet, which prevents plies from wrinkling. Because the mold is not heated, the composite sheet cools and solidifies as it comes in contact with the mold.

There are some other ways that heat and pressure can be applied. In one case, the stacked composite sheet with diaphragm is placed on the female mold and then moved inside an autoclave. In this case, the entire assembly, including the mold, is heated. Vacuum is applied between the mold and lower diaphragm, and autoclave pressure is applied above the upper diaphragm to form the composite sheet. After forming and consolidation, the assembly is cooled in ambient air inside the autoclave. Four basic mechanisms can occur during the forming and consolidation phase of composites.[50]

1. Percolation or flow of molten resin through fiber layers
2. Transverse flow (movement in thickness direction) of fibers under pressure

3. Intraply shearing, which allows fibers within each ply to move relative to each other in the axial as well as transverse directions
4. Interply slip, by which plies slide over each other while conforming to the shape of the die

The first two mechanisms help create good consolidation between plies as well as within a ply. The other two mechanisms help in forming contoured shapes without fiber wrinkling and splitting.

There is another forming process, called the hydroforming process, which is similar to the VARTM process where one open mold or die and an elastomeric diaphragm are used to form the shape. The thermoplastic sheet is laid on the die and covered by an elastomeric diaphragm, which applies pressure for consolidation using hydraulic fluid. The process is limited due to temperature limitations of the elastomeric diaphragm.

6.9.6.5 Methods of Applying Heat and Pressure

Various methods of applying heat and pressure during diaphragm forming were discussed in Section 6.9.6.4. In one case, a composite sheet with diaphragm material is heated to the melt temperature of the polymer in an oven and then placed on the female mold. The diaphragm is clamped around the edges and then formed under vacuum and pressure. Vacuum between mold and lower diaphragm, and pressure from above the upper diaphragm are used to form the shape. Many studies[49] have utilized an autoclave for processing because of easy availability. The complete process of making a part can be made faster by utilizing separate heat sources and pressure vessels.

6.9.6.6 Advantages of the Diaphragm Forming Process

This process has gained considerable attention from researchers due to its capability of making complex structural parts at high volume rates. This process has the following advantages:

1. It offers excellent structural properties because continuous fibers are used in making the part.
2. Reasonably complex shapes with uniform thickness can be produced with reasonably high production efficiencies.

6.9.6.7 Limitations of the Diaphragm Forming Process

This process has not gained much commercial importance for the following reasons:

1. The process is limited to making parts that have constant thickness.
2. Maintaining uniform fiber distribution during the manufacture of complex shapes is a challenge. In the diaphragm forming process,

composite layers float between diaphragms and are free to have all the allowable modes of deformation. This freedom necessarily results in significant reorientation of the reinforcing fibers.

6.9.7 Injection Molding

Injection molding is the predominant process for the production of thermoplastics into finished forms, and its use is increasing with fiber-filled thermoplastics. Injection molding of thermoplastics is the process of choice for a tremendous variety of parts, ranging from 5 g to 85 kg. It is estimated that approximately 25% of all thermoplastic resins are used for injection molding. Injection molding of thermoplastic composites is the same as injection molding of thermoplastics without any reinforcements. The equipment remains the same except for the change in raw material for thermoplastic composites. The use of fiber in the resin increases the mechanical strength of the part and provides better dimensional control. Injection molding is used for making complex parts at a very high rate. It is a very automated process and usually has a process cycle time of 20 to 60 s. The segment of the molding cycle that frequently requires the most time is the cooling time for the parts. Single-cavity or multiple-cavity molds are used to make the part. The process is suitable for large-volume applications such as automotive and consumer goods.

6.9.7.1 Major Applications

Injection molded, unreinforced thermoplastics are very common in household items such as buckets, mugs, soap casings, toys, housing, and enclosures for various units, etc. Reinforced composite parts include equipment housing, sprockets, computer parts, automotive parts, and more.

6.9.7.2 Basic Raw Materials

Initial thermoplastic composite materials used in this process are in pellet or granular form. Various types of pellets for injection molding are discussed in Chapter 2. These pellets are formed by pultruding composite rods and then cutting them into small pieces about 10 mm in length. Another way to make fiber reinforced pellets is by passing a continuous strand through a coating die. Coated strands are then chopped, typically to a length of 10 mm. The final molded parts contain fibers that range from 0.2 to 6 mm in length. Fiber breaks when it passes through a screw barrel, nozzle, or other part of the equipment and mold. Primarily, glass fibers are used with various types of thermoplastics such as PP, nylon, PET, polyester, etc. Moldable pellets with carbon and Kevlar fibers are also available on the market.

6.9.7.3 Tooling

Steel molds with single or multiple cavities are used to make injection molded parts. Air-hardened tool steels are generally chosen in order to maximize the life and performance of the mold. A mold made with tool steel typically costs 5 to 10% more than steel tool but it is important to buy quality material to increase the life of the mold. Multiple-cavity molds are used to increase the production rate. Each cavity is connected with a passage called a runner for raw material flow. The details of the tool design for injection molding are given in Section 6.4.

6.9.7.4 Making of the Part

Injection molding is a process that forces a measured amount (shot) of liquid fiber-filled resin into heated mold cavities. As shown in Figure 6.77, pellets of thermoplastic resins or fiber-filled thermoplastics flow from a funnel-shaped feed hopper into a heated compression cylinder. Most injection molding machines have a reciprocating screw-type barrel that transports pellets through heating stages before the material is injected. Other systems use a plunger (called a torpedo) that forces the stock around a heated mandrel. The cylinder temperature increases to the melt temperature of the resin as the material flows toward the nozzle. The purpose of heating the cylinder that surrounds the barrel is to transform the solid pellets into a viscous liquid or melt that can be forced through the connecting nozzle, sprue, and runners to the gates that lead into the mold cavities. The plunger forces a controlled quantity of material through the injection molding nozzle into the closed mold. When the viscous liquid is injected into the mold cavity, it forms a part of the desired shape. The mold is tightly clamped against injection pressure and is cooled well below the melt temperature of the resin by water running through cooling channels. After the part cools, the mold halves are opened at the mold parting plane and the parts are ejected by a knockout system. In most cases, parts ejected from the mold require no finishing other than trimming off sprues and runners.

The molding cycle is usually completed in about 20 to 60 s. In most cases, approximately 50% of the process cycle time is taken by the cooling time for the part. The cooling time depends on the part size and shape, part wall thickness, and temperature settings on the machine. The distribution of time for various mold operations in a typical mold cycle is given below for a cooling time of 15 s.

Mold close	2 s
Material injection (shot)	3 s
Hold pressure (screw dwell)	2 s
Part cooling	15 s
Mold opening	3 s
Part ejection and mold opening dwell	3 s
Total cycle time	28 s
Cycles per hour	128
Cycles/24 hr	3085
Parts made/24 hr for four mold cavities = 3085 × 4 = 12,340	

In the above example, the screw recovery time is 11 s, which is less than the part cooling time and therefore does not affect the cycle rate. The quality of injection molded parts depends on several process variables, including injection pressure, back-pressure, melt temperature, mold temperature, and shot size. Judicious selection of these process variables can avoid problems related to excess flash on parting lines, short shots, weld lines, and part warpage and shrinkage. Warpage and shrinkage may result in parts being out of tolerance. Other process-related problems include discoloration (visible flow pattern, faded color near inserts), bad appearance (bubbles and voids beneath the surface), sprue adhering to the part, or part adhering to the cavities.

6.9.7.5 Basic Processing Steps

The following are the basic steps for the fabrication of injection molded parts:

1. Pellets/molding compounds are placed in the feeder and mixed with additives and colorants.
2. The machine is switched on and desired temperature and pressure are set.
3. The mold is closed using oil pressure and the desired clamping force is applied.
4. The screw is turned at a speed of 40 to 80 rpm and fiber-filled resin is moved forward through a heated chamber where the resin is melted. As it moves toward the nozzle, polymer is pressurized because of the screw mehanism.
5. The filled polymer is injected into the mold. The rate of injection is adjustable up to 230 cm^3/s. Maximum injection pressures attainable are between 100 and 200 MPa but commonly set at 170 MPa. As the screw moves forward, a non-return or check valve in the screw tip prevents back-flow of the melt into the barrel.
6. The polymer melt flows in the cavity elastically. The melt front forces the air out from the cavity through air vents primarily located at parting lines.
7. High pressure is maintained in the cavity during the packing phase. The holding pressure can vary from material to material, but most are in the range of 50 to 100 MPa. The holding pressure prevents the back-flow of filled resin. The duration to maintain this pressure depends on part size, melt temperature, and mold temperature.
8. The melted material solidifies in the mold. The mold temperature is kept around 50 to 100°C. The mold temperature is controlled by circulating liquid coolant such as soft water or a water/ethylene glycol mixture, which is arranged to flow turbulently through drilled channels in the mold cavity.
9. The part is removed after it is sufficiently cooled to be ejected without distortion.
10. After removal of the part, the mold is closed to prepare it for the next injection.

Manufacturing Techniques

6.9.7.6 Methods of Applying Heat and Pressure

In injection molding, the pellets/molding compounds are first dried and then fed into the hopper. The feedstock moves through a heated chamber where it starts softening. The movement of stock material takes place by the screw mechanism as shown in the Figure 6.77. By the time the feedstock gets close to the nozzle, the resin is completely melted. The temperature near the nozzle is close to the melt temperature of the resin material. The melted material is injected into the mold cavity at a pressure of 100 to 200 MPa. This pressure allows the resin to flow and fill the cavity. Once the resin fills the cavity, a holding pressure of about 50 to 100 MPa, depending on material type, is maintained. The holding pressure prevents the back-flow of filled resin. The duration to maintain this pressure depends on part size, melt temperature, and mold temperature. The mold is kept at 50 to 100°C so that the resin does not immediately solidify as it nears the cavity. The mold is maintained at that temperature by circulating coolant. With thermoplastic composites, the resin material does not have to come to room temperature for parts to be removed from the mold.

6.9.7.7 Advantages of the Injection Molding Process

Injection molding is a high-volume manufacturing process with the following advantages:

1. This process allows production of complex shapes in one shot. Inserts and core materials can be used in part fabrication.
2. Part repeatability is much better in injection molding than in any other molding process. It offers tight dimensional control (±0.002 in.).
3. The process is a high-volume production method with a mold cycle ranging from 20 to 60 s. Because of this high production rate, the process is very suitable for making automotive, sporting, and consumer goods parts. The process can be completely automated to achieve the highest volume rate.
4. The process allows fabrication of low-cost parts because of its capacity for high-volume production rates. The process has very low labor costs.
5. Small (5 g) to large (85 kg) parts can be made using this process.
6. The process allows for production of net-shape or near-net-shape parts. It eliminates finishing operations such as trimming and sanding. The quality of the surface finish is very good.
7. The process has very low scrap loss. Runners, gates, and scrap are recyclable.

6.9.7.8 Limitations of the Injection Molding Process

Although injection molding is the process of choice for the production of most thermoplastic parts, the process suffers from the following drawbacks:

1. The process requires significant capital investment. An injection molding machine with 181-tonne clamping capacity and 397-g shot size costs about $150,000. Lack of expertise in product design, manufacturing, and machine maintenance can cause high start-up and running costs.
2. The process is not suitable for the fabrication of low-volume parts because of high tooling costs. The mold usually costs between $20,000 and $100,000; for this reason, changes in the design are not frequently allowed.
3. The process is not suitable for making prototype parts. To get an idea of the design, rapid prototyping is preferred for visualization of the part before going for final production.
4. The process requires a longer lead time because of the time involved in mold design, mold making, computer simulation of the manufacturing process, debugging, trial and error, etc.
5. Because there are so many process variables (e.g., injection pressure, back-pressure, melt temperature, mold temperature, shot size, etc.), the quality of the part is difficult to determine immediately.

References

1. Fisher, K., Resin flow control is the key to RTM success, *High Performance Composites*, Ray Publishing, p. 34, January/February 1997.
2. Gutowski, T.G., A resin flow/fiber deformation model for composites, *SAMPE Quarterly*, 16, 4, 1985.
3. Gutowski, T.G., Morigaki, T., and Cai, Z., The consolidation of laminate composites, *J. Composite Materials*, 21, 172, 1987.
4. Springer, G.S., Resin flow during the cure of fiber reinforced composites, *J. Composite Materials*, 16, September 1982.
5. Loos, A.C. and Springer, G.S., Curing of epoxy matrix composites, *J. Composite Materials*, 17, March 1983.
6. Hudson, A., Crafting a high-end yacht, *Composites Technology*, Ray Publishing, p. 27, July/August 1996.
7. Dawson, D., FRP top choice in tub and shower market, *Composites Technology*, p. 23, January/February 1997.
8. Bassett, S., Large-pipe fabrication spirals forward, *Composites Technology*, p. 23, September/October 1996.
9. Munro, M., Review of manufacturing of fiber composite components by filament winding, *Polymer Composites*, 9(5), 352, 1988.
10. Evans, D.O., Simulation of filament winding, *30th Int. SAMPE Symp.*, March 1985, 1255.
11. Roser, R.R., Computer graphics streamline the programming of the filament winding machine, *30th Int. SAMPE Symp.*, March 1985, 1231.
12. Roser, R.R., New generation computer controlled filament winding, *31st Int. SAMPE Symp.*, April 1986, 810.

13. Larson, D.L. et al., Advancements in control systems for filament winding, *31st Int. SAMPE Symp.*, April 1986, 222.
14. Menges, G. and Effing, M. CADFIBER — A program system for design and production of composite parts, *43rd Annu. Conf., Composites Institute*, The Society of Plastic Industry, Inc., Paper 20-D, February 1–5, 1988.
15. Wells, G.M. and McAnulty, K.F. Computer aided filament winding using nongeodesic trajectories, in *ICCM-VI and ECCM-2, Proc. 6th Int. Conf. Composite Materials*, London, U.K., 1, 1.161, 1987.
16. Steiner, K.V., Development of a robotic filament winding workstation for complex geometries, *35th Int. SAMPE Symp.*, April 1990, 765.
17. Bernard, E., Fahim, A., and Munro, M., A CAD/CAM approach to robotic filament winding, *CANCOM'91*, Montreal, Quebec, Canada, 1991.
18. Mazumdar, S.K. and Hoa, S.V., On the kinematics of filament winding on non-axisymmetric cylindrical mandrels. I. A generalized model, *Composites Manufacturing*, 2(1), 23, 1991.
19. Mazumdar, S.K. and Hoa, S.V., On the kinematics of filament winding on non-axisymmetric cylindrical mandrels. II. For convex polygonal cross-sections, *Composites Manufacturing*, 2(1), 31, 1991.
20. Mazumdar, S.K. and Hoa, S.V., Analytical models for low cost manufacturing of composite components by filament winding. I. Direct kinematics, *J. Composite Materials*, 29(11), 1515, 1995.
21. Mazumdar, S.K. and Hoa, S.V., Analytical models for low cost manufacturing of composite components by filament winding. II. Inverse kinematics, *J. Composite Materials*, 29(13), 1762, 1995.
22. Mazumdar, S.K. and Hoa, S.V., Algorithm for filament winding of non-axisymmetric tapered composite components having polygonal cross-sections in a two axes filament winding machine, *Composites Engineering*, 4(3), 343, 1994.
23. Mazumdar, S.K. and Hoa, S.V., Kinematics of filament winding during starting and reversal process for complex composite components, *Trans. Canadian Soc. Mechanical Engineering*, 17(4A), 671, 1993.
24. *SMC Design Manual*, SMC Automotive Alliance of the Society of the Plastics Industry's (SPI) Composites Institute, 1991.
25. Benjamin, B. and Bassett, S., Casting carbon as fly rod reinforcement, *High Performance Composites*, Ray Publishing, July/August 1996, 27.
26. Thermoset injection molding, *Composites Technology*, Ray Publishing, p. 23, September/October 1995.
27. Mazumdar, S.K. and Hoa, S.V., Determination of manufacturing conditions for processing PEEK/carbon thermoplastic composites using hot nitrogen gas by tape winding tehnique, *J. Thermoplastic Composite Materials*, 9, 35, January 1996.
28. Saint-Royre, D., Gueugnant, D., and Reveret, D., Test methodology for the determination of optimum fusion welding conditions of polyethylene, *J. Appl. Polymer Sci.*, 38, 147, 1989.
29. Mazumdar, S.K. and Hoa, S.V., Application of Taguchi method for process enhancement of on-line consolidation technique, *Composites*, 26(9), 669, 1995.
30. Mazumdar, S.K. and Hoa, S.V., Manufacturing of non-axisymmetric thermoplastic composite parts by tape winding technique, *Materials and Manufacturing Processes*, 10(1), 47, 1995.
31. Werdermann, C., Friedrich, K., Cirino, M. and Pipes, R.B., Design and fabrication of an on-line consolidation facility for thermoplastic composites, *J. Thermoplastic Composite Materials*, 2, 293, 1989.

32. Beyeler, E., Phillips, W., and Guceri, S.I., Experimental investigation of laser assisted thermoplastic tape consolidation, *J. Thermoplastic Composite Materials,* 1, 107, 1988.
33. Mazumdar, S.K. and Hoa, S.V., Experimental determination of process parameters for laser assisted processing of PEEK/carbon thermoplastic composites, *38th Int. SAMPE Symp.,* Anaheim, CA, 1993.
34. Seferis, J.C., Polyetheretherketone (PEEK): processing, structure and properties studies for a matrix in high performance composites, *Polymer Composites,* 7, 158, 1986.
35. Manson, J.A.E. and Seferis, J.C., Autoclave processing of PEEK/carbon fiber composites, *J. Thermoplastic Composite Materials,* 2, 34, 1989.
36. Anderson, B.J. and Colton, J.S., A study in the lay-up and consolidation of high performance thermoplastic composites, *SAMPE J.,* 25(5), 22, 1989.
37. Kim, T.W., Jun, E.J., and Lee, W.I., The effect of pressure on the impregnation of fibers with thermoplastic resins, *34th Int. SAMPE Symp.,* 1989, 323–328.
38. Fabricating with Aromatic Polymer Composites APC-2, Data Sheet No. 5, Fiberite Corporation, 1986.
39. Lee, W.I. and Springer, G.S., A model of the manufacturing process of thermoplastic matrix composites, *J. Composite Materials,* 21(11), 1017, 1987.
40. Loos, A.C. and Li, M.C., Heat transfer analysis of compression molded thermoplastic composites, *Advanced Materials: The Challenge for the Next Decade,* SAMPE, 1990, 557–570.
41. Li, M.C. and Loos, A.C., Autohesion Model for Thermoplastic Composites, Center for Composite Materials and Structures, Report No. CCMS-90-03, Virginia Polytechnic Institute and State University, Blacksburg, VA, 1990.
42. Dara, P.H. and Loos, A.C., Thermoplastic Matrix Composite Processing Model, Center for Composite Materials and Structures, Report No. CCMS-85-10, Virginia Polytechnic Institute and State University, Blacksburg, VA, 1985.
43. Wool, R.P. and O'Connor, K.M., Theory of crack healing in polymers, *J. Appl. Phys.,* 52, 5953, 1981.
44. Jud, K., Kausch, H.H., amd Williams, J.G., Fracture mechanics studies of crack healing and welding of polymers, *J. Material Sci.,* 16, 204, 1981.
45. Mallon, P.J. and O'Bradaigh, C.M., Development of a pilot autoclave for polymeric diaphragm forming of continuous fiber-reinforced thermoplastics, *Composites,* 19(1), 37, 1988.
46. Smiley, A.J. and Pipes, R.B., Simulation of the diaphragm forming of carbon fiber/thermoplastic composite laminates, *American Society for Composites, 2nd Technical Conference,* Delaware, 1987.
47. O'Bradaigh, C.M. and Mallon, P.J., Effect of forming temperature on the properties of polymeric diaphragm formed APC-2 components, *American Society for Composites, 2nd Technical Conference,* Delaware, 1987.
48. Mallon, P.J., O'Bradaigh, C.M., and Pipes, R.B., Polymeric diaphragm forming of continuous fiber reinforced thermoplastic matrix composites, *Composites,* 20(1), 48, 1989.
49. Monaghan, M.R. and Mallon, P.J., Development of a computer controlled autoclave for forming thermoplastic composites, *Composites Manufacturing,* 1(1), 8, 1990.
50. Cogswell, F.N., The processing science of thermoplastic structural composites, *Int. Polymer Processing,* 1, 157, 1987.

Bibliography

1. Cogswell, F.N., *Thermoplastic Aromatic Polymer Composites*, Butterworth-Heinemann Ltd., 1992.

Questions

1. Why is processing of thermoset composites easier than that of thermoplastic composites?
2. What are the four common processing steps in thermoset and thermoplastic composites?
3. What are the four major steps in making a mold from a solid board?
4. What are the guidelines for gate locations in closed molding operations?
5. How do you define an ideal manufacturing process and why?
6. Why is higher processing temperature required in thermoplastic tape winding as compared to the autoclave or hot press technique?
7. Write down important processing steps in making a bathtub?
8. Why is it difficult to get a void-free tube using helical thermoplastic tape winding?
9. Why are SMC compression molding and the SRIM process not used for making prototype parts?
10. What are the limitations of filament winding?
11. Under what circumstances is SMC a process of choice?
12. Write down some of the applications of filament winding?
13. What are the major differences in the die for the thermoset pultrusion process and the thermoplastic pultrusion process?
14. What are the major processing steps in an RTM process?
15. Write down the differences between closed molding and open molding processes?
16. What are the process selection criteria?

7
Process Models

7.1 Introduction

Models are used to simulate real-world situations in a mathematical and graphical form to analyze a problem or to demonstrate the behavior and characteristics of the real world under various conditions. Engineers and researchers are using models to solve various types of physical, chemical, and engineering problems. Chemical plants control various chemical reactions and operations with the help of models. Physicists use various atomic models and quantum mechanics to understand a phenomenon. Material scientists use body-centered cubic (BCC), face-centered cubic (FCC), and other crystal structures to understand the behavior of metals and polymers. Metallurgists use phase diagrams, mechanical engineers use free body diagrams and other models to describe or predict a phenomenon or real-world situation. Models are very useful for solving real engineering or physical problems without performing any experiment. In this chapter, models for composites manufacturing processes are described to gain more insight into these manufacturing processes.

7.2 The Importance of Models in Composites Manufacturing

A process model is very helpful in analyzing a manufacturing process, providing a tool for the production of high-quality parts at low cost. It can eliminate several processing problems before a manufacturing process begins or before part design is finalized. It estimates optimum process parameters for a manufacturing process to get a high-quality product. It can save a significant amount of time and money for product fabrication.

In an RTM process, a simulation model can be used to predict the flow pattern and evaluate any dry areas in the product. It can determine optimum process parameters such as injection pressure, resin viscosity, and gate location to get complete impregnation of the reinforcements with the resin. The main problems associated with the fabrication of parts using an RTM process are:

1. Resin cures before it completely fills the cavity
2. Dry areas
3. Warpage of the part because of residual stresses
4. Change in fiber orientation (fiber wash away) because of resin flow

The above problems can be solved using process models for RTM without even making the part. The model can predict optimum gate location, fiber permeability, wall thickness, injection pressure, viscosity, flow rate, and other process parameters for better resin impregnation of reinforcements and better part quality with minimum residual stresses. Similarly, several problems of injection molding, such as weld line, flash, short shot size, and part distortion, can be solved using models. Typical problems associated with autoclave processes include insufficient cure, delamination, part warpage, and voids, all of which can be solved using process models. Process variables in a filament winding process are fiber tension, resin viscosity, cure temperature, etc.; these variables can be optimized using models for making of good quality parts. Similarly, a pultrusion process model can provide pull force, resin viscosity, and die temperature for better part quality.

In general, process models are very advantageous for predicting optimum process parameters for the fabrication of good-quality composite parts. Process models are used for part design, tool design, and for the selection of optimum processing conditions.

7.3 Composites Processing

There are various processing techniques (e.g., filament winding, autoclave, RTM, injection molding, and thermoplastic tape winding) available for the fabrication of composite parts. Composite parts are made using either thermoset resins or thermoplastic resins with some form of reinforcements. In these processes, pressure and heat are applied or removed at certain rates to get well-consolidated parts with minimum flaws. In thermoset composites, heating and cooling are applied to initiate and control the degree of cure whereas in thermoplastic composites, heating and cooling are applied to melt and solidify the resin to obtain the desired crystallinity. Pressure during cure or solidification is applied to get good compaction between adjacent layers as well as to obtain good fill-out in the mold or tool surface. The magnitude and duration of the temperatures and pressures applied during the curing/solidification process significantly affect the performance of the finished product. Therefore, a processing cycle must be selected carefully for each application. Some important considerations in selecting a proper cure/process cycle are:

- The processed part should have the lowest possible void content.
- In the case of thermoset composites, the resin should be cured uniformly and completely. Uneven curing generates residual stress in the part. Similarly, for thermoplastic composites, uniform cooling should take place to avoid residual stress build-up.
- The finished part should have minimum residual stress to minimize part distortion. Part distortion or uneven shrinkage may cause the part to go out of tolerance.
- The complete process should be achieved in the shortest possible time. A shorter process cycle time provides higher production rate and lower production cost.
- During the curing process of thermoset composites, the excess resin must be squeezed out from every ply and resin distribution must be uniform.

Methods of applying heat and pressure are different in each manufacturing process. In the autoclave process, heat is supplied with the help of hot air or hot nitrogen gas inside the autoclave, and pressure is applied by vacuum bagging and by pressurizing the chamber with air. In the RTM process, pressure is controlled by injection pressure and heat is supplied by heating the mold. In the thermoplastic tape winding process, heat is supplied by a laser or hot gas, and pressure is applied with the help of a compaction roller. In the pultrusion process, heat is supplied by a heated die and pressure is caused by narrowing the die opening.

During the solidification or curing process, a number of physical and mechanical phenomena occur in the composite material,[1] including:

- Formation of a polymer structure, change in physical and mechanical properties, change in degree of cure or crystallinity, relaxational transitions (gelation and vitrification)[2]
- Chemical and temperature shrinkage[3-5]
- Stress and strain development[4-7]
- Formation of kogesive defects and adhesive debonding[6,7]
- Release of volatiles and void growth[8]

7.4 Process Models for Selected Thermosets and Thermoplastics Processing

For composites processing, various analytical models have been developed to simulate the resin flow, degree of cure (for thermosets) or crystallinity (for thermoplastics), degree of compaction and consolidation, residual stress, etc.

To analyze the complete phenomenon, the entire model is generally divided into four sub-models.[8–12] The sub-models are coupled and solved simultaneously. George Springer and his team from Stanford University have done significant work in developing process models for the various manufacturing processes (e.g., autoclave, filament winding, pultrusion, RTM, etc.).[8–12] They developed models for understanding the cure kinetics of various thermoset resin systems.[13–15] There are several other researchers who have developed models for cure/crystalline kinetics, as well as models for the various manufacturing processes.[16–26] This chapter focuses on process models for some of the selected manufacturing processes, including autoclave, filament winding, hot press, thermoplastic tape winding, and RTM. The models presented for autoclave, filament winding, hot press technique, and thermoplastic tape winding are based on the work performed by Springer and colleagues[8,10,12,20] and is explained with kind permission from Technomic Publishing Company, Inc.[8,10,12,20] The four sub-models typically used to divide complete manufacturing process are described below.

1. **Thermochemical sub-model.** This sub-model provides instantaneous local temperature, viscosity, and either degree of cure or crystallinity, depending on the type of resin. Determining the temperature distribution inside the laminated structure is very important in understanding the viscosity and degree of cure at at any time and point. An uneven degree of cure or solidification results in residual stress of the part. The temperature distribution can be easily determined using finite element analysis (FEA) software or by developing computer models and solving them using numerical techniques such as the finite difference method or finite element method. The temperature distribution inside the structure depends on the boundary and initial conditions, as well as material properties. Boundary and initial conditions for the autoclave, filament winding, and thermoplastic tape winding processes are different and are defined below in respective process sections.

2. **Flow sub-model.** This model determines resin flow, fiber and resin distribution, compaction, and in the case of thermoplastics, it estimates degree of consolidation and bonding. Resin flow takes place in a different manner in different manufacturing processes. In an RTM process, dry fiber is used inside the mold and is impregnated with the resin by injecting the resin from single or multiple inlets. In the RTM process, resin flows through porous media. In an autoclave process, prepregs are used to make the part. In prepregs, reinforcements are already pre-impregnated with the resin. In the autoclave process, resin flow takes place between the layers to remove air and excess resins. Resin flow helps in creating well-compacted parts. In the filament winding process, incoming layers of fibers and resins are laid down over pre-laminated composites.

The resin flow in the filament winding process takes place to remove air spaces and excess resins and thus to create compacted parts. In the thermoplastic tape winding process, resin flow takes place in a small volume, where heat is supplied between incoming tape and pre-consolidated laminates. In this process, resin flow is responsible for creating a good bond between the layers. Overall, resin flow analysis in a manufacturing process provides guidance on whether the parts produced will have excess resin or dry fiber areas, unfilled mold areas, or unconsolidated areas.

3. **Stress sub-model.** This model predicts the residual stress and strain in the part resulting from a manufacturing process. Residual stress generation is of considerable concern when making thick parts or large parts. Residual stress causes the part to distort and makes the part go out of tolerance. For parts made with unsymmetric laminates or dissimilar materials, residual stress is dominant. In symmetric parts, residual stress is generated by uneven curing and solidifications. The primary causes of residual stress are chemical changes in the resin, uneven curing/solidification, and thermal strains in each ply due to differential coefficients of thermal expansion.

4. **Void sub-model.** The goal of any manufacturing process is to minimize the amount of voids because the presence of voids deteriorates the performance of the composite part. Voids are formed during composite part fabrication, either by mechanical means (e.g., air or gas bubble entrapment, broken fibers) or by homogeneous or heterogeneous nucleation. This model determines the void sizes in the part.

The above sub-models are discussed in detail in the following sections.

7.4.1 Thermochemical Sub-Model

The temperature distribution, resin viscosity, and degree of cure inside the composite depend on the rate at which heat is transmitted from the environment into the material. The purpose of this model is to determine the temperature T, viscosity μ, and degree of cure α for thermosets (or crystallinity c for thermoplastics) at any time using the following energy equation[8-10]:

$$\rho C \frac{\partial T}{\partial t} = \nabla \cdot (K \cdot \nabla T) + \rho \dot{Q} \qquad (7.1)$$

where ρ is the density, C is the specific heat, t is the time, ∇ is the Laplacian operator, K is thermal conductivity of the composite, and $\dot{Q}$ is the rate at which heat is generated or absorbed by the chemical reaction. Equation (7.1) neglects the energy transfer by convection.

Equation (7.1) is shown in general form and can be written for the autoclave process, filament winding process, etc. For example, in the filament winding process, where heat transfer takes place in both radial and axial directions, the above equation takes the following form[10]:

$$\rho C \frac{\partial T}{\partial t} = \frac{1}{r} \cdot \frac{\partial}{\partial r} \left(K_r \cdot r \frac{\partial T}{\partial r} \right) + \frac{\partial}{\partial z} \left(K_z \cdot r \frac{\partial T}{\partial z} \right) + \rho \dot{Q} \qquad (7.2)$$

where r and z are the radial and axial coordinates, and K_r and K_z are the thermal conductivities in the r and z directions.

The last term in Equations (7.1) and (7.2) can be written in the following form because the chemical reaction takes place only in the composite:

$$\rho \dot{Q} = \rho_m v_m \dot{Q}_m + \rho_f v_f \dot{Q}_f \qquad (7.3)$$

where ρ_m and ρ_f are densities, v_m and v_f are volume fractions, and $\dot{Q}_m$ and $\dot{Q}_f$ are heating rates of matrix and fiber, respectively. The last term in Equation (7.3) would be zero because there is no chemical reaction in fibers. Therefore, the heat generated can be given by:

$$\rho \dot{Q} = \rho_m v_m \dot{Q}_m \qquad (7.4)$$

The degree of cure α of the resin is interpreted as[8,11]:

$$\alpha = \frac{Q_m}{H_u} \qquad (7.5)$$

where Q_m is the heat evolved from time t = 0 to time t, and H_u is the total heat of reaction of the matrix. For semi-crystalline thermoplastic material, Q_m is related to the crystallinity c as c = Q_m/H_u.[10,12] Differentiation of Equation (7.5) can give the expression for $\dot{Q}_m$ as:

$$\dot{Q}_m = \left(\frac{d\alpha}{dt} \right) H_u \qquad (7.6)$$

For semi-crystalline thermoplastics, $\dot{Q}_m$ can be related to the crystallinity c as follows:[12]

$$\dot{Q}_m = \left(\frac{dc}{dt} \right) H_u \qquad (7.7)$$

where H_u is the theoretical ultimate heat of crystallization of the polymer at 100% crystallinity.

The rate of degree of cure and crystallinity are expressed in the following forms:

$$\left(\frac{d\alpha}{dt}\right) = f_1(\alpha, t) \tag{7.8}$$

$$\left(\frac{dc}{dt}\right) = f_1\left(c, T, \frac{dT}{dt}\right) \tag{7.9}$$

Expressions for f_1 for the various thermoset composites such as Fiberite 976, Hercules 3501-6, and Hercules HBRF-55 epoxy resins are available in the literature.[13-16] Similarly, expressions for dc/dt are available in the literature[17-20] for PEEK thermoplastic composites.

Equations (7.1), (7.6), (7.7), and (7.8) or (7.9) are solved to determine the temperature and degree of cure or crystallinity inside the material at various times for any specific initial and boundary conditions. The initial conditions for this case are temperature and degree of cure/crystallinity inside the composite before or at the start of the cure ($t \leq 0$). The boundary conditions are the temperatures on the top and bottom surfaces of the composite as a function of time during cure ($t > 0$). These equations are solved by numerical techniques such as finite element or finite difference methods for given initial and boundary conditions. Following are the initial and boundary conditions for some of the thermoset and thermoplastic manufacturing processes.

7.4.1.1 Autoclave or Hot Press Process for Thermoset Composites

For making a flat plate using an autoclave or hot press process, the initial conditions ($t \leq 0$) are[8]:

$$\begin{gathered} T = T_i(z)\{0 \leq z \leq L\} \\ \alpha = \alpha_0 \end{gathered} \tag{7.10}$$

where T_i is the initial temperature in the composite, and α_0 is the initial degree of cure, which is zero in most cases.

The boundary conditions are as shown in Figure 7.1; that is:

$$\begin{gathered} T = T_L(t) \quad at \quad z = 0 \\ T = T_u(t) \quad at \quad z = L \end{gathered} \tag{7.11}$$

where T_L and T_u are the temperatures on the bottom and upper surfaces of the composite, respectively.

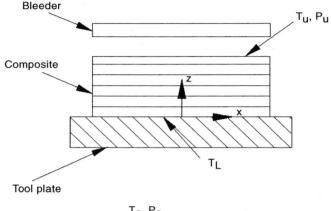

FIGURE 7.1
Boundary conditions for the autoclave process.

By solving Equations (7.1), (7.6), (7.7), (7.8), (7.10), and (7.11), values of temperature T, the degree of cure, and cure rate can be determined as functions of position and time inside the composite.

Once these parameters are obtained, the viscosity is readily determined from rheological data. For example, the viscosity of a thermoset resin is given by:

$$\mu = f_2(\alpha, T) \tag{7.12}$$

Expressions for f_2 are available in the literature[13–16] for selected resins (e.g., Fiberite 976, Hercules 3501-6, and Hercules HBRF-55 epoxy resins).

7.4.1.2 Filament Winding of Thermoset Composites

The initial conditions, such as the temperature T_0 and degree of cure α_0 at which resin-impregnated fibers are laid down on the mandrel, are specified as:

$$T = T_0 \quad \alpha = \alpha_0 \quad at \quad t = t_0 \tag{7.13}$$

In the filament winding process, time t_0 is different for each layer because filament winding is a continuous process and thickness is built up gradually.

The temperature boundary conditions for $t > t_0$ are written as follows and as shown in Figure 7.2[10]:

$$\begin{aligned} T &= T_c \quad at \quad r = R_{mo} + h \quad 0 \leq z \leq L \\ T &= T_m \quad at \quad r = R_{mi} \quad 0 \leq z \leq L \end{aligned} \tag{7.14}$$

Process Models

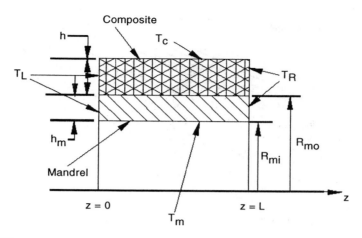

FIGURE 7.2
Boundary conditions for the filament winding process.

$$T = T_L \quad at \quad z = 0 \qquad R_{mi} \leq r \leq R_{mo} + h$$

$$T = T_R \quad at \quad z = L \qquad R_{mi} \leq r \leq R_{mo} + h$$

where T_c and T_m are the temperatures on the outer surface of the composite cylinder and on the inner surface of the mandrel, respectively; R_{mi} and R_{mo} are the inner and outer radii of the mandrel, respectively; and h, which varies with time, is the thickness of the cylinder. T_L and T_R are the temperatures at the left and right edges of the composite cylinder, respectively.

7.4.1.3 Tape Winding of Thermoplastic Composites

A typical tape winding process is shown in Figure 7.3. In a tape winding process, composite cylinders are made by online consolidation of thermoplastic tape on a cylindrical mandrel. Refer to Chapter 6 for details on the thermoplastic tape winding process. In a tape winding process, heat is supplied at the contact point of the incoming tape and preconsolidated laminate for melting the resin and a pressure P is applied using a roller for consolidation. The thermal conditions are shown in Figure 7.4. The thermochemical model will be same as Equation (7.2) with a θ term instead of z term as follows[20]:

$$\rho C \frac{\partial T}{\partial t} = \frac{1}{r} \cdot \frac{\partial}{\partial r} \left(K_r \cdot r \frac{\partial T}{\partial r} \right) + \frac{1}{r^2} \frac{\partial}{\partial \theta} \left(K_\theta \cdot r \frac{\partial T}{\partial \theta} \right) + \rho \dot{Q} \qquad (7.15)$$

The initial conditions (t ≤ 0) are given as follows:

$$T = T_{in}^m \text{ in mandrel}$$

$$T = T_{in}^c \text{ and } c = c_{in} \text{ in composite} \qquad (7.16)$$

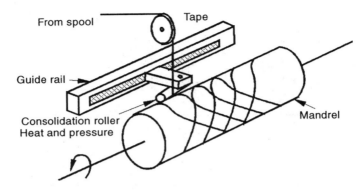

FIGURE 7.3
Schematic diagram of a tape winding process.

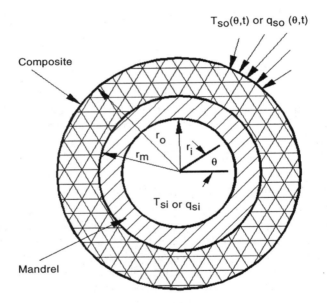

FIGURE 7.4
Thermal conditions during a tape winding process.

In the tape winding process, boundary conditions such as temperature or heat flux at the mandrel inside surface and at the composite cylinder outside surface are specified at time t > 0. At the inside surface of the mandrel, the temperature T_{si} or heat flux q_{si} is uniform and constant as shown below for t > 0,

$$T = T_{si} \quad \text{or} \quad q = q_{si} \quad \text{at} \quad r = r_i \tag{7.17}$$

where r_i is inside radius of the mandrel. At the outer surface of the cylinder, either the surface temperature T_{so} or the heat flux q_{so} can be specified. Either of these may vary in the circumferential θ direction and with time as follows for $t > 0$,[20]

$$T = T_{so}(\theta,t) \quad or \quad q = q_{so}(\theta,t) \quad at \quad r = r_o \qquad (7.18)$$

where r_o is the composite outside surface radius. The above conditions can be applied after one or more layers have been wound or at the completion of winding. In the former case, composite cylinder thickness varies with time.

Mazumdar and Hoa[27-30] performed analytical and experimental studies for the processing of thermoplastic composites (Carbon/PEEK, APC-2) during tape winding and tape laying processes. During the study, hoop-wound composite rings with various plies were produced as shown in Figure 7.5. Laser and hot nitrogen gas were used as heat sources. Based on the finite element analysis, temperature profiles for various laser powers, tape speeds, and ply thickness are plotted in Figures 7.6 through 7.9 for the tape winding process. The experimental temperature histories during laser and hot gas processing are plotted in Figures 7.10 through 7.15. For the measurement of the temperature history, a thin (0.125-mm diameter) K type thermocouple was inserted between the fifth and sixth ply without interrupting the tape winding process for making hoop-wound composite rings. Because of the high rate of temperature change near the consolidation point, a thermocouple with a small response time was selected for greater accuracy in measuring the temperature history.[30] The thermocouple was connected to a digital process indicator (DP-86, Omega Eng., Inc.). The reading rate of the indicator was 0.5 s. The temperature history was recorded on a 125-MHz digital

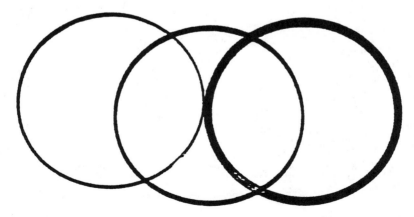

FIGURE 7.5
Hoop-wound composite rings having 10, 15, and 45 plies produced by laser-assisted processing. (Courtesy of S.K. Mazumdar.[30])

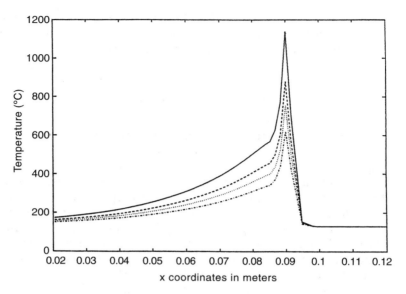

FIGURE 7.6
Numerical prediction of temperature profile for 80-W (—), 60-W (--), 50-W (···), and 40-W (-·) laser powers at 10 mm/s tape speed. (Courtesy of S.K. Mazumdar.[30])

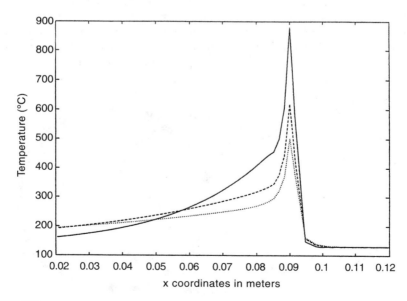

FIGURE 7.7
Temperature profile obtained by computer simulation for 60-W laser power at 10 mm/s (—), 20-mm/s (--), and 30-mm/s (···) tape speeds. (Courtesy of S.K. Mazumdar.[30])

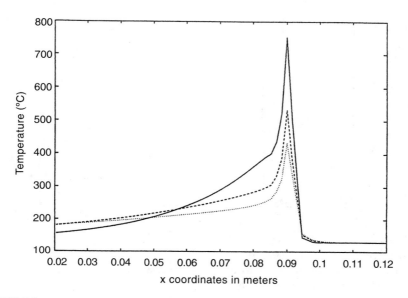

FIGURE 7.8
Effect of tape speed on temperature distribution for a 5-ply thick laminate with 50-W laser power and 10-mm/s (—), 20-mm/s (--), and 30-mm/s (···) tape speeds. (Courtesy of S.K. Mazumdar.[30])

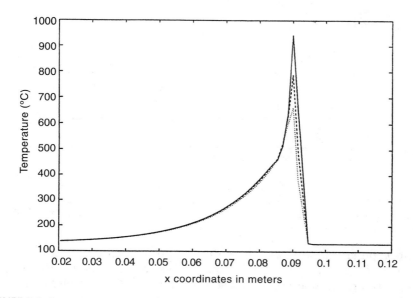

FIGURE 7.9
Temperature distribution at different ply interfaces for a 5-ply thick laminate at 50-W laser power and 6.28-mm/s tape speed. (—), (--) and (···) represent the fifth, fourth, and third ply, respectively. (Courtesy of S.K. Mazumdar.[30])

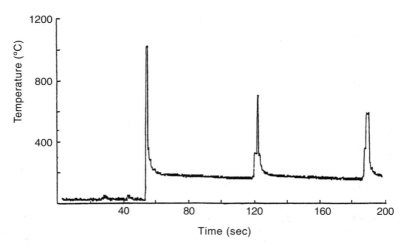

FIGURE 7.10
Experimental temperature profile for 50-W laser power, 6.28-mm/s tape speed, and 67.2-kN/m consolidation pressure. (Courtesy of S.K. Mazumdar.[30])

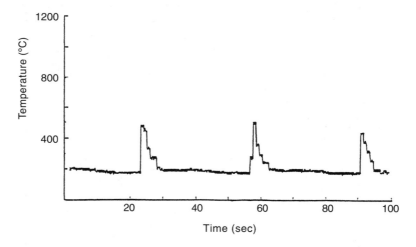

FIGURE 7.11
Experimental temperature history for 50-W laser power, 13.3-mm/s tape speed, and 67.2-kN/m consolidation pressure. (Courtesy of S.K. Mazumdar.[30])

oscilloscope and plotted as shown in Figures 7.10 through 7.15 for three consecutive heating cycles for various processing conditions. It is evident from these figures that only in a small region near the consolidation point was the matrix material heated above its melting temperature. Also, the maximum temperature reached in a ply decreases with an increase in distance from the consolidation point. Good agreements between theoretical predictions and experimental results was found.

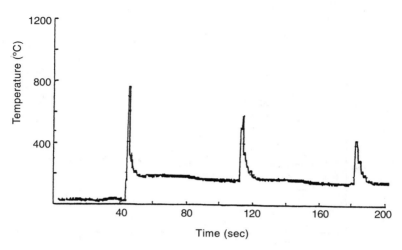

FIGURE 7.12
Experimental temperature profile for 35-W laser power, 6.28-mm/s tape speed, and 67.2-kN/m consolidation pressure. (Courtesy of S.K. Mazumdar.[30])

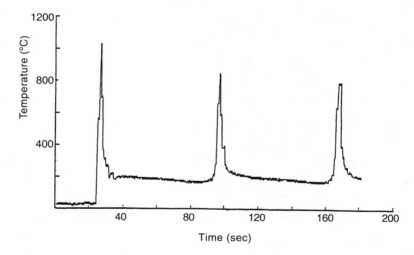

FIGURE 7.13
Temperature history during hot gas processing for a nitrogen flow rate of 146 SCFH, 6.28-mm/s tape speed, and 67.2-kN/m consolidation pressure. (From Mazumdar, S.K. and Hoa, S.V., J. *Thermoplastic Composite Mater.*, 9, 35, January 1966. Reprinted by permission of Sage Publications Ltd.)

The heating and cooling rates were calculated using the numerical model for laser processing and plotted (see Figures 7.16 through 7.18). With the increase in laser power, the heating rate was found to be increasing. The heating rate for 40-W laser power was calculated to be 31,108°C/min, whereas for 80-W laser power, it was 63,896°C/min. The melt time (i.e., the duration for which the temperature during processing remains higher than the melt temperature (343°C) of PEEK thermoplastics) was calculated to be in the range of 0.04 to

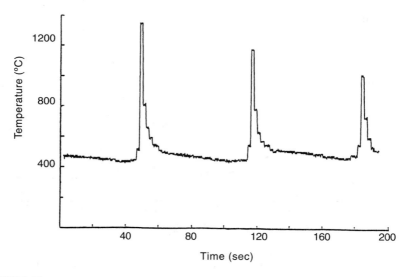

FIGURE 7.14
Temperature profile during hot gas processing for a nitrogen flow rate of 146 SCFH, 13.3-mm/s tape speed, and 67.2-kN/m consolidation pressure. (From Mazumdar, S.K. and Hoa, S.V., J. *Thermoplastic Composite Mater.*, 9, 35, January 1966. Reprinted by permission of Sage Publications Ltd.)

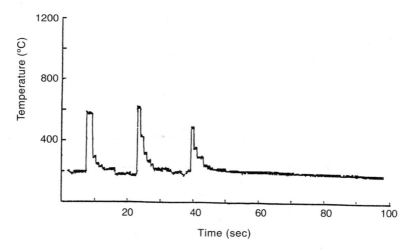

FIGURE 7.15
Temperature history during hot gas processing for a nitrogen flow rate of 264 SCFH, 28.0-mm/s tape speed, and 67.2-kN/m consolidation pressure. (From Mazumdar, S.K. and Hoa, S.V., J. *Thermoplastic Composite Mater.*, 9, 35, January 1966. Reprinted by permission of Sage Publications Ltd.)

1.75 s for 50-W and 60-W laser powers and for 10 to 30 mm/s tape speeds. Shorter melt times and higher heating rates during the tape winding process limit the amount of diffusion and therefore a higher processing temperature is needed for sufficient molecular interdiffusion. Higher processing temperatures would result in lower viscosity and higher diffusivity, which would cause a

Process Models

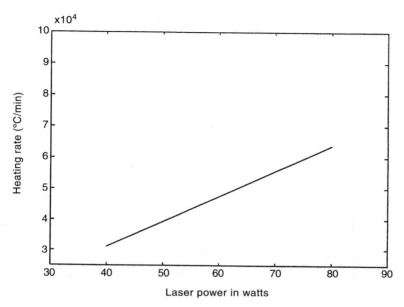

FIGURE 7.16
Effect of laser power on heating rate during tape winding process. (Courtesy of S.K. Mazumdar.[30])

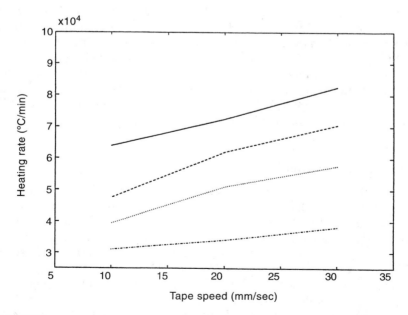

FIGURE 7.17
Effect of tape speed on heating rate for 80-W (—), 60-W (- -), 50-W (···), and 40-W (-·) laser powers. (Courtesy of S.K. Mazumdar.[30])

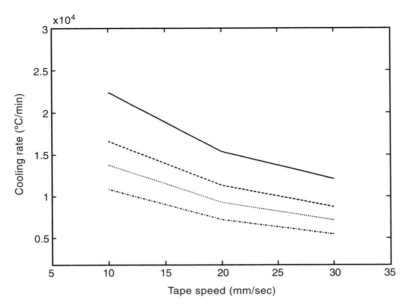

FIGURE 7.18
Effect of tape speed on cooling rate for the first 2 seconds after the maximum temperature for 80-W (—), 60-W (--), 50-W (···), and 40-W (-·) laser powers. (Courtesy of S.K. Mazumdar.[30])

greater degree of resin flow and interdiffusion and thus better interply bond properties. Saint-Royre and co-workers[31] found that the minimum temperature required for good weld conditions for polyethylene layers increases from 145 to 214°C with the increase in heating rate from 20 to 400°C/min. Mazumdar and Hoa suggested more than 500°C processing temperature for carbon/PEEK (APC-2) thermoplastics for better interply bond properties during the tape winding process.[27] Carbon/PEEK is processed in the range of 380 to 400°C for autoclave, hot press, and diaphragm molding processes.

7.4.2 Flow Sub-Model

In composites processing, resin flows inside the fiber architecture and mold to form a well-consolidated laminate. In an RTM process, resin is injected from the inlet port and flows through the fiber architecture to form a composite product. Resin flow analysis provides guidance on whether the part produced by the manufacturing process has excess resin or dry fiber area, unfilled mold area, or unconsolidated area. In an autoclave process, resin flow analysis determines whether or not all the layers (prepregs) are fully compacted. In an autoclave process, after forming a good interface between the layers, excess resin is squeezed out. In this section, resin flow models for autoclave curing, filament winding, and thermoplastic tape winding are described. Resin flow during an RTM process is described later in this chapter.

Process Models

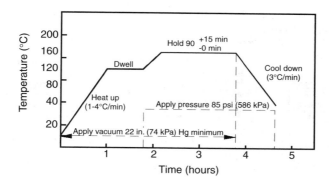

FIGURE 7.19
Typical cure cycle during an autoclave process.

7.4.2.1 Compaction and Resin Flow during Autoclave Cure

In an autoclave process, prepregs are stacked on a tool surface and then the bleeder and bagging materials are applied. The prepregs are then pressurized and temperature is raised for resin flow and proper consolidation. A typical pressure and temperature curve during autoclave process is shown in Figure 7.19. The purpose of supplying heat during the process is to:

- Decrease the resin viscosity for easy resin flow.
- Start curing action.

The purpose of pressure during the curing operation is to:

- Remove excess resin from the space between adjacent plies as well as from spaces between individual fibers. The resin flow between the individual fibers is relatively small due to the proximity of fibers.[21]
- Make a uniform, well-consolidated part.

A well-consolidated part will not have a resin-rich area between plies, whereas a poorly consolidated part will have a resin-rich area between plies. In the presence of temperature and pressure, resin flows normal to the plane of laminate, parallel to the laminate, or some combination of the two, as shown in Figure 7.20 for the autoclave process. The amount of resin flow in these directions depends on the width-to-thickness ratio of the part, edge constraints, and bleeder arrangements. Figure 7.21 shows prepreg motion when there is resin flow only in the normal direction. In this case, the top layer starts moving first, squeezing excess resin from the first and second prepregs, as shown in the figure. In the next phase, the first and second layers move in unison toward the third layer, removing excess resin between the second and third layers. This sequence is repeated for the subsequent

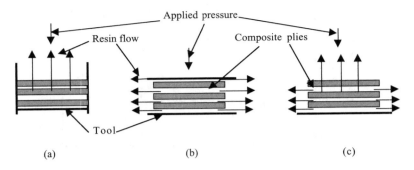

FIGURE 7.20
Resin flow illustration for three cases in an autoclave process: (a) resin flow normal to the plies, (b) resin flow parallel to the plies, and (c) resin flow in both directions.

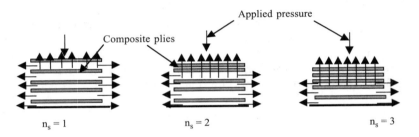

FIGURE 7.21
Resin flow motion for both normal and parallel to the tool plate directions.

layers. Thus, the layers are compacted in a wave-like, cascading manner.[8,21] When the resin flow is only in the parallel direction, compaction and ply movement take place uniformly. Springer and Loos[8,11] developed models to represent these three modes of resin flow with the assumption that fibers do not touch each other. Gutowski and co-workers[22,23] included the effect of contact between fibers in their model. It was found that the effects of fiber contact become appreciable only at fiber volume fractions above 65 to 70%. The present analysis is based on the Loos and Springer model.[8]

7.4.2.1.1 Resin Flow Normal to the Tool Plate

The present analysis of resin flow during autoclave curing is done for the case when resin flow is only normal to the tool surface. Section 7.4.2.1.2 describes the compaction behavior when resin flow is parallel to the tool surface. The flow analysis is done separately for both directions. Kardos et al.[24] have developed a model for the case when resin flow takes place in both directions simultaneously.

The present model determines the amount of resin flows to the bleeder due to the action of resin squeezing out between plies. From that, it determines the number of layers of prepregs compacted during the process.

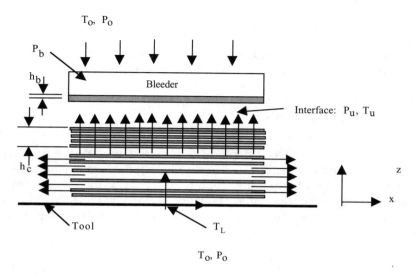

FIGURE 7.22
Resin flow model for autoclave cure. (Adapted from Loos, A.C. and Springer, G.S., *J. Composite Mater.*, 17, 135, 1983.)

To carry out the analysis, it is assumed that the resin velocities in either prepreg or bleeder follow Darcy's law (flow in a porous medium) at any instant in time. According to Darcy's law, the velocity V is represented by:

$$V = -\frac{S}{\mu}\frac{dP}{dz} \quad (7.19)$$

where S is the apparent permeability of the bleeder or prepreg, μ is the viscosity of the resin, and dP/dz is the pressure gradient. The rate of change of mass M in the composite can be determined using the law of conservation of mass and Equation (7.19):

$$\frac{dM}{dt} = -\rho_m A_z V_z = -\rho_m A_z \frac{P_c - P_u}{\int_0^{h_c} \mu dz} \quad (7.20)$$

where ρ_m is the matrix density, A_z is the cross-sectional area normal to the z-axis, h_c is the thickness of the compacted plies through which resin flow takes place (Figure 7.22). The pressure between the bleeder and composite interface is represented by P_u. P_c is the pressure in the composite at position hc. Pressure P_c in this case is equal to the applied pressure P_o.

During compaction, excess resin comes out from the prepreg and flows into the bleeder. The resin flow rate through the composite at any instant in time is equal to the resin flow rate into the bleeder:

$$\rho_m A_z V_z = \rho_m A_z V_b \tag{7.21}$$

It is assumed that the viscosity of the resin inside the bleeder is independent of position but not of time. From Equations (7.19) and (7.21):

$$\rho_m A_z V_z = \rho_m A_z \frac{S_b}{\mu_b} \frac{P_u - P_b}{h_b} \tag{7.22}$$

where h_b is the instantaneous depth of resin in the bleeder. The subscript b refers to conditions in the bleeder. In developing the above equations, the pressure drop across the porous teflon sheet between the bleeder and the prepreg was neglected. Combining Equations (7.20) and (7.22) yields the rate of change of resin mass in the composite by noting that the mass of fibers remains constant.[8]

$$\frac{dM_m}{dt} = \frac{\rho_m A_z S_c}{\int_0^{h_c} \mu \, dz} \left[\frac{P_o - P_b}{1 + G(t)} \right] \tag{7.23}$$

where the parameter $G(t)$ is defined as:

$$G(t) = \frac{S_c}{S_b} \frac{\mu_b h_b}{\int_0^{h_c} \mu \, dz} \tag{7.24}$$

where S_c and S_b are the permeabilities of the prepreg and bleeder, respectively. The mass of resin that leaves the composite and enters the bleeder in time t is given by:

$$M_T = \int_0^t \frac{dM_m}{dt} \, dt \tag{7.25}$$

The instantaneous resin depth in the bleeder is related to the mass of the resin that enters the bleeder by the following equation:

$$h_b = \frac{1}{\rho_m \phi_b A_z} \int_0^t \frac{dM_m}{dt} \, dt \tag{7.26}$$

where ϕ_b is the porosity of the bleeder and represents the volume (per unit volume) that can be filled by resin. The thickness of the compacted plies is:

$$h_c = n_p h_1 \tag{7.27}$$

where h_1 is the thickness of one compacted prepreg ply and n_p is the number of compacted prepreg plies. The value of n_p varies with time, depending on the amount of resin that has been squeezed out of the composite.

Equations (7.20) to (7.27) are used to calculate the resin flow normal to the tool plate.

7.4.2.1.2 Resin Flow Parallel to the Tool Plate

In the plane parallel to the tool plate, resin flows in two directions (along fibers and perpendicular to fibers). The resin flow perpendicular to the fibers is small because of the resistance created by the fibers and the restraints placed around the edges of the composite.[8] If such restraints are not provided, then fiber spreading (wash-out) would occur, resulting in a nonuniform distribution of fibers in the composite. This scenerio occurs most often during compression molding of unidirectional prepregs with no edge support, and must be avoided to obtain good laminated composites.

This section focuses on resin flow along the fibers because it is the most prominent resin flow mechanism in flow parallel to the tool plate. Resin flow along the fibers and parallel to the tool plate is characterized as viscous flow between two parallel plates (channel flow) separated by a distance d_n, as shown in Figure 7.23.

The distance, d_n, separating the plates (Figure 7.23) is small compared to the thickness of the composite ($d_n < L$). It is assumed that the variation in resin properties across and along the channel is constant. The pressure drop between the center of any given channel and the channel exit can be expressed as[8,25]:

$$\frac{2(P_H - P_L)}{\rho_m (V_x^2)_n} = \lambda \frac{X_L}{d_n} \tag{7.28}$$

where $(V_x)_n$ is the average resin velocity in the channel, X_L is the channel length as shown in the Figure 7.23. The subscript n represents the channel located between the n and n – 1 prepreg plies. The thickness of the nth channel is determined assuming there is one channel in each ply and all excess resin is contained in the channel. With this assumption, the thickness of the nth channel can be calculated as:

$$d_n = \frac{M_n}{\rho_n A_Z} - \frac{M_{com}}{\rho_{com} A_Z} \tag{7.29}$$

The mass M_n and density ρ_n of the nth prepreg ply is calculated by the rule of mixture.[8] M_{com} and ρ_{com} are the mass and the density of a compacted prepreg ply, respectively. To determine the experimental values of M_{com} and

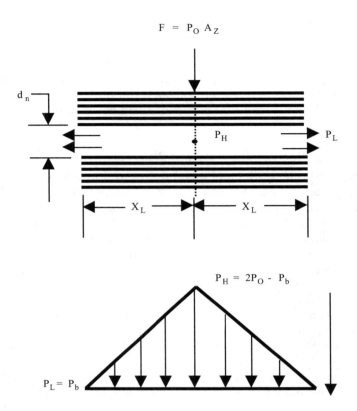

FIGURE 7.23
Resin flow model parallel to the tool plate along the fiber direction. (Adapted from Loos, A.C. and Springer, G.S., *J. Composite Mater.*, 17, 135, 1983.)

ρ_{com}, consider making a thin (4 to 16 ply) composite panel.[8] The panel is cured, employing a cure cycle that will ensure that all the excess resin is squeezed out of every ply in the composite (i.e., all plies are consolidated, $n_s = N$). The total mass of the composite M is experimentally determined after the cure. The resin content of one compacted prepreg ply $(M_m)_{com}$ is related to the composite mass by following relation. By knowing M, value of M_{com} is determined from the following relation:

$$\left(M_m\right)_{com} = \frac{M}{N} - M_f \tag{7.30}$$

where M_f is the mass of fiber in one prepreg ply and N is the total number of plies in the composite. The compacted prepreg ply thickness h_1 is given by:

$$h_1 = \frac{M/N}{\rho_{com} A_z} \tag{7.31}$$

where ρ_{com} is the density of compacted ply thickness that can be determined by the rule of mixture.[8] For laminar flow between parallel plates, λ is given by:

$$\lambda \equiv \frac{(1/B)\mu_n}{\rho_m (V_x)_n d_n} \tag{7.32}$$

where μ_n is the viscosity of the resin in the nth channel. By substituting Equation (7.32) into (7.28), the average velocity in the channel is determined by:

$$(V_x)_n = B \frac{d_n^2}{\mu_n} \frac{(P_H - P_L)}{X_L} \tag{7.33}$$

where B is a constant that is determined experimentally. The resin mass flow rate is:

$$(\dot{m}_{mx})_n = \rho_m A_x (V_x)_n \tag{7.34}$$

where A_x is the cross-sectional area defined as the product of the channel width W and thickness d_n. The law of conservation of mass, together with Equations (7.33) and (7.34), provides the following expression for the rate of change of mass in the nth prepreg ply:

$$\frac{d(M_m)_n}{dt} = -2(\dot{m}_{mx})_n = -2B \frac{d_n^3}{\mu} \rho_r W \frac{(P_H - P_L)}{X_L} \tag{7.35}$$

The amount of resin leaving the nth prepreg ply in time t is:

$$(M_E)_n = \int_0^t \frac{d(M_m)_n}{dt} dt \tag{7.36}$$

The total resin flow in all plies containing excess resin can be calculated by summing Equation (7.36):

$$M_E = \sum_{n=1}^{N-n_s} (M_E)_n \tag{7.37}$$

where N is the total number of prepreg plies.

The pressure distribution of each channel is determined as follows by assuming that the pressure gradient in the x-direction is linear and that the centerline pressure P_H is the same in each ply:

$$P = \left(\frac{P_L - P_H}{X_L}\right)x + P_H \tag{7.38}$$

where the centerline pressure P_H is estimated from the force balance applied along the boundaries of the channel. P_L is the pressure at the exit of the channel and is assumed to be equal to the environment pressure P_b surrounding the composite. A force balance applied along the channel surface gives (see Figure 7.23):

$$F = \int_A P dA = 2W \int_0^{X_L} P dx \tag{7.39}$$

F is the applied force that can be related to the cure pressure P_o as:

$$F = P_o A_z = 2P_o W X_L \tag{7.40}$$

Equations (7.38) to (7.40) yield the centerline pressure as:

$$P_H = 2P_o - P_b \tag{7.41}$$

Using Equations (7.28) to (7.41), the resin flow along the fibers is calculated.

7.4.2.1.3 Total Resin Flow

The total resin flow at any time t is determined by adding the resin flow normal and parallel to the tool plate. Using the law of conservation of mass:

$$\frac{dM}{dt} = -\left[\dot{m}_{mz} + 2\sum_{n=1}^{N-n_s}(\dot{m}_{mx})_n\right] \tag{7.42}$$

where $\dot{m}_{mz}$ and $\dot{m}_{mx}$ are the resin flow rates normal (z-direction) and parallel (x-direction) to the tool plate, respectively. The total mass of the composite at any time t is given by:

$$M = M_i - M_T - M_E \tag{7.43}$$

where M_i is the initial mass of the composite, and M_T and M_E are calculated from Equations (7.19) and (7.29), respectively. Due to the flow of resin, the

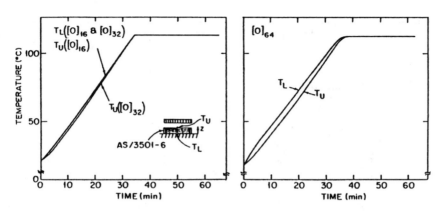

FIGURE 7.24
The temperatures measured by thermocouples on the surfaces of 16-, 32-, and 64-ply composites during cure. (Reprinted from Loos, A.C. and Springer, G.S., *J. Composite Mater.*, 17, 135, 1983.)

composite gets compacted and the composite thickness decreases. The thickness of the composite at any time t is calculated from:

$$L = \frac{M}{\rho 2 X_L W} \tag{7.44}$$

where ρ is the density of the composite.

The above thermochemical and resin flow models for autoclave processing were experimentally verified by Loos and Springer.[8] The experiments were performed using graphite/epoxy (Hercules AS/3501-6) prepreg tapes. Temperature distributions and resin flow perpendicular and parallel to the tool plate were measured during the experiments. Cure temperatures for all the tests were maintained the same. The cure temperatures of the lower and upper composite surfaces, as measured by thermocouples, are shown in Figure 7.24.

Pressure was applied at the beginning of the cure cycle and maintained constant during the cure process. Tests were conducted for various cure pressures, ranging from 103 kPa (15 psi) to 724 kPa (105 psi). The pressure in the bleeder was taken to be equal to the ambient pressure of 101.35 kPa (14.7 psi). The temperature as a function of time was measured at three locations inside a 64-ply composite and results are shown in Figure 7.25. Experimental results and theoretical model predictions were found to match very well, as shown in Figure 7.25.

Resin flow measurements were performed for different cure pressures [103 kPa (15 psi), 345 kPa (50 psi), 586 kPa (85 psi), and 724 kPa (105 psi)], different ply thicknesses (16, 32, and 64 plies), and different initial resin contents (39 and 42%) in the prepreg. The results are plotted in Figures 7.26 through 7.28, taking time t as the abscissa and mass loss due to resin flow as the ordinate. The mass losses shown in Figures 7.26 to 7.28 represent the mass loss with respect to the initial mass of the composite.

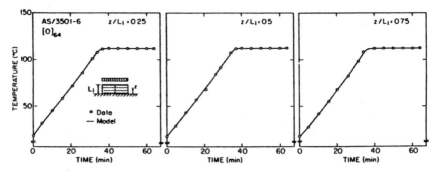

FIGURE 7.25
Theoretical and experimental plots of temperature as a function of time at three positions inside a 64-ply composite. The temperature cure cycle is shown in Figure 7.29. The cure and bleeder pressures were constant at 586 kPa (85 psi) and 101 kPa (14.7 psi), respectively. (Reprinted from Loos, A.C. and Springer, G.S., *J. Composite Mater.*, 17, 135, 1983. With permission from Technomic Publishing Company.)

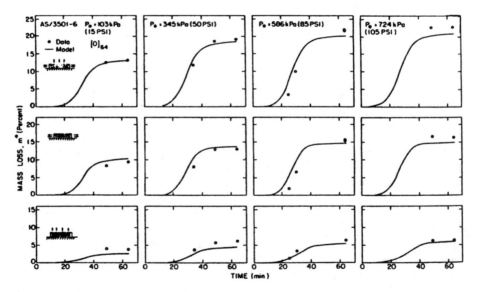

FIGURE 7.26
The mass loss, normal to the tool plate (bottom), parallel to the tool plate (center), and the total mass loss (top) as a function of time for a 64-ply composite. The temperature cure cycle is shown in Figure 7.29. The bleeder pressure was constant at 101 kPa (14.7 psi). The initial resin content was 42%. (Reprinted from Loos, A.C. and Springer, G.S., *J. Composite Mater.*, 17, 135, 1983. With permission from Technomic Publishing Company, Inc.)

The temperature distributions and resin flows were determined by models using the same cure temperature and pressures as employed during the test. The results of the models are shown as solid lines in Figures 7.25 to 7.28. It is evident from the figures that the calculated and measured temperatures

Process Models

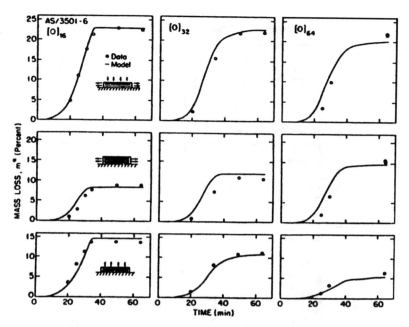

FIGURE 7.27
The mass loss, normal to the tool plate (bottom), parallel to the tool plate (center), and the total mass loss (top) as a function of time. Comparisons between the data and the results computed by the model for 16-, 32-, and 64-ply composites. The temperature cure cycle is shown in Figure 7.29. The cure and bleeder pressures were constant at 586 kPa (85 psi) and 101 kPa (14.7 psi), respectively. The initial resin content was 42%. (Reprinted from Loos, A.C. and Springer, G.S., *J. Composite Mater.*, 17, 135, 1983. With permission from Technomic Publishing Company, Inc.)

and the calculated and measured resin mass losses agree well. These agreements, which exist for a wide range of applied cure pressures, composite thicknesses, and initial prepreg resin contents, tend to validate the thermochemical and resin flow models.

The pressure applied during the cure must be high enough to squeeze out any excess resin in each ply before it starts to gel. The magnitude of pressure applied depends on the composite thickness and to a lesser extent on the cure temperature and heating rate. This is illustrated in Figure 7.29, where the number of compacted plies n_s is plotted against time. The cure cycle is shown in Figure 7.29a.

7.4.2.2 Compaction and Resin Flow during Filament Winding

The compaction model presented here was developed by Lee and Springer for filament winding of thermoset composites.[10] During the filament winding process, resin-impregnated fiber bands with bandwidth b and thickness Δh are deposited on a mandrel with an initial tension F_o. The initial winding angle (i.e., the angle between the fibers and the axial direction) is ϕ_o as shown

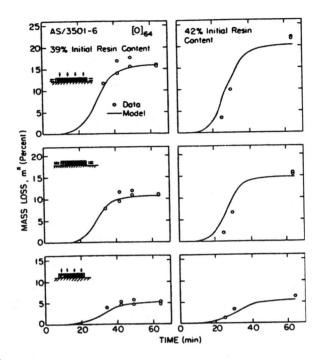

FIGURE 7.28
The mass loss, normal to the tool plate (bottom), parallel to the tool plate (center), and the total mass loss (top) as a function of time for a 64-ply composite. Comparisons between the data and the results computed by the model for different initial resin contents (39 and 42%). The temperature cure cycle is shown in Figure 7.29. The cure and bleeder pressures were constant at 586 kPa (85 psi) and 101 kPa (14.7 psi), respectively. The initial resin content was 42%. (Reprinted from Loos, A.C. and Springer, G.S., *J. Composite Mater.*, 17, 135, 1983. With permission from Technomic Publishing Company, Inc.)

in Figures 7.30 and 7.31. During consolidation, the fiber may move, causing a change in fiber tension and in fiber position. At any time t, the fiber tension is F (Figure 7.30).

A fiber layer consists of a fiber sheet of thickness $\Delta\xi$ surrounded by the resin as shown in Figure 7.32 and deposited on the mandrel. The cross-sectional area of the fiber sheet is $A_f = v_f A$, where v_f is the fiber volume fraction and A is the cross-sectional area of the entire layer ($A = b \cdot \Delta h$). As mentioned, the fiber sheet can move in a radial direction and thus change the value of the fiber tension. The motion of the fiber sheet in the hoop and axial directions is not considered because the axial and hoop components of the fiber tension are in equilibrium. The two major reasons for changes in the radial position r_f of the fiber sheet are:

1. During the filament winding process, the fiber tension in curved fibers causes the fibers to move when the viscosity of the resin is low (Figures 7.33 and 7.34). Once the resin gels or the viscosity

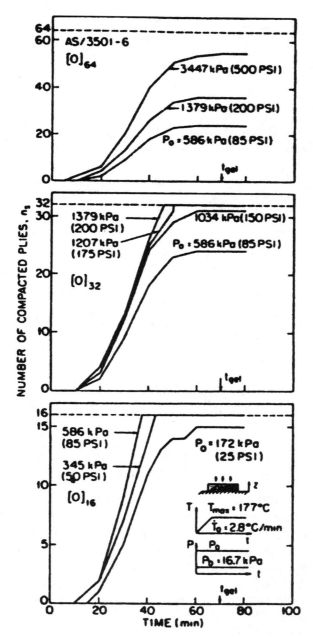

FIGURE 7.29
Number of compacted plies as a function of time for different cure pressures and different composite thicknesses: (a) 16-ply (bottom), (b) 32-ply (center), and (c) 64-ply (top). Results obtained by the model. (Reprinted from Loos, A.C. and Springer, G.S., *J. Composite Mater.*, 17, 135, 1983. With permission from Technomic Publishing Company, Inc.)

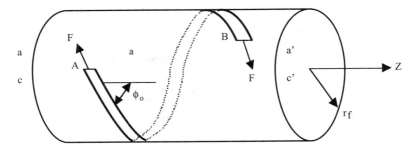

FIGURE 7.30
Fiber band at r_f position during filament winding operation. (Adapted from Lee, S.Y. and Springer, G.S., *J. Composite Mater.*, 24, 1270, 1990.)

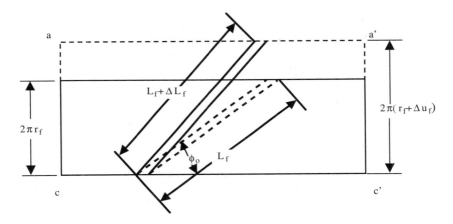

FIGURE 7.31
The original and deformed lengths of a fiber band in a filament winding operation. (Adapted from Lee, S.Y. and Springer, G.S., *J. Composite Mater.*, 24, 1270, 1990.)

becomes sufficiently high, the fibers get locked into the resin and do not move relative to the resin under fiber tension. The radial displacement of a fiber sheet relative to the resin is denoted by u_f.

2. The chemical changes (shrinkage) of the resin and the mismatch of coefficients of thermal expansion (CTE) of the mandrel and composite can cause the change in the fiber sheet position. The radial displacement in fiber sheet caused by the expansion and contraction of the composite and mandrel is denoted by u_{mc}, where $u_{mc} = u_m + u_c$ (Figure 7.34).

Therefore, the instantaneous fiber position relative to the axis of the cylinder is given by,

$$r_f = R_f^0 + u_f + u_{mc} \tag{7.45}$$

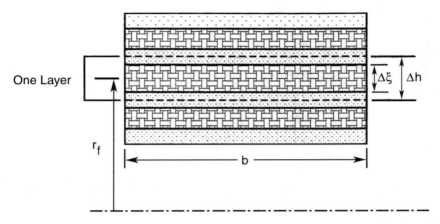

FIGURE 7.32
Demonstration of fiber sheet in a filament winding operation. (Adapted from Lee, S.Y. and Springer, G.S., *J. Composite Mater.*, 24, 1270, 1990.)

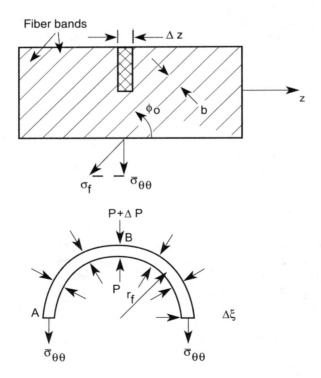

FIGURE 7.33
Pressures at the inner and outer surfaces of the fiber sheet during the filament winding process. (Adapted from Lee, S.Y. and Springer, G.S., *J. Composite Mater.*, 24, 1270, 1990.)

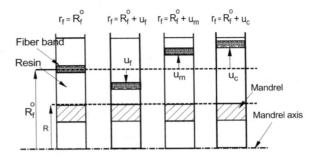

FIGURE 7.34
Illustration of changes in fiber position relative to cylinder axis. (Adapted from Lee, S.Y. and Springer, G.S., *J. Composite Mater.*, 24, 1270, 1990.)

where R_f is the radial position of fiber sheet at time t_0. The value of u_{mc} caused by CTE and chemical shrinkage is determined by performing a stress analysis[10] which is briefly discussed in Section 7.4.4. The value of u_f is calculated by assuming that an entire layer is deposited instantaneously at time t_0 so that the radial position of the fiber sheet is the same at every point in a given layer. Here, analysis is performed for the cylindrical segment A–B of the fiber sheet and the end effects associated with end enclosures of the mandrel and composite cylinder are neglected (Figure 7.30). The stress in the fiber direction is calculated as:

$$\sigma_f = \frac{F}{A_f} \tag{7.46}$$

where A_f is the cross-sectional area of fiber sheet and F is the fiber tension used during the filament winding process. The circumferential component of the fiber tension is given by:

$$\bar{\sigma}_{\theta\theta} = \sigma_f \sin^2 \phi_0 \tag{7.47}$$

where ϕ_0 is the fiber winding angle. Because the fibers are applied under tension, there is a pressure difference in the resin across the fiber sheet, as shown in Figure 7.33. The pressure at the inner surface of the fiber sheet is denoted by p, and at the outer surface by p + Δp. Assume that the fiber band is contiguous and completely covers the surface along a small Δz length of the cylinder. In this case, the force equilibrium equation for the layer of length Δz can be written as follows assuming the absence of inertia effects:

$$2\bar{\sigma}_{\theta\theta}\Delta\xi\Delta z + 2(p+\Delta p)\left(r_f + \frac{\Delta\xi}{2}\right)\Delta z - 2p\left(r_f - \frac{\Delta\xi}{2}\right)\Delta z = 0 \tag{7.48}$$

In the above equation, $\Delta\xi$ can be neglected because it is small compared to the radial position r_f. Therefore, Equation (7.48) can be simplified to:

$$\frac{\Delta p}{\Delta\xi} = \frac{dp}{dr} = -\frac{\overline{\sigma}_{\theta\theta}}{r_f} \tag{7.49}$$

The fiber velocity relative to the resin can be determined using Darcy's law by considering the fiber sheet as a porous medium[10]:

$$\dot{u}_f = \frac{S}{\mu}\frac{dp}{dr} \tag{7.50}$$

where S is the apparent permeability of the porous fiber sheet and μ is the viscosity of the resin. Utilizing Equations (7.47) and (7.49), Equation (7.50) can be rewritten as:

$$\dot{u}_f = -\frac{S}{\mu}\frac{\sigma_f}{r_f}\sin^2\phi_0 \tag{7.51}$$

In Equation (7.51), the fiber tension σ_f and viscosity μ can vary with time t and position r and therefore solutions to this equation must be obtained by numerical means. To facilitate the computation, a change in fiber position during a small time step Δt can be written as:

$$\dot{u}_f = \frac{\Delta u_f}{\Delta t} = \frac{u_f^{t+\Delta t} - u_f^t}{\Delta t} = -\frac{S}{\mu}\frac{\sigma_f}{r_f}\sin^2\phi_0 \tag{7.52}$$

Equation (7.52) can be written as:

$$\Delta u_f = -\frac{S\Delta t}{\mu}\frac{\sigma_f}{r_f}\sin^2\phi_0 \tag{7.53}$$

From Figure 7.31, we have,

$$\Delta L_f = L_f\sqrt{1 + \frac{8\pi^2 r_f u_f}{L_f^2}} - L_f \cong \frac{u_f L_f}{r_f}\sin^2\phi_0 \tag{7.54}$$

Equation (7.54) reduces to:

$$\frac{u_f}{r_f}\sin^2\phi_0 = \frac{\Delta L_f}{L_f} = \Delta\varepsilon_f \tag{7.55}$$

where ΔL_f is the elongation, L_f is the original length of the fiber sheet, and $\Delta\varepsilon_f$ is the change in fiber strain during time Δt as the fiber length changes from L_f to $L_f + \Delta L_f$. The change in fiber stress corresponding to this change in strain can be given as:

$$\Delta\sigma_f = \sigma_f^{t+\Delta t} - \sigma_f^t = E_f \Delta\varepsilon_f \tag{7.56}$$

where E_f is the longitudinal fiber modulus. By combining Equations (7.53), (7.55), and (7.56), the expression for fiber stress at time $t + \Delta t$ can be given as:

$$\sigma_f^{t+\Delta t} = \sigma_f^t \left[1 - \frac{E_f S \Delta t}{\mu} \frac{\sin^4 \phi_0}{r_f^2} \right] \tag{7.57}$$

By solving Equations (7.53) and (7.57), the fiber position and fiber stress (fiber tension) at time $t + \Delta t$ can be calculated. The initial conditions for these equations are that at time $t = t_0$, the radial displacement u_f is zero and the fiber stress is σ_f^0. Thus, initial conditions at time $t = t_0$ can be written as:

$$u_f = 0$$

$$\sigma_f = \sigma_f^0 = \frac{F_0}{A_f} \tag{7.58}$$

It is to be noted here that the $\sigma_f^{t+\Delta t}$ is the tension in the fiber caused only by the fiber motion through the resin. The tension caused by expansions and contractions of the mandrel and the composite can be determined by stress analysis, as previously mentioned.

7.4.2.3 Consolidation of Thermoplastic Composites during Autoclave or Hot Press Processing

Thermoplastic composites require more heat and presure for processing than thermoset composites. More heat is required because thermoplastics must be melted from the solid state, and high pressure is required because thermoplastics have a higher viscosity than thermosets during processing. During processing of thermoplastic composites, first the individual plies consolidate and come into intimate contact and then bonding between the layers takes place at the contact surfaces. The bonding between the layers is caused by diffusion of molecular chains across the interface, a process called autohesion. The autohesion process is described in a later section. The model presented here for consolidation of thermoplastic composites during the autoclave or hot press method was proposed by Lee and Springer.[12] This model is then derived for thermoplastic tape winding by Mantell and Springer.[20]

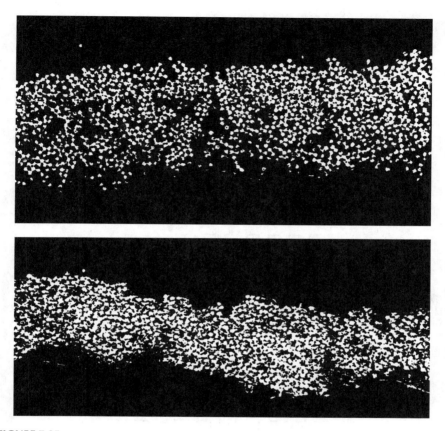

FIGURE 7.35
Cross-sectional micrograph of unprocessed APC-2 tape (200×). (Courtesy of S.K. Mazumdar.[30])

For consolidation, the ply surfaces must first deform to produce intimate contact at the interface. A typical prepreg surface is shown in Figure 7.35. Lee and Springer[12] modeled the irregular surface of the ply as a series of rectangles, as shown in Figure 7.36. This concept was originally developed by Dara and Loos[26] to describe the surface of a ply. Lee and Springer,[12] however, used the viscosity of the prepreg material (APC-2 tape), which includes the effect of fibers on the resin viscosity. Due to the applied force (consolidation pressure), the rectangular elements spread along the interface. With reference to Figure 7.37, the degree of intimate contact is defined as[12]:

$$D_{ic} = \frac{b}{w_0 + b_0} \tag{7.59}$$

where b_0 and b are the initial (t < 0) and instantaneous (at time t) widths of each rectangular element, respectively, and w_0 is the initial distance between two adjacent elements. Physically, it denotes the ratio of the base width of

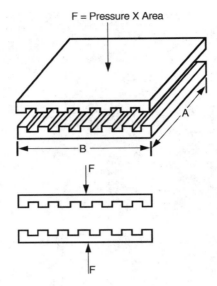

FIGURE 7.36
Schematic diagram of an idealized interface for intimate contact model. (Adapted from Lee, W.I. and Springer, G.S., *J. Composite Mater.*, 21, 1017, 1987.)

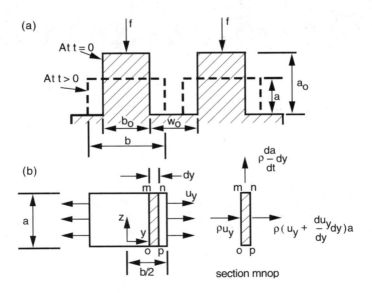

FIGURE 7.37
Rectangular elements for intimate contact model: (a) uneven surface at time $t = 0$; (b) one element at time t, and the control volume used in calculating mass. (Adapted from Mantell, S.C. and Springer, G.S., *J. Composite Mater.*, 26(16), 2348, 1992.)

an asperity to the wavelength of an idealized periodic arrangement of asperities. A degree of intimate contact is defined such that it takes a value of unity in the case of complete interfacial contact. During processing, the volume of each element remains constant; therefore,

$$V_0 = a_0 b_0 = ab \tag{7.60}$$

where a_0 and a are the initial and instantaneous heights of each rectangular element, respectively. Equations (7.59) and (7.60) give the following expression for the degree of intimate contact:

$$D_{ic} = \frac{a_0/a}{1+w_0/b_0} \tag{7.61}$$

Applying the law of conservation of mass for a control volume of width $d\xi$ for the drawing shown in Figure 7.37[20] results in:

$$a \frac{du_\xi}{d\xi} + \frac{da}{dt} = 0 \tag{7.62}$$

where t is the time and ξ is the coordinate along the interface. By assuming the flow is laminar, the average velocity u_ξ can be written as:

$$u_\xi = -\frac{a^2}{12\mu_{mf}} \frac{dP}{d\xi} \tag{7.63}$$

where μ_{mf} is the rheometric viscosity of the matrix fiber mixture measured between two parallel platens in shear under adiabatic conditions. P is the absolute pressure at a location inside an element. Here, P is the function of ξ and time t. In the space between two adjacent elements, the pressure is P_e, which is taken to be equal to the ambient pressure. The edge of the element ($\xi = b/2$) moves with a speed db/dt. Therefore, the boundary conditions for Equation (7.63) can be written for $t > 0$ as:

$$P = P_e \text{ and } u_\xi = \frac{db}{dt} \text{ at } \xi = \frac{b}{2} \tag{7.64}$$

By combining Equations (7.62) to (7.64), and by algebraic manipulations, one can get:

$$P - P_e = \frac{6\mu_{mf}}{a^3} \frac{da}{dt} \left[\xi^2 - \left(\frac{b}{2}\right)^2 \right] \tag{7.65}$$

As shown in Figure 7.36, the force F is applied to a ply of length L and width W. The force applied per unit length to one element can be determined as:

$$f = \frac{F}{L}\frac{(b_0 + w_0)}{W} = (P_{app})(b_0 + w_0) \tag{7.66}$$

where P_{app} is the applied gauge pressure. The force applied to an element must be balanced by the pressure inside the element and can be written as:

$$f = \int_{-b/2}^{b/2} (P - P_e) d\xi \tag{7.67}$$

By combining Equations (7.65) through (7.67), one obtains

$$P_{app}(b_0 + w_0) = \int_{-b/2}^{b/2} \frac{6\mu_{mf}}{a^3} \frac{da}{dt}\left[\xi^2 - \left(\frac{b}{2}\right)^2\right] d\xi \tag{7.68}$$

The integration of Equation (7.68) gives:

$$P_{app}(b_0 + w_0) = -\mu_{mf} \frac{da}{dt} \frac{b^3}{a^3} \tag{7.69}$$

By substituting Equation (7.60) into (7.69) and integrating with respect to the instantaneous element height a, one obtains[20]:

$$(b_0 + w_0)\int_0^{t_c} \frac{P_{app}}{\mu_{mf}} dt = \frac{(a_0 b_0)^3}{5}\left(\frac{1}{a^5} - \frac{1}{a_0^5}\right) \tag{7.70}$$

where t_c is the contact time during which pressure is applied. By combining Equations (7.61) and (7.70), one can derive the expression for the degree of intimate contact as:

$$D_{ic} = \frac{1}{1+\frac{w_0}{b_0}}\left[1 + 5\left(1 + \frac{w_0}{b_0}\right)\left(\frac{a_0}{b_0}\right)^2 \int_0^{t_c} \frac{P_{app}}{\mu_{mf}} dt\right]^{1/5} \tag{7.71}$$

In the event that P_{app} and μ_{mf} are independent of time t, Equation (7.71) can be written as:

$$D_{ic} = \frac{1}{1+\frac{w_0}{b_0}}\left[1 + \frac{5P_{app}}{\mu_{mf}}\left(1 + \frac{w_0}{b_0}\right)\left(\frac{a_0}{b_0}\right)^2\right]^{1/5} \tag{7.72}$$

Complete intimate contact is achieved when D_{ic} becomes unity. The height "a" corresponding to complete intimate contact can be determined from Equation (7.61) as:

$$a_{D_{ic}=1} = \frac{a_0}{1+w_0/b_0} \tag{7.73}$$

By substituting Equation (7.73) into (7.72), one can get an expression for the time required to get complete intimate contact:

$$t_{ic} = \frac{\mu_{mf}}{5P_{app}} \frac{1}{1+w_0/b_0} \left(\frac{b_0}{a_0}\right)^2 \left[\left(1+\frac{w_0}{b_0}\right)^5 - 1\right] \tag{7.74}$$

Equation (7.72) can be used to calculate the degree of intimate contact for an applied pressure P_{app} and processing time t. Equation (7.74) is used to determine the time needed to achieve complete intimate contact. Lee and Springer[12] performed experiments to verify the intimate contact model. To perform the test, they created a simple flat plate mold and then placed one ply of PEEK/carbon (APC-2) thermoplastic composite between the top and bottom aluminum plates of the mold. The assembly was then placed in a hot press at specified temperatures and pressures. After a certain time, the pressure was released and the assembly cooled to room temperature. The APC-2 ply was then taken out of the mold and areas of the uncompressed surfaces were measured. From these measurements, the degree of intimate contact was calculated as follows.

$$D_{ic} = \frac{\text{Total surface area} - \text{Uncompressed surface area}}{\text{Total surface area}} \tag{7.75}$$

The test results are shown in Figures 7.38 and 7.39. Tests were conducted at 40 psig and at 350, 360, and 370°C (662, 680, and 698°F); and at 40, 96, and 227 psig at 350°C (662°F). There is good correlation found between experimental results and theoretical predictions.

7.4.2.4 Consolidation and Bonding Models for Thermoplastic Tape Laying and Tape Winding

The thermoplastic tape winding process is similar to the filament winding of thermoset composites with the difference being that in thermoplastic tape winding, heat is continuously supplied at the contact point of incoming tape and preconsolidated laminate, and pressure is applied using a consolidation roller as shown in Figure 7.3. The heat source could be a laser, hot nitrogen gas, or a heated roller as discussed in Chapter 6.

The consolidation model for the thermoplastic tape winding process presented here was developed by Mantell and Springer[20] and is an extension

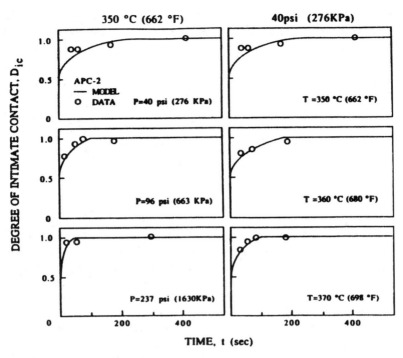

FIGURE 7.38
Comparison of experimental test results with the intimate contact model. The degree of intimate contact vs. time is plotted for the various applied temperatures and pressures. (Reprinted from Lee, W.I. and Springer, G.S., *J. Composite Mater.*, 21, 1017, 1987. With permission from Technomic Publishing Company, Inc.)

of the intimate contact model described in Section 7.4.2.3. Mazumdar and Hoa[27–30] conducted various analytical and experimental studies for processing of thermoplastic composites using laser and hot nitrogen gas as heat sources for tape laying and tape winding processes. Results of theoretical predictions and experimental data for temperature distribution, heating rates, and cooling rates are shown in Figures 7.5 through 7.18.

To derive the model for the tape winding process, Equation (7.71) is considered the starting point for the derivation. This equation can be simplified to:

$$D_{ic} = \frac{1}{w^*}\left[1 + a^* \int_0^{t_c} \frac{P_{app}}{\mu_{mf}} dt\right]^{1/5} \qquad (7.76)$$

where w^* and a^* are defined as follows:

$$w^* = 1 + \frac{w_0}{b_0} \quad \text{and} \quad a^* = 5w^*\left(\frac{a_0}{b_0}\right)^2 \qquad (7.77)$$

Process Models   277

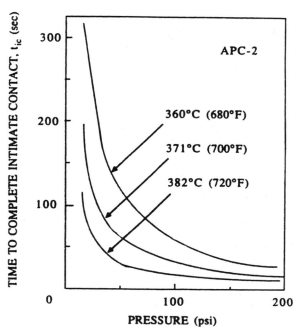

FIGURE 7.39
Time required for complete intimate contact is plotted against pressure for various temperatures. (Reprinted from Lee, W.I. and Springer, G.S., *J. Composite Mater.*, 21, 1017, 1987. With permission from Technomic Publishing Company, Inc.)

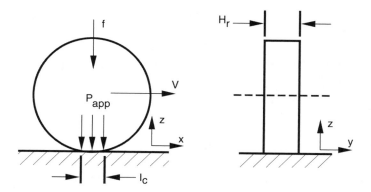

FIGURE 7.40
The contact point between the roller and the composite in a tape winding/tape laying process.

For the tape laying and tape winding processes, P_{app} is the pressure applied by the roller to the composite (Figure 7.40). The roller moves at constant velocity V and presses the incoming tape against the preconsolidated laminate. Under an applied force f, the roller touches an arc length g–h of the composite. The linear length between point g and h is l_c. Here, it can be assumed that when the roller moves the small distance l_c, the temperature is nearly constant in the region g to h. With this assumption, the viscosity μ_{mf} can be considered as constant and the integral term in Equation (7.71) can be written as:

$$\int_0^{t_c} \frac{P_{app}}{\mu_{mf}} dt = \frac{1}{\mu_{mf}} \int_0^{t_c} P_{app} dt \tag{7.78}$$

where t_c is the contact time, during which the roller travels the distance l_c, and t_c is given by:

$$t_c = \frac{l_c}{V} \tag{7.79}$$

where V is the roller speed. By assuming that the arc between g and h is shallow, the force balance in the vertical direction can be written as:

$$f \approx \int_0^{H_r} \int_0^{l_c} P_{app} dx dy \tag{7.80}$$

where H_r is the width of the roller. Because x = Vt, Equation (7.78) can be integrated to give:

$$\frac{1}{\mu_{mf}} \int_0^{t_c} P_{app} dt = \frac{f}{\mu_{mf} V H_r} \tag{7.81}$$

Using Equation (7.81), the degree of intimate contact in Equation (7.76) can be written for the tape laying process as:

$$D_{ic} = \frac{1}{w^*} \left[1 + a^* \frac{f}{\mu_{mf} V H_r} \right]^{1/5} \tag{7.82}$$

For the tape winding process, Equation (7.82) can be rewritten by replacing $V = wr_c$ as follows:

$$D_{ic} = \frac{1}{w^*} \left[1 + a^* \frac{f}{\mu_{mf} \omega r_c H_r} \right]^{1/5} \tag{7.83}$$

where w is the mandrel's rotational speed in rpm (revolutions per minute) and r_c is the radius of the composite cylinder at the interface at which the

Process Models

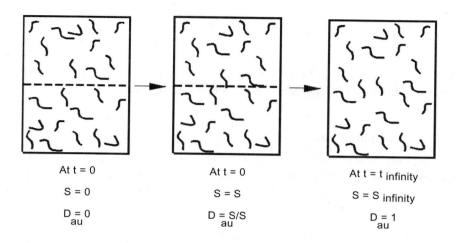

FIGURE 7.41
Illustration of the autohesion process.

degree of intimate contact is calculated. It should be noted that Equations (7.82) and (7.83) do not require values of contact length l_c and contact time t_c.

Bonding Sub-Model

During thermoplastic processing, the individual plies first consolidate and come into intimate contact and then bonding takes place at the contact surfaces. It is suggested that bonding is caused primarily by autohesion.[12,26] During the autohesion process, the segments of chain-like molecules diffuse across the interface as shown in Figure 7.41. The amount of molecular diffusion depends on the temperature at the interface as well as the duration during which the interface was kept at the processing temperature. The autohesion process can be modeled as follows[12,26,32-34]:

$$D_{au} = kt_a^{n_a} \tag{7.84}$$

where t_a is the time elapsed from the start of the autohesion process and exponent n_a is a constant (1/4 for amorphous polymers). k is a constant which is related to the temperature through Arrhenius relation as follows:

$$k = k_0 \exp(-E/RT) \tag{7.85}$$

In Equation (7.85), k_0 is a constant, E is the activation energy, R is the universal gas constant, and T is the temperature in Kelvin. To determine the experimental value of the degree of autohesion, the following relation is used[26]:

$$D_{au} = \frac{S}{S_U} \qquad (7.86)$$

where S is the bond strength at time t and S_U is the ultimate bond strength, which is the bond strength of a completely bonded surface.

7.4.3 Void Sub-Model

Voids are formed during composite part fabrication, either by mechanical means (e.g., air or gas bubble entrapment, broken fibers) or by homogeneous or heterogeneous nucleation.[8,35] Once a void is established, its size and shape change, due primarily to:

1. Changes in inside pressure of the void due to pressure and temperature changes in the prepreg
2. Changes in vapor mass inside the void caused by vapor transfer through the void/prepreg interface
3. Thermal expansion or shrinking due to temperature gradients in the resin

The model presented here is based on that of Loos and Springer[8] and takes into consideration only the first two effects.

Assume that there is a spherical vapor nucleus of diameter d_i at any given location in the prepreg. The nucleus is filled with water vapor resulting from the humid air surrounding the prepreg. The partial pressure of water vapor inside the nucleus PP_{wi} can be written in terms of the relative humidity ϕ_a as follows[8]:

$$PP_{wi} = \phi_a P_{wga} \qquad (7.87)$$

where P_{wga} is the saturation pressure of water vapor at ambient temperature. From the known values of the initial nucleus volume and initial partial pressure PP_{wi}, the initial mass m_{wi} and the initial concentration $(c_{vw})_i$ of the water vapor in the nucleus can be calculated.

During the consolidation and curing process, the volume of the void changes because:

- The cure pressure increases the pressure at the location of the void.
- Water, air, gas, and other types of molecules are transported across the void/prepreg interface.

For a spherical void of diameter d, the total pressure inside the void P_v is related to the pressure P in the prepreg surrounding the void by:

$$P_v - P = \frac{4\sigma}{d} \tag{7.88}$$

where σ is the surface tension between the void and the resin, and P_v is the sum total of partial pressures of the water, air, and other types of vapors present inside the void. For simplicity, the model below assumes that only water vapor is transported through the void/prepreg interface. If necessary, other types of vapors can be included in this model.[11] After the migration of water vapor, the pressure inside the void becomes:

$$P_v = PP_w + PP_{air} \tag{7.89}$$

where PP_w and PP_{air} are the partial pressures of water vapor and air inside the void, respectively. The partial pressures are related to mass, temperature, and void diameter as follows:

$$PP_{air} = f(T, m_{air}, d) \tag{7.90}$$

$$PP_w = f(T, m_w, d) \tag{7.91}$$

Using Equations (7.88) through (7.91), one can calculate the partial pressure, the total pressure, and the void diameter if one knows the pressure in the prepreg around the void, the temperature inside the void, and the mass of vapor in the void. The temperature inside the void can be taken as the prepreg temperature at the void location. The temperature and pressure at any point and at any time t can be determined by thermochemical and resin flow models, as described in previous sections. The air mass in the void can be considered as constant. The water vapor mass in the void changes with time t. The change in water vapor mass is calculated by assuming that the vapor molecules are transported through the prepreg by Fickian diffusion (Fick's law) as follows:

$$\frac{dc}{dt} = D\left(\frac{\partial^2 c}{\partial r^2} + \frac{2}{r}\frac{\partial c}{\partial r}\right) \tag{7.92}$$

where c is the water vapor concentration at a radial coordinate r with r = 0 at the center of void; and D is the diffusivity of the water vapor through the resin in the r direction.

At the beginning (t < 0) the vapor is assumed to be uniformly distributed in the prepreg at the known concentration c; that is:

$$c = c_i \text{ at } r \geq d_i/2, \ t < 0 \tag{7.93}$$

As time increases (t ≥ 0), the vapor concentration at the void/prepreg interface must be specified. By denoting the concentration at the prepreg surface by the subscript p, the vapor concentration can be written as:

$$c = c_p \text{ at } r = d/2, \quad t \geq 0 \tag{7.94}$$

$$c = c_i \text{ at } r \to \infty, \quad t \geq 0 \tag{7.95}$$

Equation (7.95) reflects that the concentration remains unchanged at a distance far from the void. The surface concentration is related to the maximum saturation level M_p in the prepreg by:

$$c_p = \rho M_p \tag{7.96}$$

The value of M_p can be experimentally determined for each vapor-resin system.[36]

By solving Equations (7.92) through (7.96), vapor concentration as a function of position and time c = f(r,t) can be determined. The mass of vapor transported in time t through the surface of the void can be expressed as:

$$m_T = -\int_0^t \pi d^2 D \left(\frac{dc}{dr}\right)_{r=d/2} dt \tag{7.97}$$

The mass of vapor in the void at time t can be written as:

$$m = m_i - m_T \tag{7.98}$$

where m_i, the initial mass of water vapor, is known, as discussed earlier. By solving Equations (7.87) to (7.98), the void size and pressure inside the void as a function of time can be determined for a known void location and initial size.

The void sub-model requires that both the location and the initial size of the void be known. This information is generally not available because void nuclei formation is a random process. However, the sub-model presented here can still be used in selecting a cure cycle that results in a low void content. It has been found that the void content is reduced significantly when pressure is applied in the prepreg, a pressure high enough to collapse vapor bubbles.[37] Vapor bubbles can be collapsed by applying a pressure that is equal to or greater than the saturation pressure inside the void at that local temperature. It is to be noted here that the pressure is applied just before the gel point is reached (t < t_{gel}). In a typical autoclave cure cycle, a pressure in the range of 80 to 100 psi at 270°F is applied to drive out the volatiles.

The temperature is then increased to 350°F for curing the resin for about 1 to 2 hours. The gel time and local temperature are determined by thermochemical and resin flow models. The appropriate thermodynamic relationship between the temperature and the saturation pressure provides the required pressure.

7.4.4 Stress Sub-Model

Residual stress in the laminate is generated when laminate is cooled from an elevated temperature (cure temperature) to room temperature. Two main reasons for residual stress are chemical changes in the resin as well as thermal strains in each ply due to the differential coefficient of thermal expansion. During the curing process, as the cross-linking takes place, the epoxy shrinks and chemical changes occur. These chemical changes cause more deformation in the transverse direction than in the longitudinal direction of a unidirectional laminate. Therefore, within the laminate, the deformation of one ply is constrained by the other plies with different fiber orientations, and hence residual stresses are built up in each ply. During composites processing, cross-linking usually takes place at elevated temperature, which is called the cure temperature. At this temperature, the resin can still be sufficiently viscous to allow complete relaxation of the residual stresses. Thus, the cure temperature can be considered as the stress-free temperature.[38] In the model described here, residual stress developed by chemical shrinkage is not considered.

The second cause of residual stress generation involves thermal strain. When a composite laminate is cooled from the cure temperature to room temperature, significant residual stress may develop, owing to the thermal mismatch of various laminas. In some cases, these curing stresses may be sufficiently high to cause interlaminar cracks.[39]

For example, consider a $[0/90]_s$ laminate being cooled from the cure temperature to room temperature. Suppose the plies were not joined and could contract freely; the 0° ply will contract much less in the x-direction than the 90° ply, while the reverse is true in the y-direction. Because the plies are joined and must deform together, internal stresses are generated to maintain the geometric compatibility between the plies. In the case of $[0/90]_s$ laminate, residual stresses are compressive in the fiber direction but tensile in the transverse direction in both 0° and 90° plies.

For a symmetric laminate, the residual stress in any ply is given by[8,38]:

$$\sigma_i = Q_{ij}\left(e_{oj} - e_j\right) \tag{7.99}$$

where Q_{ij} is the modulus of a ply as defined by Tsai and Hahn.[38] The strain e_j is defined as:

$$e_j = \alpha_j(T - T_a) \tag{7.100}$$

where α_j is the thermal expansion coefficient, T_a is the ambient temperature, and T is the temperature in the ply at the end of cure given by the thermochemical model. The laminate curing strain is:

$$e_{oj} = a_{ij} \int_0^L Q_{ij} e_j dz \qquad (7.101)$$

where a_{ij} is the in-plane compliance of a symmetric laminate as defined by Tsai and Hahn. By solving the above equations, residual stresses in each ply can be calculated.

7.5 Process Model for RTM

In an RTM process, resin is injected from single or multiple inlet ports to the mold cavity and it flows inside the cavity through a porous fiber architecture to completely impregnate the fibers. In an RTM process, various simulation models have been developed to determine the position of the resin flow front at any given time, to calculate mold filling time and other parameters for processing variables such as mold location, applied pressure, and flow rate. Optimum processing variables are determined to make sure that the mold filling is accomplished in a minimum amount of time without any dry spots. In the RTM process, the area that is not saturated with resin is called the dry spot. Dry spots are the most serious defects during manufacturing of composite parts by the RTM process. The mechanical performance of the part is seriously affected by dry spots. Local permeability variation of a fiber preform and inappropriate inlet (gate) and/or outlet (vent) positions are the cause of dry spot formation. Local variation in the permeability occurs during manufacture of the preform or during fiber mat/preform placement. During preform placement, the preform might be bent, rolled, or stretched. Resin flows faster around the low-permeability region and may cause dry spot formation around the high-permeability region. Although the preform permeability is uniform, inappropriate placement of gates and/or vents may cause dry spot formation during mold filling. In industry, gate and vent locations are often determined from experience or by trial and error. If the part shape is simple, only a few tests are needed to determine the inlet and outlet positions. However, for large or complicated shapes, the tests must be repeated many times. This significantly increases the cost of running tests. To solve such problems, mold filling simulations are performed to come up with optimum gate and vent locations.

The resin flow through a preform can be considered a flow through porous media, and Darcy's law can be used to perform flow analysis. According to Darcy's law for one-dimensional flow analysis:

$$v = -\frac{k}{\mu} \cdot \frac{\partial p}{\partial x} \tag{7.102}$$

where v is the velocity, k is the permeability of fiber preform, μ is the resin viscosity, and p is the resin pressure.

For a two-dimesional flow, the velocity vector v consists of two components (u and v) in the x- and y-directions, respectively. Equation (7.102) can be rewritten for two-dimensional flow as:

$$\begin{bmatrix} u(x,y) \\ v(x,y) \end{bmatrix} = -\frac{1}{\mu} \begin{bmatrix} k_{xx} & k_{xy} \\ k_{yx} & k_{yy} \end{bmatrix} \begin{bmatrix} \frac{\partial p(x,y)}{\partial x} \\ \frac{\partial p(x,y)}{\partial y} \end{bmatrix} \tag{7.103}$$

where k_{ij} (i,j = x,y) are the components of the permeability tensor. For an incompressible fluid, the continuity equation can be reduced to the following form.

$$\frac{\partial u}{\partial x} + \frac{\partial v}{\partial y} = 0 \tag{7.104}$$

Substituting the value from Equation (7.104) into (7.103) and aligning the equations along the principal directions of the fiber preform/mat, one can obtain:

$$\frac{\partial}{\partial x}\left(\frac{k_{xx}}{\mu}\frac{\partial p}{\partial x}\right) + \frac{\partial}{\partial y}\left(\frac{k_{yy}}{\mu}\frac{\partial p}{\partial y}\right) = 0 \tag{7.105}$$

The pressure field during mold filling can be determined by solving Equation (7.105). The finite element method can be used to solve Equation (7.105). Once the pressure field is known, the velocity field and flow front pattern can be calculated from Equation (7.103).

References

1. Gorovaya, T.A. and Korotkov, V.N., Quick cure of thermosetting composites, *Composites*, 27A(10), 953, 1996.
2. Gillham, S.K. and Enns, J.B., On the cure and properties of thermosetting polymers using torsional braid analysis, *Trends Polymer*, 2(12), 406, 1994.
3. Penn, L.S., Chou, R.C.T., Wang, A.S.D., and Binienda, W.K., The effect of matrix shrinkage on damage accumulation in composites, *J. Composite Mater.*, 23, 570, 1989.

4. Ochi, M., Yamashita, K., and Shimbo, M., The mechanism for occurrence of internal stress during curing epoxide resins, *J. Appl. Polymer Sci.*, 43, 2013, 1991.
5. Ghasemi Nejhad, M.N., Cope, R.D., and Guceri, S.I., Thermal analysis of *in-situ* thermoplastic composite tape laying, *J. Thermoplastic Composite Mater.*, 4, 20, 1991.
6. Korotkov, V.N., Chekanov, Y.A., and Rozenberg, B.A., The flaws of shrinkage formed in the course of curing of polymer based composites (Review), *Polymer Sci.*, 36(4), 684, 1994.
7. Plepys, A.R. and Farris, R.J., Evolution of residual stresses in three-dimensional constrained epoxy resins, *Polymer*, 31, 1932, 1990.
8. Loos, A.C. and Springer, G.S., Curing of epoxy matrix composites, *J. Composite Mater.*, 17, 135, 1983.
9. Calius, E.P. and Springer, G.S., A model of filament wound thin cylinders, *Int. J. Solids and Structures*, December 1988.
10. Lee, S.Y. and Springer, G.S., Filament winding cylinders. I. Process models, *J. Composite Mater.*, 24, 1270, 1990.
11. Springer, G.S., A model of the curing process of epoxy matrix composites, in *Progress in Science and Engineering of Composites*, T. Hayashi, K. Kawaka, and S. Umekawa, Eds., Japan Society for Composite Materials, 1982, 23–35.
12. Lee, W.I. and Springer, G.S., A model of the manufacturing process of thermoplastic matrix composites, *J. Composite Mater.*, 21, 1017, 1987.
13. Lee, W.I., Loos, A.C., and Springer, G.S., Heat of reaction, degree of cure and viscosity of Hercules 3501-6 resin, *J. Composite Mater.*, 16, 510, 1982.
14. Dusi, M.R., Lee, W.I., Ciriscioli, P.R., and Springer, G.S., Cure kinetics and viscosity of Fiberite 976 resin, *J. Composite Mater.*, 21, 243, 1987.
15. Bhi, S.T., Hansen, R.S., Wilson, B.A., Calius, E.P., and Springer, G.S., Degree of cure and viscosity of Hercules HBRF-55 Resin, in *Advanced Materials Technology*, Vol. 32, Society for the Advancement of Materials and Process Engineering, 1987, 1114.
16. Calius, E.P., Lee, S.Y., and Springer, G.S., Filament winding cylinders. II. Validation of models, *J. Composite Mater.*, 24, 1299, 1990.
17. Seferis, J.C. and Velisaris, C.N., Modeling-processing-structure relationships of PEEK based composites, *Materials Sciences for the Future*, Society for the Advancement of Materials and Process Engineering, 1986, 1236.
18. Velisaris, C.N. and Seferis, J.C., Crystallisation kinetics of polyetheretherketone (PEEK) matrices, *Polymer Eng. Sci.*, 26, 1574, 1986.
19. Blundell, D.J., Chalmers, J.M., Mackenzie M.W., and Gaskin, W.F., Crystalline morphology of the matrix of PEEK-carbon fiber aromatic polymer composites. I. Assessment of crystallinity, *SAMPE Q.*, 16, 22, 1985.
20. Mantell, S.C. and Springer, G.S., Manufacturing process models for thermoplastic composites, *J. Composite Mater.*, 26(16), 2348, 1992.
21. Springer, G.S., Compaction and Consolidation of Thermoset and Thermoplastic Composites, Rapra Technology Limited, U.K., 1990.
22. Gutowski, T.G., A resin flow/fiber deformation model for composites, *SAMPE Q.*, 16, 58, 1985.
23. Gutowski, T.G., Morigaki, T., and Cai, Z., The consolidation of laminate composites, *J. Composite Mater.*, 21, 172, 1987.
24. Kardos, J.L., Dave, R., and Dudukovic, M.P., A model for resin flow during composite processing. I. General mathematical development, *Polymer Composites*, 8, 29, 1987.

25. White, F.M., *Viscous Fluid Flow*, McGraw-Hill, New York, 1974, 336–337.
26. Dara, P.H. and Loos, A.C., Thermoplastic Matrix Composite Processing Model, Center for Composite Materials and Structures, Report CCMS-85-10, VPI-E-85-21, Virginia Polytechnic Institute and State University, Blacksburg, 1985.
27. Mazumdar, S.K. and Hoa, S.V., Determination of manufacturing conditions for hot gas aided thermoplastic tape winding technique, *J. Thermoplastic Composite Mater.*, 9, 35, January 1996.
28. Mazumdar, S.K. and Hoa, S.V., Application of Taguchi method for process enhancement of on-line consolidation technique, *Composites*, 26(9), 669, 1995.
29. Mazumdar, S.K. and Hoa, S.V., Manufacturing of non-axisymmetric thermoplastic composite parts by tape winding technique, *Mater. Manuf. Processes*, 10(1), 47, 1995.
30. Mazumdar, S.K., Automated Manufacturing of Composite Components by Thermoplastic Tape Winding and Filament Winding, Ph.D. thesis, Concordia University, Montreal, 1994.
31. Saint-Royre, D., Gueugnant, D., and Reveret, D., Test methodology for the determination of optimum fusion welding conditions of polyethylene, *J. Appl. Polymer Sci.*, 38, 147, 1989.
32. Wool, R.P. and O'Connor, K.M., Theory of crack healing in polymers, *J. Appl. Phys.*, 52, 5953, 1981.
33. Wool, R.P., Relations for Healing, Fracture, Self-diffusion and Fatigue of Random Cool Polymers, *ACS Polymer Prepr.*, 23(2), 62, 1982.
34. Jud, K., Kausch, H.H., and Williams, J.G. Fracture mechanics studies of crack healing and welding of polymers, *J. Mater. Sci.*, 16, 204, 1981.
35. Kardos, J.L., Dudukovic, J.P., McKague, E.L., and Lehman, M.W., Void formation and transport during composite laminate processing, in *Composite Materials, Quality Assurance and Processing*, ASTM STP 797, 1983, 96–109.
36. Springer, G.S., *Environmental Effects on Composite Materials*, Technomic Publishing Co., 1981.
37. Brown, G.G. and McKague, E.L., Processing Science of Epoxy Resin Composites, Technical Orientation, General Dynamics, Convair Division, San Diego, CA, August 1982.
38. Tsai, S.W. and Hahn, H.T., *Introduction to Composite Materials*, Technomic Publishing Co., 1980.
39. Hahn, H.T., Residual stresses in polymer matrix composite laminates, *J. Composite Mater.*, 10, 226, 1976.

Questions

1. Why is it important to develop a process model before making a product?
2. How can a simulation model for RTM help the manufacturing engineer?
3. What are the process parameters in an RTM process?
4. What are the common process-related defects?

5. What are the four submodels commonly used to derive the complete manufacturing process model? Write the purpose of each submodel.
6. Why is it necessary to have higher processing temperatures in a thermoplastic tape winding process than in an autoclave or hot press process for thermoplastic composites?
7. Write down initial conditions and boundary conditions for the filament winding process.
8. How do you define degree of intimate contact?
9. What is an autohesion process? Give an illustration of it.
10. What is the principle used to determine total number of compacted plies during an autoclave process?

8
Production Planning and Manufacturing Instructions

8.1 Introduction

This chapter discusses the various procedures that industry uses for making successful composite products. Mere knowledge of a manufacturing or a design process would not help in translating the concept into a final product. The technicians who make the product are different from the designers and manufacturing engineers who are involved in the product development process. Therefore, the design and manufacturing instructions need to be clearly communicated to the technicians, the people who are actually involved in making the product. Moreover, projects are usually large and not completed by one person. There are designers, manufacturing engineers, purchasing personnel, technicians, quality assurance personnel, and others involved in project completion. Because there are so many departments and groups involved in making a product, it is important that each department knows what it is supposed to do for successful completion of the product. This chapter shows procedures to write the bill of materials so that the purchasing department can buy the material for product fabrication, and describes methods of writing manufacturing instructions so that technicians can make the part consistently and repeatedly.

Production planning starts when a decision is made for the fabrication of the product. Production planning covers all stages of production, from procurement of raw materials to shipping the final product. At this stage, the manufacturing department works with the design engineering team to understand the requirements of the product, and then prepares the bill of materials (parts and materials listing) and writes manufacturing instructions for the production of various assembly and sub-assembly parts. Many tasks are performed during the production planning stage. These include bill of materials preparation, manufacturing instructions (process sheets) preparation, capacity planning, demand planning, layout planning, raw materials procurement planning, and more. The objectives of production planning are described below.

8.2 Objectives of Production Planning

Production planning is performed to ensure that the task of delivering the product is done smoothly and in a timely manner. Production planning involves the following tasks:

1. It estimates the total time needed for fabrication of the product. It determines the time needed for procurement of raw materials, inspection of raw materials, storage, manufacturing operations, delays, quality control, packaging, and shipping.
2. It calculates the capacities of manufacturing equipment, raw material storage, lay-up area, and more.
3. It prepares the bill of materials and identifies methods for procuring all the materials and parts listed in the bill of materials.
4. It lists all the major activities and sub-activities of the entire project.
5. It prepares manufacturing instructions for the fabrication of sub-assembly and final part assembly.
6. It estimates manufacturing lead times and prepares manufacturing schedules.

This chapter focuses on the following three production planning tasks:

1. Bill of materials
2. Preparation of manufacturing instructions
3. Capacity planning

8.3 Bill of Materials

The bill of materials (BOM) is a list of materials/parts, by quantity and size, that are required for the fabrication of a product. The manufacturing department handles the make-or-buy decisions on the various components needed for product fabrication. The BOM is prepared well in advance so that all the materials are procured by the time fabrication starts. Once the BOM is prepared, it is given to the purchasing department to find the right supplier and then to buy the material under certain shipping terms and conditions.

A sample BOM is shown in Table 8.1 for the fabrication of a composite sandwich panel of dimensions $0.5 \times 46 \times 95$ in. The panel has a 0.03-in. thick fiberglass skin on both sides and a Nomex honeycomb core in between, as shown in Figure 8.1. The edges of the panel are open. These kinds of sandwich panels are used as tooling panels or as platforms for various applications such as aircraft floor panel.

TABLE 8.1
Bill of Materials for Making a Sandwich Panel

Part No.	Part Description	Size	Quantity
1131	Fiberglass sheet, Grade A, white	0.03 × 46 × 95 in.	2
21562	Nomex honeycomb, industrial grade, type 4.0 (lb/ft^3) × 1/4 in. (cell size)	0.44 × 46 × 95 in.	1
1281	Film adhesive, high temperature, L312, Supplier: XYZ, 50 in. wide roll	60.69 sq. ft.	1

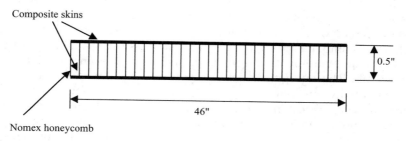

FIGURE 8.1
Side view of a composite sandwich panel with dimensions 0.5 × 46 × 95 in.

In Table 8.1, the part numbers are unique and are assigned for each item according to company policy. Part numbers could be three-digit or ten-digit numbers, or anything with combinations of digits and characters. Companies keep a record, such as a drawing, MSDS sheet, and/or material specification, of each part/item number. In this particular bill of material, there are only three materials needed to make the final sandwich panel. In most cases, the number of parts needed is more than three. However, in this example, two fiberglass sheets are needed as the bottom and top layers of the sandwich panel. Composite sheets are available in the marketplace with different thicknesses and sizes. The standard size close to the present requirement is 0.03 × 48 × 96 in. This size is cut to the proper length and width as directed in the manufacturing instructions (Section 8.4). In some BOMs, standard sizes are also written in one column because the purchasing department orders only the standard sizes. Instead of fiberglass sheets, fiberglass prepregs can also be used, but the use of fiberglass sheets is shown in this example. When fiberglass prepregs are used, the use of adhesives can be avoided because of the presence of resin in the prepreg material. Resin works as an adhesive for bonding prepregs with the core material. The designer makes the choice as to when to use prepregs and when to use precured composite sheets. Prepregs have some advantages because one can eliminate the need for adhesive and lower the cost, but the surface finish obtained with prepregs usually has voids and undulations due to nonuniform support by the honeycomb core. (Note that the adhesive needed in this case is to bond two surface areas.)

The number of fiberglass sheets, Nomex cores, and adhesive rolls are determined by the number of sandwich panels to be made during this project. If the project requires manufacture of only one sandwich panel, then the cost of making that panel is very high because of higher material and shipping and handling costs. Shipping and handling costs per piece are less if the quantities of parts ordered are more. Moreover, the adhesive manufacturer sells the adhesive in roll form and not in small, square-foot quantities. If the panel manufacturing company does not have the particular adhesive (as listed in the BOM) in its inventory, then the company may have to order the whole roll of adhesive for this project. This increases the cost of making the panel. Similarly, the cost of ordering one Nomex honeycomb of size 0.44 × 48 × 96 in. is very high because the honeycomb manufacturer usually makes the honeycomb core in 12-in.-thick sections. When someone orders 0.44-in. thick honeycomb, it is cut to the desired thickness using a high-speed rotating saw. The equipment set-up cost is included in each cut piece of the honeycomb. The equipment set-up cost is a constant number and gets divided equally into number of pieces cut during that set-up time. These factors are very critical in determining both project and panel costs. If these factors are not taken into consideration, the company may lose money.

Another BOM example presented here is for making flaps for aerospace applications. Instead of using four columns in a table, all the material information is provided in three columns in Table 8.2. Flaps are attached to the trailing edge of the wing by hinged joints, they create lift and drag during takeoff and landing operations. There is more than one flap attached to each wing. They come in a variety of sizes, ranging from 3 to 8 ft wide, 5 to 12 ft long, and 3 to 12 in. thick, depending on the size of the aircraft. Flaps are tapered in width and thickness from one end to the other. Figure 8.2 is a location of flaps in an aircraft whereas Figure 8.3 shows the photograph of a flap. One can see the movement of flaps by sitting close to the wing section of a passenger aircraft.

TABLE 8.2

Bill of Materials for Making One Flap of Size 3 × 6 ft

Part No.	Part Description	Quantity
10-1278	Fiberglass prepreg 50 in. wide, type 7781 (4 filler plies on each side)	144 ft^2
10-1502	Fiberglass prepreg 50 in. wide, type 220 (2 skin plies on each side)	72 ft^2
12-2001-10	Tedlar film, 26-in. width, 1-mil shell, white (for outer appearance)	18 ft^2
123A53-2	Aluminum core	1
DAB12N	Bolt	4
DAB15K	Nut	4
DAB18A	Washer	4
DAB14B	Rivets	80

Production Planning and Manufacturing Instructions

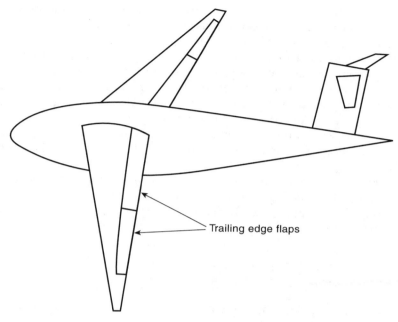

FIGURE 8.2
Location of flaps in an aircraft wing.

FIGURE 8.3
Photograph of a flap. (Courtesy of Marion Composites.)

Flaps are made by two methods. In the first method, honeycomb is used as a core material and then prepregs are laid down on the top and bottom of the core material. After lay-up of all the necessary materials, it is vacuum bagged and cured in an autoclave. In the second method, aluminum mandrels (rectangular cross section) are used to make the flap. The number of mandrels depends on the number of stiffeners to be created on a single flap. Prepregs are laid down on four sides of the mandrel and then butted together to form I-beam-like sections. After all the mandrels are laid down, prepregs are applied on the top and bottom of the assembly as skin materials and then cured in an autoclave. Mandrels are removed once curing is complete. The bill of materials presented in Table 8.2 is for making flaps using the first method where honeycomb core is used. The size of the flap is assumed to be 3 × 6 ft. Manufacturing instructions for making flaps is discussed in Section 8.4.

The number of rivets, washers, bolts, and other components depends on the size of the flap. The BOM in Table 8.2 does not represent the true number of components for flaps; instead, it shows how to prepare a bill of materials for an application.

8.4 Manufacturing Instructions

In industry, the manufacturing department writes step-by-step instructions for technicians and workers to follow — a set of procedures for the fabrication of finished products. This sheet of instructions is called a manufacturing instructions sheet, process sheet, or routine sheet. Writing this sequence of operations is a very important element of manufacturing engineering because it involves a good knowledge of manufacturing processes as well as skill in analyzing engineering drawings. In the aerospace industry, quality personnel check and sign the process sheet after completion of a certain number of steps, and then the next step begins. For example, after laying one graphite/epoxy prepreg layer at 45°, the quality inspector may come and verify whether proper material and lay-up angle was used to lay-up that prepreg sheet. In the aerospace and space industries, missing a step or performing a step incorrectly can result in serious loss or injury. For example, in September 1999, the $125 million orbiter for the Mars project burned up in the Martian atmosphere because of a failure by engineers to convert figures to metric. Two months later, a software glitch caused the Polar Lander spacecraft's descent engines to shut off prematurely, destroying the $165 million spacecraft. In the space industry, a small error can lead to a big failure; thus, extra precautions are taken in making various components. Each manufacturing step is documented so that, in the event of failure, engineers and scientists can trace the cause of failure and learn from it for future projects.

In product fabrication, the material/part flows from one area to another until all the operations are performed. This is shown in Figure 8.4 as an example. There are many steps performed at each station to include all the functionality and requirements of the product.

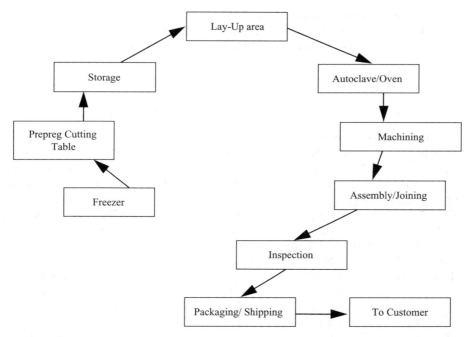

FIGURE 8.4
Process flowchart for a typical application.

Manufacturing instructions provide a step-by-step procedure for proper fabrication of the product. It provides the following benefits:

1. Clear instructions for technicians and quality control personnel to complete the product fabrication.
2. A written document, which is available to everybody in the organization.
3. The complete manufacturing process is broken down into small steps that are easy to understand and follow.
4. A chance for an engineer to think and come up with better way to fabricate the part before manufacturing begins. (While writing manufacturing instructions, an engineer must think of every small step involved in the process and that helps in clarifying the whole process.)
5. If anything should go wrong with the product in the future, manufacturing instructions work as a reference document to conduct investigations to find out the mistake.

Sections 8.4.1 and 8.4.2 provide two examples of manufacturing instructions: one for making tooling panels and the other for making flaps. The bill of materials for both of these cases were described in Tables 8.1 and 8.2. The format for writing manufacturing instructions can vary from company to company; but essentially, all the details of manufacturing operations are covered.

8.4.1 Manufacturing Instructions for Making Tooling Panels

In this example of manufacturing instructions, assume that the company XYZ Inc. gets a project for making 100 sandwich panels for a tooling company called Tooling Enterprise. Company XYZ has already discussed product specifications with Tooling Enterprise. Company XYZ then decides to prepare the panel by laying fiberglass skins and core as shown in Figure 8.1. The BOM for this project is shown in Table 8.1 for making one panel. To make 100 panels, 100 times of the materials listed in Table 8.1 are ordered. To be on the safe side, companies usually order more than the required number of parts so that if any part/product gets damaged during production, the company can remake it without delay. The manufacturing process selected for this project is the compression molding/hot press method where skins with adhesives and a honeycomb core are co-cured between two heated platens for a certain time duration under pressure. After completion of the cure, the assembly is taken from the press and slowly cooled. The panel then goes to the machining station for trimming to the desired size. Quality control (QC) personnel verify the panel dimensions and finish requirements. Once the panel is found to be in the tolerance range as described in the work order, quality personnel okay the panel and then it goes for packaging and shipping. Manufacturing date and operation time shown on the right two columns of Table 8.3 are kept blank and are filled by technicians and supervisors in charge of various operations. This is filled when the operation is performed. The complete procedure is described in detail in Table 8.3.

8.4.2 Manufacturing Instructions for Making Flaps

To make actual airplane flaps, there are many operations (e.g., lamination, vacuum bagging, autoclave processing, machining, and shipping) a company must perform. Each operation involves several steps. Many times companies use standard procedures to perform repetitive jobs such as vacuum bagging, curing, and machining. These repetitive jobs are written as standard procedures and are referred to in the manufacturing instructions as shown in Table 8.4. Just by referring to standard procedures, many steps in the manufacturing instruction can be avoided. The manufacturing instructions in Table 8.4 cannot be used for making actual flaps for aircraft applications, however, the table shows how the various steps in making the flaps can be written down using such a method. The format for numbering the operation number has been changed in this example as compared to a

TABLE 8.3

Manufacturing Instructions for Making the Tooling Panel for Tooling Enterprise

Customer Name: Tooling Enterprise, San Diego, CA
Product Name: Tooling Panel
Work Order No.: 10561
P.O. Number: TE15632
Manufacturing Processes: Material preparation, Lamination, Pressing, Finishing, Inspection, Packaging and Shipping

Operation No.	Operation Name	Process Description	Manufacturing Date	Operation Time
10-0005	Issue material from store and inspect	Issue material from store Skin specification: fiberglass sheet, Grade A, Part No. 1131 Size: see bill of material (BOM), Table 8.1 Core Specification: Nomex honeycomb, industrial grade, type: 4.0 (lb/ft^3) x 1/4 in. (cell size) Part No.: 21562 Size: See BOM Adhesive specification: Film adhesive, high temperature, L312 Part No.: 1281, Size: See BOM Inspect Inspect the skin, core, and adhesive for any defect. QC Employee No. _____ For any defective material, fill out form QA-ABC-05 and submit to Quality Department for disposition.		
10-0010	Material preparation	(Procedure is shown for making tooling panels in a batch of five.) 10-0010-01: Obtain ten fiberglass sheets and cut to a size 1 in. larger in the length and width of the panel per BOM. 10-0010-02: Sand the bonding surface of the skin with 60–80 grit paper. 10-0010-03: Clean the bonding surfaces with methyl ethyl ketone (MEK) followed by wiping with isopropyl alcohol (IPA).		

TABLE 8.3 (continued)
Manufacturing Instructions for Making the Tooling Panel for Tooling Enterprise

Customer Name: Tooling Enterprise, San Diego, CA
Product Name: Tooling Panel
Work Order No.: 10561
P.O. Number: TE15632
Manufacturing Processes: Material preparation, Lamination, Pressing, Finishing, Inspection, Packaging and Shipping

Operation No.	Operation Name	Process Description	Manufacturing Date	Operation Time
		10-0010-04: Obtain five cores and ensure that it is of the proper size and thickness. Cut to a size 1 in. larger in the length and width of the panel per BOM.		
		10-0010-05: Obtain adhesive roll and defreeze it for 1 hour. Cut adhesive to a size 1 in. larger in the length and width of the panel as per BOM.		
10-0015	Lamination	10-0015-01: Lay down one fiberglass sheet on a lamination table. Make sure sanded surface is facing up.		
		10-0015-02: Lay the adhesive film on top of the sheet.		
		10-0015-03: Remove any entrapped air. Remove the release film from the adhesive.		
		10-0015-04: Lay the Nomex core on top of the adhesive film. Make sure the edges are aligned.		
		10-0015-05: On another lamination table, lay the top skin facing sanded surface up.		
		10-0015-06: Apply adhesive film on top of the skin. Remove any entrapped air. Remove the release film from the adhesive.		
		10-0015-07: Place the top skin (pre-applied with adhesive) on top of the Nomex core, facing adhesive film down.		
		10-0015-08: Move the assembly to another table and similarly prepare other four panels.		

10-0020	Hot press processing	10-0020-01: Put five assembled panels inside the heated press. (XYZ Inc. has one press with six moving platens of size 5 × 12 ft. All five panels are cured in one batch.) 10-0020-02: Press panels at 250°F for 60 min at 10 to 15 psi. 10-0020-03: Record the following actual readings: Cure temperature: _____ °F Cure pressure: _____ psi Cycle time: _____ min (Follow the above steps 10-0010 to 10-0015 for preparing another five panels during the period that the above five panels are getting cured.)
10-0025	Machining	10-0025-01: Cut the panel to the finished size as per Work Order No. 10561 with a tolerance of ±1/32 in. Refer MIS-2500 for standard trimming procedure (companies have policies to document various standard procedures for reference).
10-0030	Panel inspection	10-0030-01: Manufacturing personnel shall inspect each finished panel and verify the following: 1. Finished panel dimensions match Work Order No. 10561. 2. There are no more than 6 defects, scratches, and dents combined, with none being visible at 3 ft from the panel, and no scratches longer than 1.5 in.
10-0035	Quality Control	10-0035-01: Quality control shall verify 15% of all the panels and shall record the following measurements: 1. Panel size: ___ × ___ × ___ 2. Thickness at three places: ___ ___ ___
10-0040	Packaging	10-0040-01: Label each panel with XYZ Inc. label with Work Order 10561. 10-0040-02: Clean, package, and ship the panel. Refer to MIS-0540 for standard packaging procedures.

TABLE 8.4
Manufacturing Instructions for Making Flaps

Customer Name: ABC Aerospace, London, U.K.
Product Name: Flaps, 3 × 6 ft.
Work Order No.: 10570
P.O. Number: ABC15632
Manufacturing Processes: Prepreg cutting, lay-up, vacuum bagging, autoclave processing, machining, surface finishing, fastening, quality control, packaging and shipping.

Operation Number	Operation Name	Process Description	Manufacturing Date	Operation Time
0010	Issue material from store and inspect	Issue material from store and record prepreg batch and roll numbers for each type of materials used Inspect material Employee No. _____		

0020	Prepreg cutting	Layout prepregs and cut the following:
		Trim filler plies type 7781
		  3 layers at 0°, 36 × 72 in. warp short
		  3 layers at 90°, 36 × 72 in. warp long
		Trim filler plies type 7781
		  2 layers at +45° 36 × 72 in.
		Trim skin plies type 220
		  2 layers at 0°, 36 × 72 in. warp short
		  2 layers at 90°, 36 × 72 in. warp long		

0030	Lay-up	0030: Lay-up one skin ply type 220 at 0° Employee No. _____ 0040: Lay-up one skin ply type 220 at 90° Employee No. _____ 0050: Lay-up one filler ply type 7781 at 0° Employee No. _____		

0060:
　Lay-up one filler ply type 7781 at 90°
　Employee No. _____
0070:
　Lay-up one filler ply type 7781 at 0°
　Employee No. _____
0080:
　Lay-up one filler ply type 7781 at 90°
　Employee No. _____
0090:
　Apply honeycomb core. Mark the location and position.
　Employee No. _____
0100:
　Inspect
　Employee No. _____
0110:
　Lay-up one partial filler ply type 7781 at −45°
　Employee No. _____
0120:
　Lay-up one partial filler ply type 7781 at +45°
　Employee No. _____
0130:
　Lay-up one filler ply type 7781 at 90°
　Employee No. _____
0140:
　Lay-up one filler ply type 7781 at 0°
　Employee No. _____
0150:
　Lay-up one skin ply type 220 at 90°
　Employee No. _____
0160:
　Lay-up one skin ply type 220 at 0°
　Employee No. _____

TABLE 8.4 (continued)
Manufacturing Instructions for Making Flaps

Customer Name: ABC Aerospace, London, U.K.
Product Name: Flaps, 3 × 6 ft.
Work Order No.: 10570
P.O. Number: ABC15632
Manufacturing Processes: Prepreg cutting, lay-up, vacuum bagging, autoclave processing, machining, surface finishing, fastening, quality control, packaging and shipping.

Operation Number	Operation Name	Process Description	Manufacturing Date	Operation Time
		0165: Apply white tedlar Employee No. _____		
		0170: Apply part number and serial number label. Also apply "Made in U.S.A" label. Employee No. _____		
0180	Vacuum bagging	0180: Prepare vacuum bagging as per procedure XYZ 1452-98. (Vacuum bagging procedure is a standard procedure and instead of writing the comlplete procedure in detail, it is referred to the procedure number, whenever there is a vacuum bagging process required. Similarly, there are other standard procedures for cleaning, trimming, curing, and more, which are referred to during writing of manufacturing instructions.) Employee No. _____		
		0190: Inspect the bagging. Employee No. _____		
0200	Autoclave loading and curing	0200: Load the panel inside the autoclave and cure the assembly as per procedure XYZ 1632-99. Autoclave load no.: _____ Employee No. _____		

0210	Autoclave offload and debagging	0210: Offload the panel. 0220: Debag as per procedure XYZ 1652-99. 0230: Remove excess resin and resin ridges.
0240	Machining	0240: Machine panel as per CNC program #5021. The program includes drilling holes, trimming of all edges, chamfering the aft edge including cutouts. 0250: Deburr all edges and corners. Remove tabs.
0260	Quality control	0260: Perform dimensional inspection on the panel. Use 100% visual inspection as per procedure XYZ 1236-97. Use backlighting as per requirement. Employee No. _____
0270	Surface finishing	0270: Prepare the surface as per procedure XYZ 1036-98. Employee No. _____
0280	Fastening	0280: Fasten nuts, bolts and other components as per procedure XYZ 1656-99. Employee No. _____
0290	Quality control, assembly	0290: Inspect the entire assembly. Ensure fastener flushness as per XYZ-233. Employee No. _____
0300	Packaging/shipping	0300: Package the panel as per procedure XYZ-355 and ship to the shipping department. Employee No. _____

previous example (Table 8.3) to show you various possibilities for numbering an operation. Operation numbers are a must in manufacturing instructions for easy reference to various steps.

The manufacturing instructions written in Table 8.4 are for manual operations. If the operation is performed by machine (e.g., machining operations using the CNC machine), the steps are programmed into the machine as a standard procedure; and by calling the standard procedure, the operation is completed as shown in Table 8.4. In the automobile industry, most of the processes for part fabrication are automated because of the high-volume requirements. However, assembly operations are still widely performed by manual operations, and manufacturing instructions are written down for those operations. In the automobile industry, there is typically no sign-up sheet by employees after each step because it is a time-consuming process. Moreover, in the automobile industry, one complete assembly operation is divided into small groups of operations and then each group operation is performed by a team of five to ten technicians, where each technician is responsible for one or a few steps. After the technician completes his/her task of bolting or putting an insert or applying lubricants or adhesive, the partial assembly/sub-assembly is passed to the next person in the team. That person performs his/her task and hands it over to the next person. Either the partial assembly is handed over to the next person by placing it on his/her workbench, or it is moved automatically by a carriage unit. There are human errors; for example, an operator misses one or two steps or places the wrong insert, etc. These human errors can be minimized by error proofing the operation, and are discussed in Chapter 5.

8.5 Capacity Planning

After the completion of a product design, capacity planning is done to ensure that the plant is capable of producing the desired number of parts per month (or per week) according to customer requirements. Capacity planning takes into account various aspects, including raw materials storage capacity, fabrication equipment capacity, raw materials cutting capacity, etc. A freezer capacity study is performed for parts requiring prepregs and film adhesives, and a rack storage analysis is performed for raw materials such as fabrics and fiber tows. Based on capacity analysis, total amounts of processing equipment, tools, floor space, cutting capacity, etc. are determined. For example, to make 10,000 golf shafts per day by the roll wrapping process, companies perform capacity planning to determine the total number of mandrels, total freezer capacity, total lay-up area, total number of ovens, total number of roll wrapping equipment, and total man-hours required to meet the objective of making 10,000 golf shafts per day.

Capacity planning proceeds through the following analysis.

1. It determines the output (production rate) of manufacturing equipment such as autoclave, filament winding, pultrusion, RTM, and injection molding. For example, if a typical process cycle time for an autoclave is 6 hr, then a maximum of four loads are possible per day, assuming 24 work hours per day. In each load, if five parts are made, then only 20 parts can be made in a day using one autoclave. Similarly for an RTM process, if the process cycle time is half an hour then production rate is 48 parts per day assuming three shifts per day and one mold cavity. Based on this analysis, if the production rate requirements are known, then the total equipment requirements can be determined to meet the production rate.
2. It determines the raw material storage capacity, such as freezer capacity for storing adhesives and prepregs, rack area analysis, floor space study. This analysis calculates the number of racks or freezers required to maintain a particular inventory for a fixed period of time (e.g., 1 week or 1 month).
3. It analyzes the cutting equipment capacity for cutting prepregs or preforms or cores. For pultrusion, filament winding, and other automated processes, this step is not usually required because these processes require continuous feeding of raw materials.
4. It determines the total man-hours required to meet production requirements. To do this analysis, various steps for the fabrication of composite parts are determined as shown in the manufacturing instructions, and then a conservative man-hour estimate is calculated based on the amount of time required to complete each small step. The procedure for determining time and cost to make a product is discussed in Chapter 11.
5. It determines the machining capacity to perform various tasks, including drilling, trimming, deburring, etc.

To get a clear picture of capacity planning, an example is shown below for making aircraft flaps using an autoclave process. Manufacturing instructions for making the flaps has already been discussed. The steps shown below will help in capacity planning for other manufacturing processes as well.

8.5.1 Problem Definition

Determine the number of autoclaves and freezers required to produce 2000 aircraft flaps per month. It is desired that these flaps be made by the autoclave process. The materials required for making these flaps are listed in Table 8.2 in the bill of materials. Manufacturing instructions for the fabrication of flaps are given in Table 8.4.

8.5.2 Assumptions

The following two assumptions are made for the present analysis:

1. The manufacturing facility will work 5 days a week and 3 shifts per day if required. Saturdays and Sundays are kept open as a reserve capacity to make up for unexpected shortfalls during the week's production, as well as for machine maintenance.
2. The manufacturing facility will work in "just-in-time" mode as much as possible to minimize inventory cost and floor-space requirements.

8.5.3 Capacity Analysis

This step involves capacity analyses of the autoclave and freezer capacity requirements.

8.5.3.1 Autoclave Capacity Analysis

The number of panels made by one autoclave depends on the size of the autoclave, the size of the part, the design of the rack, and the processing time. The rack is used for loading and unloading panels. For a higher production rate, the rack should be designed to maximize the use of autoclave space. For the present analysis, it is assumed that the autoclave to be used is 10 ft in diameter and 30 ft long.

The present analysis has the following objectives:

1. To determine the number of loads required per week to meet production requirements
2. To calculate the number of autoclaves required to meet the production target
3. To design the rack to maximize the use of autoclave processing space

Because panel size is 3 × 6 ft, a T-type rack as shown in Figure 8.5 is designed. The rack consists of five trays: trays A, B, C, and D are 3.5 ft wide and 28 ft long, and tray E is 5 ft by 28 ft long.

The T-rack design in Figure 8.5 can provide 20 panels per load, assuming 4 panels per tray. For the present project, to meet the target of making 2000 flaps per month, 100 loads (2000/20) per month are required. Assuming 4 working weeks per month and 5 days a week, five loads [100/(4 × 5)] are required per day.

In this analysis, it is assumed that there are enough ports to connect thermocouples and vacuum lines to make 20 panels. The connections for vacuum lines and thermocouples from rack to autoclave should be made at the open end of the autoclave. That way if there is any vacuum leak or other adjustment problem, it is easily done without completely removing the rack.

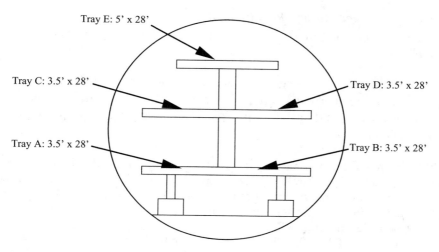

FIGURE 8.5
Rack design for a 10-ft-diameter by 30-ft-long autoclave.

TABLE 8.5

Conservative Time Estimate for Processing a Load in an Autoclave

Description of the Process	Amount of Time
Heat-up time	150 min
Dwell time	90 min
Cool-down time	90 min
Installation and removal of rack:	
Insert and position rack in the autoclave	10 min
Make vacuum line and thermocouple connection	20 min
Close door	5 min
Leak check: maintain vacuum	15 min
Leak check: check leaks	10 min
Leak check: fix leaks	10 min
Unplug thermocouples and vacuum lines	10 min
Remove panels from rack	10 min
Total cycle time	420 min ($\equiv$ 7 hr)

Table 8.5 shows a conservative time estimate for processing a load in an autoclave. The process cycle time is estimated to be 7 hr (420 min) per load. Therefore, to meet the production requirement of five loads per day, two autoclaves are required: one autoclave will have three loads per day and the other will have two loads per day. The first autoclave will provide 3 spare hours per day and the second autoclave will provide 10 spare hours per day. With the assumption of three shifts a day and 5 days a week, these two autoclaves will provide a spare capacity of 65 hr per week.

8.5.3.2 Freezer Storage Requirement

To calculate the total freezer space requirement, the following assumptions are made:

1. The nominal prepreg size is 50 in. wide and 65 yards long.
2. Prepreg rolls are delivered in $18 \times 18 \times 72$ in. cardboard containers.
3. It is assumed that 50% of the material will be wasted (scrap, angle cut, etc.).

One prepreg roll contains 812 ft^2. To make one panel, the total prepreg required is $144 + 72 = 216$ ft^2 (as shown in the BOM). Therefore, to make 2000 panels per month, the total requirement is 432,000 ft^2. To obtain 432,000 ft^2 of prepreg, the number of rolls needed will be 532 (432,000/812). It is assumed that 50% of the material will be scrapped and therefore the total number of rolls will be $2 \times 532 = 1064$. The reason for the high amount of scraps in this case is that the panel has a width of 36 in. (3 ft), and therefore, out of 50-in.-wide prepreg, 14 in. will go as scrap for a 0° fiber layer. In most aerospace applications, a typical buy-to-fly ratio is 1.7:1.0 (see Chapter 12). This implies that for every 1.7 kg of prepreg purchased, 0.7 kg are unused and scrapped.

Each roll occupies $1.5 \times 1.5 \times 6$ ft = 13.5 ft^3. Therefore, the total freezer capacity requirement for 500 rolls will be $1064 \times 13.5 = 14{,}361$ ft^3.

Companies use various types of freezers for prepreg and adhesive storage purposes. Suppose a company uses a freezer that has a capacity of $3 \times 3 = 9$ rolls. The total number of freezers required will be $1064/9 = 118.22$, or 119 freezers for 1 month of storage.

Questions

1. What is bill of material?
2. Why is it necessary to write manufacturing instructions for the making of parts?
3. Write down manufacturing instructions for the making of golf shafts. A method of making golf shafts is described in Chapter 6.
4. Determine the number of automotive door panels that can be manufactured per year using SMC compression molding equipment. Assume process cycle time is 4 min including loading and unloading. Also assume 5 days a week, 2 shifts of operation and 50 weeks in a year. Suggest what a company should do to increase the production capacity to 120,000 door panels per year per equipment.
5. Determine the number of steering rack housing using an injection molding machine that can be manufactured per week as well as per year. Assume process cycle time is 42 sec and 4 cavities per mold. Make other assumptions as necessary.

9
Joining of Composite Materials

9.1 Introduction

In any product, there are generally several parts or components joined together to make the complete assembly. For example, there are several thousands of parts in an automobile, a yacht, or an aircraft. The steering system of an automobile has more than 100 parts. Heloval 43-meter luxury yacht from CMN Shipyards is comprised of about 9000 metallic parts for hull and superstructure, and over 5000 different types of parts for outfitting. These parts are interconnected with each other to make the final product. The purpose of the joint is to transfer loads from one member to another, or to create relative motion between two members. This chapter discusses joints, which create a permanent lock between two members. These joints are primarily used to transfer a load from one member to another.

Joints are usually avoided in a structure as good design policy. In any structure, a joint is the weaker area and most failures emanate from joints. Because of this, joints are eliminated by integrating the structure. Joints have the following disadvantages:

1. A joint is a source of stress concentration. It creates discontinuity in the load transfer.
2. The creation of a joint is a labor-intensive process; a special procedure is followed to make the joint.
3. Joints add manufacturing time and cost to the structure.

In an ideal product, there is only one part. Fiber-reinforced composites provide the opportunity to create large, complicated parts in one shot and reduce the number of parts in a structure.

There are two types of joints used in the fabrication of composite products:

- Adhesive bonding
- Mechanical joints

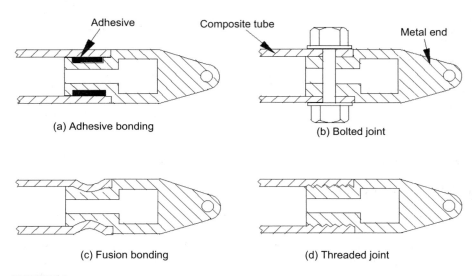

FIGURE 9.1
Various types of joints for joining a metal end with a composite tube.

Adhesive bonding is the more common type of joint used in composites manufacturing. In adhesive bonding, two substrate materials are joined by an adhesive. Mechanical joints for composites are similar to the mechanical joints of metals. In mechanical joints, rivets, bolts, and/or screws are used to form the joint. Fusion bonding is also used for joining purposes; however, this chapter focuses on adhesive bonding and mechanical joints. Fusion bonding is primarily used to join thermoplastic parts by means of heat.

Figure 9.1 depicts an application in which a composite tube is joined with a metal end by various means. Every joint has its advantages and disadvantages, as discussed later. A design engineer studies the various options for joining the two substrate materials and selects the best one for the application.

9.2 Adhesive Bonding

In adhesive bonding, two substrate materials are joined by some type of adhesive (e.g., epoxy, polyurethane, or methyl acrylate). The parts that are joined are called substrates or adherends. Various types of bonded joints are shown in Figure 9.2. The most common type of joint is a single lap joint wherein the load is transferred from one substrate to another by shear stresses in the adhesive. However, because the loads applied are off-centered (Figure 9.3) during a single lap joint, the bending action caused by the applied load creates normal stresses (cleavage stress) in the thickness direction of the adhesive. The combination of shear stress and normal stress at lap ends of

Joining of Composite Materials

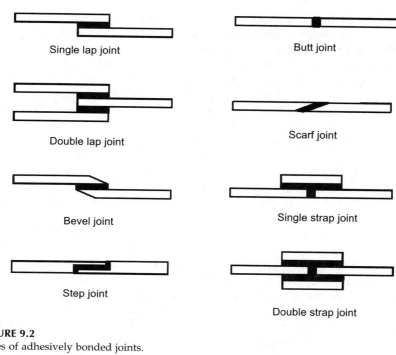

FIGURE 9.2
Types of adhesively bonded joints.

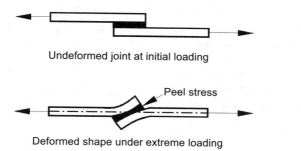

FIGURE 9.3
Illustration of joint under loading.

the adhesive reduces the joint strength in a single lap joint. To overcome the bending effect, a double lap joint is preferred. In a double lap joint, the bending force and therefore normal stresses are eliminated. The joint strength obtained by double lap joint testing is greater because of the absence of normal stresses. For adhesive selection and characterization purposes, single lap joint tests are conducted because single lap joints are very easy to manufacture. The stepped and scarf joints shown in Figure 9.2 provide more strength than single lap joints, but machining of stepped or scarf ends is difficult.

In adhesively bonded joints, the load gets transferred from one member to another by shear, and therefore shear tests are conducted on adhesives.

For this reason, the adhesive supplier seldom reports the tensile strength of the adhesive. The data reported by adhesive suppliers is usually the shear strength obtained from a single lap joint test. An exception to this is that the tensile tests are conducted by sandwich panel manufacturers to calculate the bond strength between skin and honeycomb core materials. This test is performed for quality control purposes or for selecting the correct adhesive for a sandwich structure. In honeycomb-cored sandwich structures, the bonded surface area between the core and skin material is much less than the surface area of the skin and therefore failure usually takes place at the interface.

The most common test method for shear testing is the single lap joint test (ASTM D 1002). The lap shear test measures the strength of an adhesive in most extent in shear. In this test, the specimen is prepared as shown in Figure 9.4. Substrate materials are cut into 1×4-in. test coupons from a large, flat rectangular sheet. The substrate materials are then joined in a 0.5-in. overlap area with the desired adhesive. Once the adhesive is cured, the two ends of the substrate materials are pulled under tension as shown in Figure 9.3. Because shear at the bonded surface during the lap-shear test is created by applying tensile load on the substrate material, this test is also called the tensile-shear test. This test method is used extensively because of its simplicity and low cost for evaluation purposes. However, due to non-uniform stress at the joint, the strength values obtained by this test method are of little use for engineering design purposes. However, this test method can be used to see the effects of lap length, adhesive thickness, adhesive material, etc. on bonded joints. When the bond strength test is performed for two dissimilar materials (e.g., glass/epoxy and aluminum) the thickness of the substrates is maintained such that the stiffness values of the adherend materials are the same (i.e., $E_1 t_1 = E_2 t_2$, where E_1, E_2, t_1, t_2 are the stiffnesses and thicknesses of the adherend materials 1 and 2, respectively).

Mazumdar and Mallick[1] conducted a series of tests on SMC-SMC and SRIM-SRIM joints to determine the effects of lap length, bond thickness, and joint

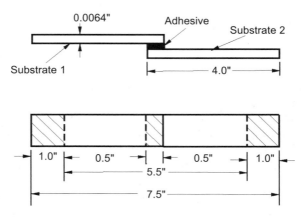

FIGURE 9.4
Standard lap shear test specimen.

Joining of Composite Materials

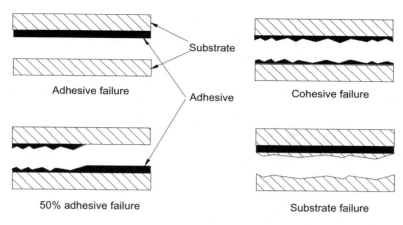

FIGURE 9.5
Common failure modes during bond test.

configuration. Single lap joints, single step lap joints, symmetric lap joints, and rounded edge single lap joints were considered and various failure modes were determined. They found that the failure load increases with increasing lap length; however, average lap shear strength value decreases. They also found that the joint strength improves with the increase in substrate stiffness.

9.2.1 Failure Modes in Adhesive Bonding

There are two major types of failure during testing of adhesively bonded joints: adhesive failure and cohesive failure, as shown in Figure 9.5. Adhesive failure is a failure at the interface between the adherend and the adhesive. Cohesive failure can occur in the adhesive or in the substrate material. Cohesive failure of the adhesive or substrate material occurs when the bond between the adhesive and the substrate material is stronger than the internal strength of the adhesive or substrate material.

The goal of any good bond design is substrate failure; that is, the bond is stronger than the joining materials themselves. In substrate failure, the parent materials fail either away from the joint or near the bond area by tearing away the parent materials. Another expected failure mode might be the cohesive failure of the adhesive, wherein the adhesive splits in the bond area but remains firmly attached to both substrates. Adhesive failure, where adhesive releases from substrate materials, is considered a weak bond and is generally unacceptable.

9.2.2 Basic Science of Adhesive Bonding

There is no single theory that explains the complete phenomena of adhesion. Some theories are more applicable for one type of application than others. However, the theories presented herein provide a general idea about the formation of good bond.

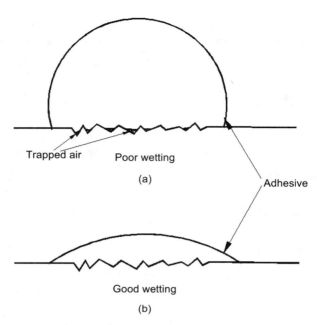

FIGURE 9.6
Demonstration of (a) poor wetting and (b) good wetting.

9.2.2.1 Adsorption Theory

According to this theory, adhesion results from molecular contact between two materials and the surface forces that develop between these materials. The surface forces are usually designated as secondary or Van der Waals forces. To develop forces of molecular attraction, there should be intimate contact between the adhesive and the substrate surfaces, and the surfaces must not be more than 5 Å apart. The process of developing intimate contact between the adhesive and substrate material is known as wetting. Figure 9.6 illustrates good and poor wetting. For an adhesive to wet a solid surface, the adhesive should have a lower surface tension than the solid's critical surface tension. Metals usually have high critical surface tension and organic surfaces usually have a lower critical surface tension. For example, epoxy has a critical surface tension of 47 dyn/cm and aluminum has a critical surface tension of about 500 dyn/cm. For this reason, epoxy wets a clean aluminum surface very well. Epoxy has poor wetting with polycarbonate, polystyrene, polyimide, polyethylene, and silicone surfaces because these substrates have critical surface tensions of 46, 33, 40, 31, and 24 dyn/cm, respectively, which are lower than epoxy's critical surface tension.

9.2.2.2 Mechanical Theory

According to this theory, bond formation is primarily due to the interlocking of adhesive and substrate surfaces. The true surface of the substrate material is never a flat, smooth surface; instead, it contains a maze of peaks and valleys. During the wetting process, adhesive flows into microcavities of substrate surfaces and fills them. When the adhesive hardens, the two substrates are held together mechanically. Application of pressure during the bonding process aids in penetrating the cavities and displacing the entrapped air from the interfaces.

During the bonding of composites or metals, sandblasting or surface roughening is performed on joining surfaces to increase joint strength. Surface roughening provides benefits such as removal of oily surface, formation of a more reactive surface, increased mechanical locking, and formation of larger surface area. The larger surface area increases the bond strength by increasing intermolecular forces (adsorption theory).

9.2.2.3 Electrostatic and Diffusion Theories

This theory is not as well regarded as the above two theories (adsorption and mechanical) on adhesion. According to this theory, electrostatic forces in the form of an electrical double layer are formed at the adhesive/substrate interface. These forces create resistance against separation. According to the diffusion theory, adhesion occurs due to the inter-diffusion of molecules on the adhesive and substrate surfaces. This theory is more applicable for the cases in which both the substrate and the adhesive material are polymer based. The key to diffusion processes is that the substrate and adhesive materials should be chemically compatible. Solvent or fusion welding of thermoplastic substrates is considered as bonding due to diffusion of molecules.

9.2.3 Types of Adhesives

For a better understanding of the types of adhesives available in the marketplace, adhesives are divided into three categories: (1) two-component mix adhesives; (2) two-component, no mix adhesives; and (3) one-component, no-mix adhesives. The majority of epoxy and polyurethane adhesives fall into two-component mix adhesives. Acrylic and anaerobic adhesives fall into two-component, no-mix adhesives. These adhesives are described as follows.

9.2.3.1 Two-Component Mix Adhesives

The adhesives falling into this category require prior mixing before being applied to the substrate surface. Epoxies and polyurethanes fall into this category. Once two components are mixed, there is a limited pot life.

9.2.3.1.1 Epoxy Adhesives

Two-component epoxy adhesives are very common in the composites industry and offer many benefits (epoxy adhesives also come as one-component and are discussed later). Two-component epoxy adhesives have a good shelf life and do not require refrigeration. They can be cured at room temperature but generally require elevated temperature curing to enhance performance.

Epoxies are sensitive to surface condition and mix ratio. Usually, surface preparation prior to bonding is required to enhance the bond quality. Epoxies have good gap-filling characteristics. Large volumes of epoxy, such as in potting, can be cured easily without the need for light, moisture, and/or absence of air to activate the curing process.

Epoxy adhesives are usually applied by hand; but for high-volume applications, automatic dispensing equipment is used. In automatic dispensing equipment, adhesives are mixed automatically prior to application on the bonding surface. This avoids human error in mixing and applying. Epoxies are mostly brittle and do not provide good peel strength. Elastomers (e.g., rubbers) are mixed to increase the toughness of epoxies. Epoxies are good for bonding stiff surfaces. For bonding flexible members, epoxies are not good because they lack peel strength.

9.2.3.1.2 Polyurethane Adhesives

Similar to epoxies, polyurethane adhesives also come as one- or two-component systems. Two-component polyurethane adhesives are available with a broad range of curing times to meet various application requirements. Polyurethane adhesives provide higher peel strength than epoxies. They provide good, durable bond strength to many substrate surfaces, although a primer may be necessary to prepare the bond surface. The primers are usually moisture reactive and require several hours to react sufficiently before parts can be used. They bond well to wood, composites, and some thermoplastics. They also have good gap-filling characteristics.

9.2.3.2 Two-Component, No-Mix Adhesives

In this category, adhesive is applied on one substrate surface and an activator, usually in a very small amount, is applied on other substrate surface. When these two surfaces are put together, the adhesive cures by the reaction of two components. There is no mixing required for the cure of the adhesive. Acrylic and anaerobic adhesives fall into this category.

9.2.3.2.1 Acrylic Adhesives

Acrylic adhesives have a polyurethane polymer backbone with acrylate end groups. They are formulated to cure through heat or the use of an activator applied on a substrate surface, but many industrial acrylic adhesives are cured by light. Light-cured adhesives are typically used for high-volume applications in which the capital investment for creating the light source can be justified. An additional need for light-cured adhesives is that bond geometry should allow light to reach the adhesives. Acrylic adhesives are relatively insensitive to minor variations in mix ratio and are mostly used in high-volume production environmenents such as automobile, speaker magnets, and consumer items. Acrylics are used for bonding thermoset composites, thermoplastic composites, wood, metals, and ceramics. They create a tough and durable bond, with a temperature resistance of up to 180°C. Some commercial acrylics have high heat generation (exothermic reaction) during cure and therefore the amount of adhesive needs to be judiciously selected for the specific application.

9.2.3.2.2 Urethane Methacrylate Ester (Anaerobic) Adhesives

Methacrylate structural adhesives are mixtures of acrylic esters that remain liquid when exposed to air but harden in the absence of air. An activator is generally required on one substrate surface to initiate the cure. These adhesives are used in a large number of industrial applications where high-reliability bonding is required. There is no mixing required, and no pot-life or waste problem with this adhesive. These adhesives provide flexible/durable bonds with good thermal cycling resistance. Anaerobic adhesives are good for bonding metals, composites, ceramics, glass, plastics, and stone.

9.2.3.3 One-Component, No-Mix Adhesives

The adhesive in this category is one component and therefore no mixing is required. Most one-component adhesives consist of two or more premixed components such as resin, curing agent, fillers, and additives. One-component epoxies, polyurethanes, cyanoacrylate adhesives, hot-melt, light-curable adhesives, and solvent-based adhesives fall into this category.

9.2.3.3.1 Epoxies

One-component epoxies are premixed and come in a bottle. They are refrigerated and have limited shelf lives. These adhesives require a high heat cure.

9.2.3.3.2 Polyurethanes

One-component polyurethanes cure by reaction with atmospheric moisture over a period of several hours or days. Usually, these adhesives are used as sealants. Heat is sometimes used to expedite the curing process.

9.2.3.3.3 Cyanoacrylates

Cyanoacrylate adhesives are one component and have rapid strength development. In a matter of few seconds, they provide good handling strength. These adhesives cure in the presence of surface moisture when confined between two substrate surfaces. Cyanoacrylate adhesives have poor heat, solvent, and water resistance as compared to structural adhesives. They provide excellent adhesion to many substrates, including metals and thermoplastics.

9.2.3.3.4 Hot-Melt Adhesives

Hot-melt adhesives are used heavily in wood, furniture, and consumer industries. Hot melts are thermoplastics and come in solid form. They are melted before being applied to a bonding surface. After application, the melt cools and solidifies, resulting in a bond between the mating surfaces. The bond is mostly a mechanical lock. The joining is achieved within a minute. Because of rapid bond formation, hot melts are used in high-volume applications. Hot melts have lower temperature resistance.

9.2.3.3.5 Solvent- or Water-Based Adhesives

In this type of adhesive, solvent or water is used as a carrier material. The purpose of the solvent or water is to lower the viscosity of the adhesive so that it can be easily dispensed and applied. Once it is applied, the solvent or water is evaporated into the air by heat or by diffusion into a porous substrate. Solvent- and water-based adhesives find application in porous substrates such as wood, paper, fabrics, and leather. Contact adhesives and pressure-sensitive adhesives fall into this category. Contact adhesives are heavily used in the wood industry. They are applied to both substrate surfaces by spray or roll coating. The substrate surfaces are then pressed under ambient or heated conditions. With water-based adhesives, the bonded assembly is typically kept at 220°F for faster evaporation of water. After evaporation, the adhesive rapidly bonds or knits to itself with the application of pressure. Pressure-sensitive adhesives usually come in a carrier film similar to a prepreg or film adhesive. The adhesive is laid down on one substrate surface and then carrier film is removed. After removal of carrier film, the other substrate surface is brought into contact with the adhesive. Very little pressure is required for mating the two substrate surfaces. Once the solvent is removed, the adhesive develops aggressive and permanent tackiness. Pressure-sensitive adhesives are applied similar to contact adhesives.

9.2.4 Advantages of Adhesive Bonding over Mechanical Joints

Joining of materials using an adhesive offers several benefits over mechanical joints. In the composites industry, adhesive bonding is much more widely used compared to the metals industry.

1. In adhesively bonded joints, the load at the joint interface is distributed over an area rather than concentrated at a point. This results in a more uniform distribution of stresses.
2. Adhesively bonded joints are more resistant to flexural, fatigue, and vibrational stresses than mechanical joints because of the uniform stress distribution.
3. The weight penalty is negligible with adhesive bonding compared to mechanical joints.
4. Adhesive not only bonds the two surfaces but also seals the joint. The seal prevents galvanic corrosion between dissimilar adherend materials.
5. Adhesive bonding can be more easily adapted to join irregular surfaces than mechanical joints.
6. Adhesive bonding provides smooth contours and creates virtually no change in part dimensions. This is very important in designing aerodynamic shapes and in creating good part aesthetics.
7. Adhesive bonding is often less expensive and faster than mechanical joining.

Joining of Composite Materials

9.2.5 Disadvantages of Adhesive Bonding

Adhesive bonding suffers from the following disadvantages:

1. Adhesive bonding usually requires surface preparation before bonding.
2. Heat and pressure may be required during the bonding operation. This may limit the part size if curing needs to be performed in an oven or autoclave.
3. With some adhesives, a long cure time may be needed.
4. Health and safety could be an issue.
5. Inspection of a bonded joint is difficult.
6. Adhesive bonding requires more training and rigid process control than mechanical joints.
7. Adhesive bonding creates a permanent bond and does not allow repeated assembly and dis-assembly.

9.2.6 Adhesive Selection Guidelines

The selection of an adhesive depends on the type of substrate material, application need, performance requirements, temperature resistance, chemical resistance, etc. A successful application requires a good joint design, good surface preparation, proper adhesive selection, and proper adhesive curing.

The first step in selecting an adhesive is to define the substrate materials and set up durability and other requirements. The following is a checklist for setting up some requirements:

- Strength requirement
- Cost requirement
- Loading type
- Impact resistance
- Temperature resistance
- Humidity, chemical, and electrical resistances
- Process requirements
- Production rate requirements

Once the above requirements are met, then the processing parameters for the adhesive bonding can be defined. These include the production rate, adhesive position, clamp time and position, surface preparation, fixture time, open time, cure parameters such as time, temeperature, and pressure, dispensing method, manual or automated assembly, and inspection method.

TABLE 9.1
Adhesive Selection Guidelines

Characteristics	Standard Epoxies	Urethane	Acrylic	Silicones	Polyolefins (Vinylics)
Adhesive type[a]	L1, L2, F	L, W, HM	L1, L2, W	L1, L2	F
Cure requirement	Heat, ambient	Heat, ambient	Heat, ambient	Heat, ambient	Hot melt
Curing speed	Poor	Very good	Best	Fair	Very good
Substrate flexibility	Very good	Very good	Good	Good	Fair
Shear strength	Best	Fair	Good	Poor	Poor
Peel strength	Poor to fair	Very good	Good	Very good	Fair
Impact resistance	Fair	Very good	Fair	Best	Fair
Humidity resistance	Poor	Fair	Fair	Best	Fair
Chemical resistance	Very good	Fair	Fair	Fair	Good
Temperature resistance (°C)	Fair	Fair	Fair	Good	Poor
Gap filling	Fair	Very good	Very good	Best	Fair
Storage (months)	6	6	6	6	12

[a] Adhesive type: L1 = Liquid one part, L2 = Liquid two part, F = Film, W = Waterborne, HM = Hot melt.

There are many types of adhesives avilable on the market. The most common adhesives are epoxies, acrylics, urethanes, silicones, and polyolefins. Table 9.1 categorizes these adhesives by their relative properties.

9.2.7 Surface Preparation Guidelines

The bond strength in an adhesively bonded joint greatly depends on the quality of the adherend surface. Therefore, surface preparation is key to the creation of successful joint. Surface preparation is performed to remove weak boundary layers and to increase wettability of the surface. Certain low-energy surfaces must be modified by plasma treatment, acid etching, flame tratment, or some other means to create attractive forces necessary for good adhesion. To prepare the surfaces, all dust, grease, oil, and foreign particles should be removed from joining surfaces. It is important for good wetting that the adherend have a higher surface tension than the adhesive.

Surface preparation can range from simple solvent wiping to sandblasting to chemical etching or combinations of these. Metals are best cleaned by vapor degreasing with trichloroethane, followed by sandblasting or, preferably, chemical etching. Chromic acid is often used as a chemical etching process for steels. Aluminum surfaces are primed to improve their bondability. Once the surfaces are cleaned, bonding is performed as soon as possible to avoid accumulation of foreign materials. If storage is necessary, special precaution is taken so that the assembly does not become contaminated. Composite surfaces very often need some form of preparation because their surfaces are often contaminated with mold release agents or other additives. These contaminants are removed, either by abrading the surface with sandpaper or by alkaline wash. With some polymeric materials, the surfaces are chemically modified to encourage wetting and to achieve acceptable bonding. Thermoplastic materials are generally difficult to bond and some kind of treatment, such as oxidation by flame treatment, plasma or corona treatment, ionized inert gas treatment, or application of primers or adhesion promoters, is usually necessary.

The amount of surface preparation needed for an application depends on the production volume, ultimate joint strength requirement, and cost. For low- to medium-strength applications, extensive surface preparation is not necessary. For applications where maximum bond strength or reliability and safety are concerns, such as in aerospace applications, surface preparation is performed in a controlled atmosphere. For consumer and automotive applications, cost and cycle time play major roles in selecting a surface preparation method. Surface preparation methods can be classified as passive surface treatment or active surface treatment. In passive surface treatment, such as sanding and solvent cleaning, the chemistry of the surface does not change. It only cleans and removes the weakly bonded outer surface layers. In active surface treatment processes, such as plasma treatment, anodizing, and etching, the surfaces are chemically modified. The following sections describe some of the surface preparation methods.

9.2.7.1 Degreasing

Degreasing implies the removal of an oily surface from the surfaces to be bonded. Degreasing is performed differently for metals and composites. To degrease metals, the surfaces are suspended in a stabilized trichloroethane vapor bath for approximately 30 seconds. If a bath is not available, the surfaces are cleaned with pieces of absorbent cotton dampened with trichloroethane. Trichloroethane is a toxic chemical, although non-flammable, and therefore the working area should be well-ventilated.

For degreasing composites or polymeric surfaces, solvents such as acetone and methyl alcohol or detergent solutions are used to remove the mold release agents or waxes.

9.2.7.2 Mechanical Abrasion

This process is quite common in the composites industry. In this process, the smooth surfaces of metals or composites are roughened to increase the surface area and to remove the contaminants and loose particles from the surface. Sandpaper, wire brush, and emery cloth are used for this purpose. For sanding big sheets of metals and composites, as in making large sandwich panels with aluminum and composite skins, the surfaces are passed between two moving rollers (mechanical sanders), one containing a sandpaper belt. The surfaces to be bonded with honeycomb or foam are sanded and the other side is covered with a protective coating to avoid scratching.

Sandblasting, vapor honing, or hand-held mechanical abraders are used for localized roughening. After abrading, the solid particles are removed by solvent wiping or by blasting with clean air.

9.2.7.3 Chemical Treatment

Chemical treatment greatly increases bond strength. In this process, strong detergent solutions are used to emulsify surface contaminants on both metallic and nonmetallic substrates. Parts are typically immersed in a well-agitated bath containing detergent solutions at 150 to 200°F for about 10 minutes. Following that, the surfaces are rinsed immediately with deionized water and then dried.

Typical alkaline detergents are combinations of alkaline salts, such as sodium metasilicate and tetrasodium pyrophosphate with surfactants included.

9.2.8 Design Guidelines for Adhesive Bonding

Structural joints are used to transfer the load from one component to the other. Because adhesives are good in shear, the joint design is made to ensure that load is transferred in shear. Adhesives are not good in peel and tensile stress, and therefore load transfer in these modes is avoided. The mechanical joints such as bolting and riveting used in the metals industry are good in transfering loads in peel. Therefore, when bolted and riveted joints need to be replaced by adhesive bonded joints for composite structures, proper care must be taken in redesigning the joint. It is easy to visualize tensile, compressive and shear stress in an application, whereas peel stresses are not so obvious. Peel stresses are created when joint edges are subjected to bending loads. Tensile and shear loads get translated to peel stress when substrates are flexible and when applied loads are not in line or parallel with each other.

Guidelines for designing adhesively bonded joints include:

1. Design the joint in such a way that load transfer predominates in shear or compressive mode.
2. Select the right adhesive material to meet the application needs (e.g., temperature resistance, chemical resistance, strength, etc.).
3. Design the joint to be production friendly. The bond area should be easily accessible and technicians should be able to perform surface preparation and bonding with minimum effort and movement.
4. Use the maximum area possible for adhesive bonding so that stress induced in the joint is minimal.
5. When joining dissimilar materials, stress caused by thermal expansion and contraction should be considered in the joint design.
6. There is an optimum adhesive thickness required to create the best bond strength. Too thin or too thick an adhesive layer provides poor bond strength. There should be enough adhesive to wet the joining surface.

9.2.9 Theoretical Stress Analysis for Bonded Joints

Theoretical stress analyses for adhesively bonded joints are derived either from classical analytical methods or finite element methods. A good review of these techniques is that by Matthews, Kilty, and Godwin.[2] The majority of the work was performed analyzing single lap joints loaded in tension.[3-7] In the classical analytical approach, linear and nonlinear analyses are performed. The pioneering work in the category of linear analysis was performed by Volkersen,[8] the so-called shear lag analysis. In this analysis, the only factors considered are the shear deformation of the adhesive and the elongation of the adherends. This scenario is more applicable for double lap joints than single lap joints because in single lap joints, bending of adherends and thus peel stresses are induced. The effect of bending in single lap joints was first considered by Goland and Reissner[9] and then by several other authors.[10,11] During a single lap joints loaded in tension test, peel stresses are induced due to eccentric loading, as shown in Figure 9.3. The magnitudes of peel and shear stresses increase with increaseing applied load, and they are higher at the edges. Several authors[12-14] have performed parametric studies to identify the factors that influence the maximum stresses in a joint. Maximum stress is induced at the ends of the overlap along the length. To minimize the maximum stresses at the joint, the use of the same adherend materials, stiff adherend material, modulus adhesive, and tapering of adherend along the length is suggested.

The work described above was performed considering that adhesive stresses remain within the elastic range. Because of this assumption, the ultimate static load of the joint can be underestimated. This assumption can be true under fatigue loads where stresses are relatively low. In reality, adhesives behave in a nonelastic mode even with so-called "brittle" adhesives, but discrepancies are much more serious for "ductile" adhesives.[2] Adams et al.[15] and Grant[16] have considered the nonlinearity of adhesives in their studies. Hart-Smith[17-20] performed extensive studies on single lap, double lap, and scarf joints, taking into account adhesive nonlinearity. Adhesive nonlinearity is based on assumptions of idealized elastic, perfectly plastic shear stress and strain curve. Various researchers[21-25] have used finite element methods for analyzing bonded joints. In finite element methods, there is no need to simplify assumptions as done in classical methods.

9.3 Mechanical Joints

Mechanical joining is most widely used in joining metal components. Examples of mechanical joints are bolting, riveting, screw, and pin joints. Similar to the mechanical joints of metal components, composite components are also joined using metallic bolts, pins, and screws; except in a few cases where

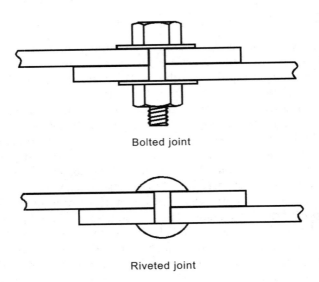

FIGURE 9.7
Schematic diagram of mechanical joints.

RFI shielding and electrical insulation are required, composite fasteners are used. For most mechanical joints, an overlap is required in two mating members and a hole is created at the overlap so that bolts or rivets can be inserted. When screws are used for fastening purposes, mostly metal inserts are used in the composites, the reason being that the threads created in the composites are not strong in shear and therefore metal inserts are used. Figure 9.7 shows examples of bolting and riveting.

In bolted joints, nuts, blots, and washers are used to create the joint. In riveting, metal rivets are used. Bolted joints can be a single lap joints, double lap joints, or butt joints, as shown in Figure 9.8.

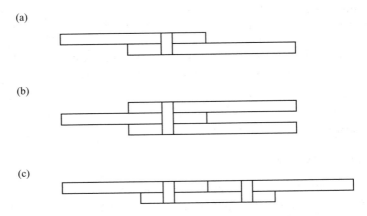

FIGURE 9.8
Types of bolted joints: (a) single lap joint, (b) double lap joint, and (c) butt joint.

Joining of Composite Materials

Sections 9.3.1 and 9.3.2 describe the advantages and disadvantages of joints using mechanical fasteners.

9.3.1 Advantages of Mechanical Joints

1. They allow repeated assembly and disassembly for repairs and maintenance without destroying the parent materials.
2. They offer easy inspection and quality control.
3. They require little or no surface preparation.

9.3.2 Disadvantages of Mechanical Joints

1. Mechanical joints add weight to the structure and thus minimize the weight-saving potential of composite structures.
2. They create stress concentration because of the presence of holes. The composite materials do not have the forgiving characteristics of ductile materials such as aluminum and steel to redistribute local high stresses by yielding. In composites, stress relief does not occur because the composites are elastic to failure.
3. They create potential galvanic corrosion problems because of the presence of dissimilar materials. For example, aluminum or steel fasteners do not work well with carbon/epoxy composites. To avoid galvanic corrosion, either metal fasteners are coated with nonconductive materials such as a polymer or composite fasteners are used.
4. They create fiber discontinuity at the location where a hole is drilled. They also expose fibers to chemicals and other environments.

9.3.3 Failure Modes in a Bolted Joint

A bolted joint is made by drilling holes in mating parts. The mating parts are then aligned and a nut is passed through it and then bolted. Failure in a bolted joint may be caused by:

1. Shearing of the substrate
2. Tensile failure of the substrate
3. Crushing failure of the substrate
4. Shearing of the bolt

The first three failure modes are shown in Figure 9.9. Among these three, crushing failure is the most desirable failure mode in the joint design. Crushing helps in relieving the stress concentration around the hole.

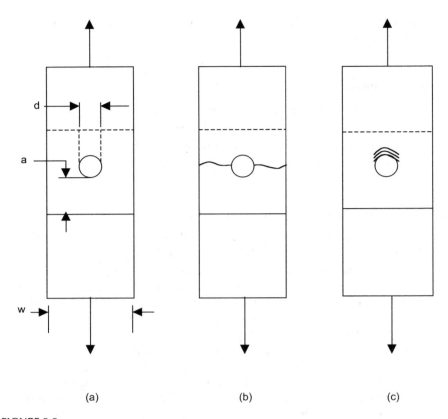

FIGURE 9.9
Failure modes in bolted joints: (a) shear failure, (b) tensile failure, and (c) bearing failure.

Bolt failures are not common because steel fasteners are very strong in shear. There is another type of failure mode called cleavage failure, which takes place in laminates having 0° fibers along the loading direction. Cleavage failure is shown in Figure 9.10.

9.3.4 Design Parameters for Bolted Joints

The following parameters affect the strength of a bolted joint:

1. Material parameters such as fiber orientation, lay-up sequence, and type of reinforcement
2. Joint parameters such as ratios of width to bolt hole diameter (w/d), edge distance to bolt hole diameter (a/d), and thickness to bolt hole diameter (t/d)
3. Quality of hole, such as delaminated edge
4. Clamping force

Joining of Composite Materials

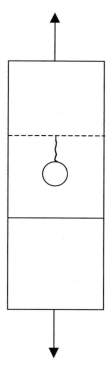

FIGURE 9.10
Cleavage failure in the 0° composite laminate.

The stress concentrations around the hole are reduced using doublers (local increase in thickness around the hole), minimizing component anisotropy, and using softening strips of lower modulus such as fiberglass plies in graphite composites. In multi-bolting joints, where more than one row of bolts is used, the spacing between holes (pitch) and the hole pattern are also important in designing bolted joints.

9.3.5 Preparation for the Bolted Joint

Bolted joints are prepared by drilling holes in mating parts at specified places. Methods of drilling holes are described in Chapter 10. Once the holes are machined, the mating parts are brought closer together and aligned to pass the bolt through the hole. A washer is placed on the other side and then the nut is placed. Bolts are often tightened by applying torque to the bolt head or nut, which causes the bolt to stretch. The stretching results in bolt tension or preload, which is the force that holds the joint together. For various applications, a prespecified torque is applied according to the design requirement using a torque wrench and is repeated for each assembly.

References

1. Mazumdar, S.K. and Mallick, P.K., Strength Properties of Adhesive Joints in SMC-SMC and SRIM-SRIM Composites, 9th Annual ACCE Conf., Detroit, MI, April 7–10, 1997, 335.
2. Matthews, F.L., Kilty, P.F., and Godwin, E.W., A review of strength of joints in fiber-reinforced plastics. 2. Adhesively bonded joints, *Composites*, 13(1), January 1982.
3. Sneddon, I.N., The distribution of stress in adhesive joints, *Adhesion*, D.D. Eley, Ed., OUP, 1961, chap. 9.
4. Benson, N.K., Influence of stress distribution on the strength of bonded joints, *Int. Conf. Adhesion: Fundamentals and Practice*, Nottingham University, September 1966.
5. Kutscha, D. and Hofer, K.E., Jr., Feasibility of Joining Advanced Composite Flight Vehicle Structures, Technical Report AFML — TR-68-391, U.S. Air Force, January 1969.
6. Niranjan, V., Bonded Joints — A Review for Engineers, UTIAS Rev. No. 28, University of Toronto, September 1970.
7. Lie, A-T, Linear Elastic and Elastoplastic Stress Analysis for Adhesive Lap Joints, Ph.D. thesis, University of Illinois, 1976.
8. Volkersen, O., Die Nietkraftoerteilung in Zubeanspruchten Nietverbindungen mit konstanten Loschonquerschnitten, *Luftfahrtforschung* 15, 41, 1938.
9. Goland, M. and Reissner, E., Stresses in cemented joints, *J. Appl. Mechanics*, p. A17, March 1944.
10. Pahoja, M.H., Stress Analysis of an Adhesive Lap Joint Subjected to Tension, Shear Force and Bending Moments, T & AM Report No. 361, University of Illinois, 1972.
11. Privics, J., Two dimensional displacement stress distributions in adhesive bonded composite structures, *J. Adhesion*, 6(3), 207, 1974.
12. Srinivas, S. Analysis of Bonded Joints, NASA TN D-7855, April 1975.
13. Renton, W.J. and Vinson, J.R., The efficient design of adhesive bonded joints, *J. Adhesion*, 7, 175, 1975.
14. Nadler, M.A. and Yoshino, S.Y., Adhesive Joint Strength as a Function of Geometry and Material Parameters, Society of Automotive Engineers (SAE), Aeronautic and Space Eng. and Manufacture Meet., Paper 670856, Los Angeles, October 1967.
15. Adams, R.D., Coppendale, J., and Peppiatt, N.A., Failure analysis of aluminum-aluminum bonded joints, *Adhesion 2*, K.W. Allen, Ed., Applied Science Publishers, 19, chap 7.
16. Grant, P.J., Strength and Stress Analysis of Bonded Joints, Report No. SOR (P) 109, British Aerospace, Warton, 1976.
17. Hart-Smith, L.J., Adhesive Bonded Double Lap Joints, NASA CR-112235, January 1973.
18. Hart-Smith, L.J., Adhesive Bonded Single Lap Joints, NASA CR-112236, January 1973.
19. Hart-Smith, L.J., Adhesive Bonded Scarf and Stepped-Lap Joints, NASA CR-112237, January 1973.

20. Hart-Smith, L.J., Analysis and Design of Advanced Composite Bonded Joints, NASA CR-2218, April 1974.
21. Wooley, G.R. and Carver, D.R., Stress concentration factors for bonded lap joints, *J. Aircraft*, p. 817, October 1971.
22. Allred, R.E. and Guess, T.R., Efficiency of double-lapped composite joints in bending, *Composites*, 9(2), 112, 1978.
23. Guess, T.R., Comparison of lap shear test specimens, *J. Testing and Evaluation*, 5(3), 84, 1977.
24. Adams, R.D. and Peppiat, N.A., Stress analysis of adhesive bonded lap joints, *J. Strain Anal.*, 9(3), 185, 1974.
25. Chan, W.W. and Sun, C.R., Interfacial stresses and strength of lap joints, in 21st Conf. Structures, Structural Dynamics and Materials, Seattle, May 1980, AIAA/ASME/ASCE/AHS.

Questions

1. What are the commonly used joining methods in the composites industry?
2. What are the advantages of adhesive bonding over mechanical joints?
3. Why should the critical surface tension of substrate material be higher than that of adhesive material?
4. How does the bond form in adhesive bonding?
5. What type of failure mode is recommended in adhesive bonding as well as in bolted joints?
6. How many types of adhesives are commonly available?
7. How would you select a right adhesive for an application?
8. What are the process parameters in adhesive bonding? Process parameters are those that affect the quality of the bond.
9. Why is there no need of surface preparation in a mechanical joint?
10. If a joint needs to be designed under peel load, which type of joint will you select and why?

10

Machining and Cutting of Composites

10.1 Introduction

Composite materials offer the benefits of part integration and thus minimize the requirement for machining operations. However, machining operations cannot be completely avoided and most of the components have some degree of machining. Machining operations are extensively used in the aerospace industry. In a typical aerospace application, assembly and sub-assembly labor costs account for as much as 50% of the total manufacturing costs of current airframes.[1] A fighter plane has between 250,000 and 400,000 holes and a bomber or transport has between 1,000,000 and 2,000,000 holes; therefore, machining cost has become major production cost factor in aerospace applications.[1] A typical wing on an aircraft may have as many as 5000 holes.[2]

There are several types of machining operations, such as cutting, drilling, routing, trimming, sanding, milling, etc., performed to achieve various objectives. The majority of these machining processes are similar to metal machining. The objectives of these machining operations are discussed below.

10.2 Objectives/Purposes of Machining

Machining of composites is done to fulfill the following objectives.

1. To create holes, slots, and other features that are not possible to obtain during manufacturing of the part. For example, if a pultruded part needs holes and other features as shown in Figure 10.1 then machining of the part is unavoidable.
2. Machining is done to create the desired tolerance in the component. For example, if a filament wound part requires the outside diameter to have a tolerance of 0.002 in., then centerless grinding is done to get that tolerance on the outer surface.

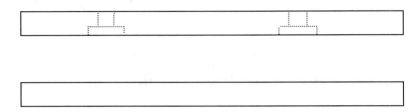

FIGURE 10.1
Machined holes in a pultruded tube.

3. Machining is performed to prepare the surface for bonding, coating, and painting purposes. In general, the outer surface is sanded to remove oils, grease, and release agents.
4. Machining is performed to create smoothness on the desired surface.
5. To make prototype parts from a big blank or sheet of material, a machining operation is performed. This process is very economical. For example, if a designer wants to test glass/nylon short fiber composites for a bushing application, he can machine a composite rod/tube to develop a prototype part instead of making an expensive mold and then injection molding the part. Similarly, test coupons for tensile and bond testing are made from big sheets of materials.

Machining of metals is very common and is easily performed; however, the machining of composites poses several challenges, as listed in Section 10.3.

10.3 Challenges during Machining of Composites

1. Machining of composite parts creates discontinuity in the fiber and thus affects the performance of the part.
2. Machining exposes fibers to chemicals and moisture.
3. The temperature during cutting should not exceed the cure temperature of the resin for thermoset composites to avoid material disintegration. Glass and Kevlar fibers have poor thermal conductivity and such high temperatures may lead to localized heating and degradation. With thermoplastic composites, if the temperature comes close to the melting temperature of the resin, it may clog the tool.
4. It is difficult to attain dimensional accuracy during the cutting of composites because of differences in the coefficients of thermal

expansion (CTE) in the matrix (highly positive CTE) and fiber (slightly negative CTE in carbon and aramid). Drilled holes are often found to be smaller than the drill used.[3]

5. There is heat build-up in the cutting zone due to the low thermal conductivity of the composite. A suitable coolant should be selected to dissipate the heat from the tool and the workpiece. In drilling metal components, chips absorb 75% of total heat, whereas the tool and workpiece absorb 18 and 7%, respectively. In the drilling of carbon/epoxy composites, the tool absorbs half of the heat and remainder is equally absorbed by workpiece and chips.[4]
6. Tool life is usually shorter because of the abrasive nature of the composite. For this reason, high-speed steel tools are coated with tungsten carbide or titanium nitride to increase the life of the tool.
7. Obtaining a smooth cut edge is difficult with composites, especially aramid composites. Aramid fibers are tough and absorb the cutting energy. Fiber kinking or burr surfaces are obtained during cutting of aramid composites.
8. The effect of coolant materials on composites is unknown and therefore any coolant material must be selected judiciously.
9. Machining of composites causes delaminations at the cut edges of continuous composites. The lay-up sequence and fiber orientations have a significant effect on the amount of delamination.

10.4 Failure Mode during Machining of Composites

Composites are predominantly made of glass, carbon, Kevlar, or combinations of these as reinforcing materials. Glass and carbon fibers show brittle fracture, whereas aramid fibers show ductile failure. With aramid fibers, axial splitting of the fiber is common because of weak molecular bonds transverse to the fiber axis, which allows relative motion of molecular chains in sliding planes.[3] This also results into the lower compressive strength of aramid fibers. For this reason, special tools are required for machining of aramid composites. During cutting of aramid composites, aramid fibers absorb a significant amount of energy due to their ductile nature and the cut surface is not smooth. Due to their low compressive strength, aramid fibers have a tendency to recede within the matrix during the machining process instead of being sheared off. Machined edges in aramid composites have burr and fiber kinking.

To obtain high quality edges and to avoid fuzzing in aramid composites, the cutting process must proceed in such a way that the fibers are being preloaded by tensile stress and then cut in a shearing action. For a rotating tool, this means that fibers must be pulled from the outside diameter toward the center. A sharp cutting edge and a comparatively high cutting speed are desirable to avoid receding of fibers into the matrix.[3]

Laminate sequencing, the type of fabric material, and the type of resin materials all have an effect on the quality of cut edges of the composite. Laminates made from angular layers are easier to machine than unidirectional composites, and plain weaves are easier to machine than satin.[3] Denser fabrics give better machined qualities than looser ones. Epoxy resins are favored over phenolic resins in terms of machinability.

10.5 Cutting Tools

Cutting tools similar to those in metal machining are used for composites as well. However, high-speed steel (HSS) tools are coated with tungsten carbide, titanium nitride, or diamond to avoid excessive wear on the tool. HSS tools without any coating performs reasonably well for a few cuts; but after a while, the tool edge becomes dull and the cut quality deteriorates.

In terms of tool life, carbide tools are superior, especially if carbide grades of fine grain size are used. However, tool cost is considerably higher. Polycrystalline diamond (PCD) tools are extensively used for machining glass- and carbon-reinforced composites due to their high wear resistance. PCD tools cost about 10 times more than carbide tools.

Figure 10.2 shows end mills and drills for machining composites and Figure 10.3 shows inserts for turning, boring, and milling operations. These tools are coated with diamond in a chemical vapor deposition (CVD) reactor and the result is called CVDD (CVD Diamond). The CVDD coating is pure diamond with no metallic binder. CVDD-coated tools are less costly than PCD tools. Usually, there is no coolant necessary while using CVDD tools because of the lubricity of the diamond. The ability to cut dry avoids contamination of the workpiece by the coolant. A coolant (such as 5% water-soluble oil) can sometimes be used to improve the surface finish and/or to enhance chip clearing.

Choosing the correct diamond coating thickness is very critical to the success of an application. For example, to machine fiberglass composites (commercially available G10), a 20- to 24-μm-thick coating works very well and it is 65 times better in sliding and 30 times better in side milling as compared to carbide tools.[5] A 10- to 14-μm-thick diamond coating does not work well for machining G10 fiberglass composites. During the slotting test, the carbide tool machined 192 linear inches before three corners were worn away, whereas the CVD tool with a 20-μm-thick coating machined 12,384 linear inches before one corner wore away. A CVD tool with 10-μm-thick coating machined 480 linear inches before one corner wore away.[5]

FIGURE 10.2
End mills and drills for machining composites. (Courtesy of sp^3 Inc., Mountain View, CA.)

FIGURE 10.3
Inserts for turning, boring, and milling operations. (Courtesy of sp^3 Inc., Mountain View, CA.)

In a side cutting operation, the carbide tool machined 2304 linear inches to the end of life. End of life was measured as 0.007 in. of flank wear. A CVD tool with a 20-μm-thick diamond coating machined 12,384 linear inches when the test was halted because all the G10 material had been consumed. When the test was halted, the CVD tool exhibited only 0.001 in. of wear.[5] To run the CVD tool to the end of life would have taken 7 days and several hundred square feet of materials.

Diamond-coated tools are not suitable for machining cast iron and steel alloys because these materials chemically interact with the diamond, causing

TABLE 10.1
Recommended Starting Parameters for CVDD Tools

Workpiece Material	Application	Turning, Facing, and Boring		Milling	
		Cutting Speed (surface ft/min, sfm)	Feed Rate (in. per revolution, ipr)	Cutting Speed (surface ft/min, sfm)	Feed Rate (in. per tooth)
Reinforced plastics/composites	Rough	500–1000	0.008–0.020	2000–4000	0.003–0.010
	Finish	1000–1500	0.003–0.010	2000–4000	0.003–0.010
Unfilled plastics, graphite, carbon	Rough	2000–3000	0.010–0.020	2000–4000	0.003–0.008
	Finish	3000–4000	0.003–0.010	2000–4000	0.003–0.008
Aluminum alloys (less than 16% Si)	Rough	2000–3000	0.010–0.020	2000–4000	0.003–0.008
	Finish	3000–3500	0.003–0.010	2000–4000	0.003–0.008
Aluminum matrix composites	Rough	1000–1500	0.010–0.015	1000–2500	0.002–0.006
	Finish	1500–2000	0.003–0.010	1000–2500	0.002–0.006
Green ceramics	All	100–500	0.002–0.010	500–1000	0.002–0.006

Source: sp^3 Inc., Mountain View, CA.

rapid tool wear. Table 10.1 shows the recommended starting parameters for various machining operations using CVDD tools.

10.6 Types of Machining Operations

Machining operations are performed to achieve various objectives, as previously discussed. The operation involves cutting, drilling, sanding, grinding, milling, and other techniques similar to metal machining. Standard machining equipment similar to metal machining is used with some modifications, mostly in the cutting tool and coolant. In all machining operations, it is important to keep the tool sharp to obtain good-quality cuts and to avoid delaminations. During the machining of composites, the proper backing material is required to avoid delamination. Two of the major machining operations — cutting and drilling — are discussed in the following sections.

10.6.1 Cutting Operation

The cutting operation is performed to get the desired dimensions or to make several parts from one part. For example, a large sheet of FRP is cut into

Machining and Cutting of Composites

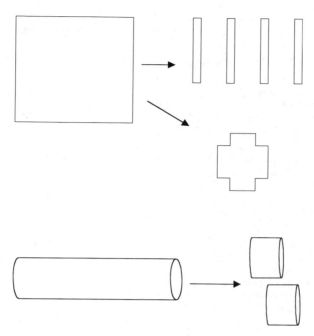

FIGURE 10.4
Conversion of a flat sheet and rod into smaller pieces.

small rectangular strips, or any other shape, as shown in Figure 10.4. Sometimes, the cutting operation is performed to fabricate net-shape parts. Flashes, runners, shear edges, etc. obtained during molding processes are removed by cutting operations. For example, during compression molding of electronic enclosures or automotive parts, shear edges are trimmed using a file while the part is still hot.

Cutting operations are performed using hand-held hacksaws, bend saws, circular saws, abrasive files, routers, and more. The tool is diamond coated for increased wear resistance. During machining, the cutting speed should be as high as the matrix can sustain for better cut quality. The higher cutting speed means lower cut forces perpendicular to the workpiece and feed direction, which consequently reduces the amount of manufacturing-induced damages.[4] Nowadays, waterjet cutting and laser cutting are gaining more importance and is discussed here.

10.6.1.1 Waterjet Cutting

Waterjet cutting is used for machining composites as well as sheet metals made of steel and aluminum. In waterjet cutting, high-velocity water is forced through a small-diameter jet. As the waterjet impinges on the surface, it cuts the material by inducing a localized stress failure and eroding the material. In waterjet cutting, water pressures up to 60,000 psi (414 MPa) are

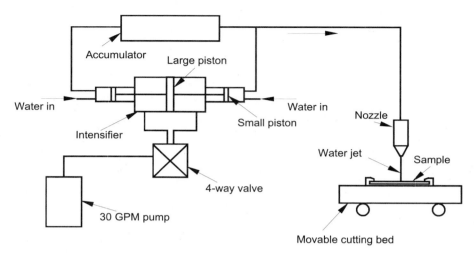

FIGURE 10.5
Schematic of waterjet cutting equipment. (Adapted from Hurlburt, G.H. and Cheung, J.B., Waterjet Cutting of Advanced Composite Materials, SME Technical Paper No. MR77-225, Society of Manufacturing Engineers, 1997.)

used to cut the material. Water speeds of 2600 ft/s (800 m/s) and nozzle diameters on the order of 0.010 in. (0.25 mm) are typical.

For most composite applications, abrasive particles are added with the water to increase the cutting speed and to cut thick composite laminates. A schematic diagram of commercially available waterjet cutting equipment is shown in Figure 10.5. As shown, the water nozzle remains stationary and the sample material travels by a hydraulic cylinder. A 30-gpm (113.5 l/min) hydraulic pump delivers hydraulic oil at up to 3000 psi (21 MPa) to an intensifier via a four-way valve. The intensifier is a differential-area, double-acting piston type in which a large piston is shuttled back and forth by the 3000 psi oil. There are two small pistons attached directly to the large piston. The small pistons have an area 1/20th of the large piston; thus it converts the 3000-psi oil to 60,000 psi water. Compressed water then flows out of the high-pressure cylinders to the nozzle through a pair of check valves.

During waterjet cutting, the process parameters that affect cutting performance include:

- Waterjet pressure
- Cutting speed
- Laminate thickness
- Nozzle orifice diameter (0.2–8 mm)

Hurlburt and Cheung[6] performed waterjet cutting on graphite/epoxy, glass/epoxy, Kevlar/epoxy, boron/epoxy, and hybrids of graphite/epoxy and boron/epoxy, graphite/epoxy and glass/epoxy, and graphite/epoxy and Kevlar/epoxy. These composites were made using bidirectional fabric layers. The purpose of these tests was to determine how well waterjet cutting can be performed on these materials. Test results for the various cutting parameters, such as waterjet pressure (P_o), orifice diameter (d_o), traverse speed (v), and material type and thickness, are shown in Table 10.2. In all these tests, the nozzle stand-off distance was constant at about 0.125 in. (3 mm). Test results showed that smaller orifice diameter produced finer cuts, but a larger orifice diameter was needed to cut thicker materials. It was also noted that a higher nozzle pressure and lower traverse speed gave a better-quality cut. The Kevlar composite gave a smooth edge, whereas a boron composite had the roughest cut quality. Hard and strong boron fibers tend to break rather than cut, leaving a short stubble of fibers protruding from the cut surface.

Figure 10.6 shows examples of waterjet cutting in a 0.125-in. thick glass/epoxy sample. The slots cut in Figure 10.6A and B show the effect of traverse speed. In both the cases, the waterjet pressure was 60,000 psi and the nozzle diameter was 0.010 in. The traverse speed in A was 0.10 ips (2.5 mm/s) and in B, it was 2.0 ips (51 mm/s). In the case of B, some of the water from the jet stagnated in the material and penetrated sideways between the layers of fibers, causing some delamination along the length of the slot. Figure 10.6C shows the exit side of Figure 10.6A. It is clear from Figure 10.6C that the exit side has very little delamination at the speed of 0.10 ips. Figure 10.6D shows the increase in intensity of delamination on the exit side for a traverse speed of 0.25 ips (6.4 mm/s). The nozzle pressure was 55,000 psi and the nozzle diameter was 0.010 in. for Figure 10.6D.

Test results show that these representative composite samples can be cut by waterjet with no airborne dust and a noise level of less than 80 dB. The possibility of water absorption must not be overlooked.

10.6.1.2 Laser Cutting

In laser cutting, a concentrated monochromatic raw light beam is focused on the workpiece into a spot size of 0.1 to 1 mm. The cutting operation takes place by local melting, vaporization, and chemical degradation. Laser operation requires expertise because of the danger of high-voltage radiation exposure and hazardous fumes. The laser beam typically damages the resin in the areas of the cut and may score the workstand. Proper ventilation is required while performing the laser cutting operation.

Cutting of unreinforced thermoplastics and thermosets is much easier than for reinforced composites. Cutting thermoplastics results in local melting, whereas laser cutting of thermosets results in local vaporization and chemical degradation. Once reinforcements are included into the resin, very high temperatures are required to vaporize the fibers. Vaporization temperature for carbon fiber is 3300°C, E-glass fiber is 2300°C, and aramid fiber is 950°C. The high-temperature requirement for cutting reinforced plastic results in local matrix degradation.

TABLE 10.2

Results of Waterjet Cutting Tests

Material Type[a]	Thickness (in.)	P_o (psi)	d_o (in.)	v (ips)	Remarks
Gr/Ep	0.067	55,000	0.008	1.0	Virtually no separation of cloth backing
	0.136	60,000	0.010	0.50	Minor separation of the cloth backing on exit side
	0.273	60,000	0.014	0.11	Good cut, no separation of cloth backing
Boron/Ep	0.058	60,000	0.012	2.0	Poor cut, loose fiber left along cut
	0.136	60,000	0.010	2.0	Poor cut, separation of the surface resin layer, top and bottom
Kevlar/Ep	0.058	55,000	0.006	2.0	Only slight separation of the cloth backing, exit side
	0.125	55,000	0.010	0.50	Good cut
Glass/Ep	0.139	60,000	0.010	0.10	Small amount of separation of the cloth backing, exit side only
Hybrid Gr/Ep and Boron/Ep	0.088	60,000	0.012	0.23	Poor cut, loose fibers left along cut
	0.154	60,000	0.012	2.0	Poor cut, loose fiber and some separation of cloth backing, bottom side
	0.321	60,000	0.014	0.15	Some exit-side separation of the cloth backing
Hybrid Gr/Ep and Kevlar/Ep	0.125	60,000	0.010	0.25	Good cut
	0.250	60,000	0.014	0.08	Very good cut
Hybrid Gr/Ep and Glass/Ep	0.068	55,000	0.012	0.15	Good cut
	0.253	60,000	0.012	0.15	Rather severe delamination of bottom fibers of composite

[a] Gr = graphite; Ep = epoxy.

Source: From Hurlburt, G.H. and Cheung, J.B., Waterjet Cutting of Advanced Composite Materials, SME Technical Paper No. MR77-225, Society of Manufacturing Engineers, 1997.

Laser cutting produces a sharp, clean edge with little discoloration at speeds unrivaled by other cutting methods for prepregs (uncured composites). Continuous-wave 250-W CO_2 lasers have produced cutting speeds up to 300 in/min (127 mm/s) in single-ply uncured boron/epoxy and up to

Machining and Cutting of Composites

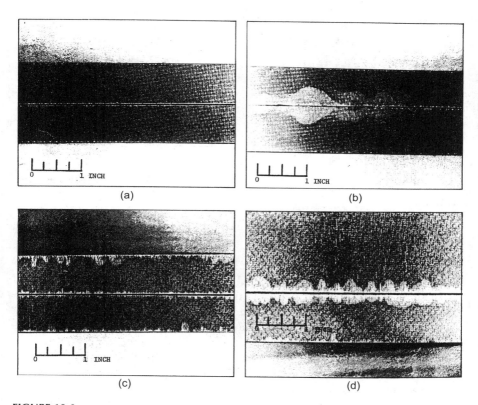

FIGURE 10.6
Waterjet cut quality in 0.125-in. thick glass/epoxy sample. (From Hurlburt, G.H. and Cheung, J.B., Waterjet Cutting of Advanced Composite Materials, SME Technical Paper No. MR77-225. With permission from the Society of Manufacturing Engineers.)

400 in./min (169 mm/s) in single-ply Kevlar prepregs. Cutting of cured composites requires higher laser power and cut edges generally reveal a charring effect. Charring can be reduced in thinner materials by increasing the speed. A 1-kW laser can cut up to 0.2 in. (5 mm) thick Kevlar/epoxy laminate and results in significant charring if 0.3-in. (7.6 mm) thick laminate is cut.

10.6.2 Drilling Operation

The drilling operation is performed to create holes in a component. Holes are created either for fastening purposes, such as riveting or bolting, or for creating special features, such as a passage for liquid injection or wire connection.

Drilling is performed similar to metal drilling but the tool used is usually a tungsten carbide tool because of the abrasive nature of composites. In metal drilling, drill tips are designed for metal-working, the tip heating the metal

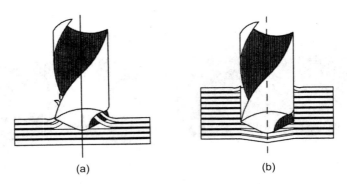

FIGURE 10.7
Schematics of delaminations caused by drilling (a) upon entry, and (b) upon exit. (*Source:* Reprinted from Persson et al., *Composites*, 28A, 141, 1997. With permission from Elsevier Science Ltd., U.K.)

to provide the plastic flow needed for efficient cutting. In composites, heat generation is kept low to avoid local matrix degradation and/or to avoid tool clogging. The chip formation in metal drilling is long; whereas in composites, chips are dry and small, and can be easily removed. If the drilling speed is high, then local heat generation makes the resin sticky and produces a lumpy chip.

Drilling creates delaminations in laminated composites, as shown in Figures 10.7 and 10.8. When the drill bit first enters the laminate, it peels up the uppermost laminae (Figure 10.7a); and when it leaves the laminate, it acts as a punch, causing delaminations on the other side of the laminate as shown in Figure 10.7b.[7] Delamination on the other side can be minimized by supporting the laminate at the back side using a plastic or wooden support, and also by lowering the feed rate at the time of exit. A more pointed drill bit tip also lowers the amount of delamination, as shown in Figure 10.8, because a pointed tip creates gradual penetration. In drilling composites, a negative or neutral rake angle in the tool is avoided. A neutral rake angle tends to push the reinforcing fibers out in front, requiring a great deal of pressure to penetrate the workpiece. This pressure causes the fibers to bend, resulting in undersized and furry holes. Moreover, this pressure produces excessive heat, which causes galling and clogging of the tool. A positive rake angle is preferred when designing the tool geometry for composites drilling. With a positive rake angle, reinforcing fibers are pulled into the workpiece and sheared or broken between the cutting edge and the uncut material. Positive rake removes more material per unit of time and per unit of pressure than negative rake, but the more positive rake at cutting edge makes the tool sensitive and fragile.

Fiber orientation and lay-up sequence affect the extent of delamination during drilling. Angle-ply laminates provide better-machined surfaces than unidirectional laminates. In unidirectional composites, fibers tend to pull out

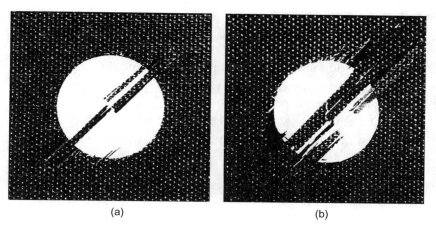

FIGURE 10.8
Photographs of delaminations on the exit side of a carbon/epoxy laminate caused by drilling with (a) pointed drill and (b) blunt drill. (*Source:* Reprinted from Persson et al., *Composites*, 28A, 141, 1997. With permission from Elsevier Science Ltd., U.K.)

of matrix when the local motion between tool and workpiece is parallel to the fibers (0°). The best surface quality is obtained when fibers are sheared off at a right angle (90°), while the worst surface is obtained when fibers are compressed and bent, which occurs at intermediate angles (20° to 45°).[4]

References

1. Micillo, C. and Huber, J., Innovative manufacturing for automated drilling operations, *Proc. Autofact West*, 2, 253, 1980.
2. Kline, G.M., Ultrasonic drilling of boron fiber composites, *Modern Plastics*, 52, 88, June 1974.
3. Konig, W., Grass, P., Heintze, A., Okcu, F., and Schmitz-Juster, Cl., New developments in drilling and contouring composites containing Kevlar aramid fiber, *Technical Symposium V, Design and Use of Kevlar Aramid fiber in Composite Structures*, 1984, 95–103.
4. Abrate, S. and Walton, D.A., Machining of composite materials. I. Traditional methods, *Composites Manufact.*, 3, 75, 1992.
5. Correspondence between author Sanjay Mazumdar and president Jim Herlinger of sp^3 Inc. on June 11, 2001. sp^3 Inc. manufactures CVD tools. sp^3 Inc. conducted a series of tests on composite materials.
6. Hurlburt, G.H. and Cheung, J.B., Waterjet Cutting of Advanced Composite Materials, SME Technical Paper No. MR77-225, Society of Manufacturing Engineers, 1997.
7. Persson, E., Eriksson, I., and Zackrisson, L., Effects of hole machining defects on strength and fatigue life of composite laminates, *Composites*, 28A, 141, 1997.

Questions

1. Why should machining of composites be avoided?
2. Why are cutting tools for composites machining coated with diamond or some other substance?
3. What are the process parameters in waterjet cutting?
4. Write down the differences between metal cutting and composites cutting.

11
Cost Estimation

11.1 Introduction

Cost estimating is an essential element in running a successful business. A design that can be manufactured at low cost will have a significant impact on the market. Success in business decisions relies on how well the cost estimating is performed.

Many times, design changes are made or new materials are used to lower the cost of a product. Many product development activities focus on how to reduce the cost of manufacturing or the overall cost of a product. In this globally competitive market, management decisions revolve around the cost of the product. Therefore, it is important to have a good understanding of the factors that affect product cost.

Cost estimating should not be left to accountants or salespeople. The manufacturing engineer should play a key role in determining the cost of a new product or an existing product. Any design changes should be discussed with the manufacturing engineer as to the feasibility of making it or achieving it at low cost.

The cost estimating job becomes vital in the area of composite materials because the composite product must compete with well-developed metal technologies. A careful consideration must be given to investigate all the possible ways to lower cost. Composite products are often not selected for an application because they are not cost-competitive. In this competitive market, where the entire world represents a single market, product cost has become a crucial factor in deciding the success or failure of a company. In the automotive industry, where the market is very cost-sensitive, product cost plays a vital role in selecting a technology. Typically, structural and nonstructural product costs range from $2 to $20/lb in the automotive field and $100 to $1000/lb in the aerospace market. Newly developing countries such as Taiwan, Korea, India, China, and Mexico, where the labor cost is low, are competing with well-established nations by acquiring technologies. Today, tooling and composite products such as golf shafts, fishing rods, tennis rackets, etc. are cost effectively made in these countries and marketed to the United States and other established countries. For labor-intensive manufacturing processes, manufacturing engineers need to consider the right country and region to produce low-cost parts. All possible ways of keeping the cost down need to be explored. The effects of automation and new technology on the fabrication process must be investigated to lower the cost. The relationship between the production volume rate and product cost for various manufacturing processes needs to be analyzed for the right selection of manufacturing process.

11.2 The Need for Cost Estimating

Cost estimating is performed to predict the cost of a product at some future point in time. This prediction can be close to the actual cost if it is prepared by an experienced professional, taking all the factors into account. Cost estimating is done for the following reasons:

- *To provide a quotation or to establish a selling price.*
- *To perform a make-or-buy decision.* Often, it is more profitable to buy a product from outside sources than to make it in-house.
- *To compare design alternatives.* Various design changes are made in the product to determine which design is more cost-effective.
- *To compare costs of materials, labor, tooling, equipment, etc. between various alternatives.* As shown in Figure 11.1, approach 2 has a higher production cost than that of either approach 1 or 3.
- *To examine the effect of changes in the production scenario.* Figure 11.2 shows the effect of production volume change on the cost of a part; it is shown that for a volume less than 120,000 parts/yr, manufacturing process 3 is more suitable, whereas for a volume above 120,000 parts/yr, manufacturing process 1 is more cost-effective. This information is used to select a cost-effective fabrication technology to meet production needs. In automotive industry, SMC (sheet molding compound) is selected when the annual volume requirement is less than 150,000 parts/year. For greater than 150,000 parts/year, steel stamping is used. This is because the initial tooling cost is less in the SMC process than in the stamping process. In the steel stamping process, to get a final shape, three to eight dies are needed, whereas in SMC, one mold is enough to transform the raw material to final shape.
- *To get an insight into the profits of launching a new product or design.*
- *To determine whether capital investments need to be committed,* based on early cost projections for advanced manufacturing processes such as braiding, RTM, automatic tow placement, and filament winding.

Cost Estimation

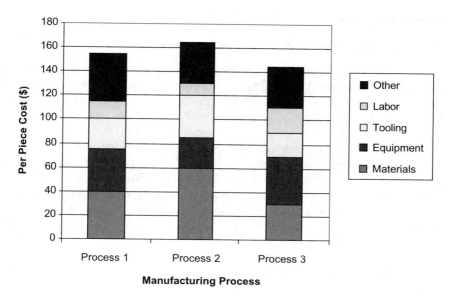

FIGURE 11.1
Product cost breakdown comparison for a fixed annual volume.

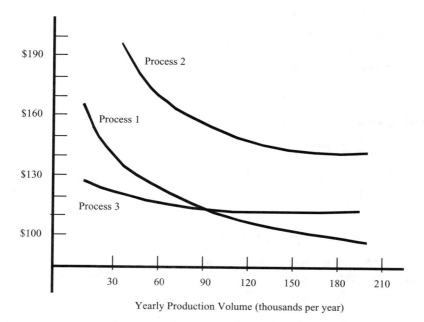

FIGURE 11.2
Effect of production volume on the price of each component.

11.3 Cost Estimating Requirements

To forecast the cost of a product, the engineer should have some basic product information. Well-sketched product drawings, detailing critical product definitions, need to be made available and should include product design and specifications. Inadequate product and material specifications may cause underestimation or over-estimation of the cost of the product. For example, if a composite shaft needs to be designed for an application in which the shaft moves back and forth in two end bushings, then surface finish and roundness requirement need to be specified. If there is any wear-resistant coating requirement for the surface, it must be noted on the design. In addition to these, fiber and matrix materials, fiber orientation, fabrication method, etc. need to be specified on the drawing. For a good estimate, the following list of parameters that influence the cost of a product is required:

- Detailed product drawing
- Product and material specifications
- Fabrication methods and sequence of manufacturing operations
- Equipment and accessory requirements
- Tooling requirements
- Equipment depreciation and tool wear knowledge
- Process cycle time and processing requirements
- Production rate requirements
- Packaging requirements
- Delivery schedules

11.4 Types of Cost

Product cost can be divided into two major categories: nonrecurring costs and recurring costs.

11.4.1 Nonrecurring (Fixed) Costs

The costs that are charged only one time are called nonrecurring costs. These include the initial cost of tooling, equipment, facility development, and engineering development. Nonrecurring costs are basically start-up costs and are independent of the production rate and therefore are also called fixed costs. These costs are included in the product cost as a fixed capital cost or depreciation cost.

Example 11.1

Filament winding equipment worth $300,000 depreciates 10% per year. This equipment is used to make 100,000 driveshafts per year. Determine the depreciation cost of the equipment per driveshaft.

SOLUTION:

Based on the given information, the depreciation cost of the equipment per year will be = 0.1 × $300,000 = $30,000. Therefore, the depreciation cost per driveshaft will be = $30,000/100,000 = $0.30 = 30 cents. This 30 cents per driveshaft will be included in the calculation of product cost. Similarly, other costs are determined and added into the product cost.

A detailed list of fixed costs include:

- *Overhead (burden) costs*: these costs are not directly related to product fabrication but include engineering and technical services, nontechnical (secretary, security) services, and office supplies.
- *Investment costs*: these costs include depreciation on capital investment such as tooling, equipment, and assembly line, and interest, tax, and insurance on the property and investment.
- *Executive and management costs.*
- *Selling expenses.*

11.4.2 Recurring (Variable) Costs

Recurring costs occur on a day-to-day basis and are also called operating costs. The costs are directly related to the fabrication process. Recurring costs depend on the number of parts produced and therefore are also called variable costs. Variable costs include:

- *Direct material costs*: the materials we see in the product are direct materials, and costs associated with it are direct material costs, including the cost of fiber, resin, insert, coatings, etc.
- *Indirect material costs*: materials such as release agent and vacuum bagging materials are used to support the manufacturing process and are called indirect materials. The costs associated with such materials are called indirect material costs.
- *Direct labor costs*: the labor costs associated with manufacturing, machining, assembly, and inspection are included in this category. For a hand lay-up process, direct labor would involve cutting the prepreg, cleaning the tool, applying the mold release, laying the prepreg on the tool, inspecting the lay-up sequence, preparing the vacuum bagging, putting into autoclave, operating the autoclave, removing the material and tool, etc.

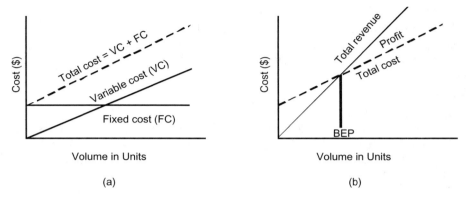

FIGURE 11.3
Break-even curve showing the relationship between fixed and variable costs: (a) fixed, variable, and total costs; (b) revenue and profit.

- *Indirect labor costs*: the costs associated with supervision and process trouble-shooting come in this category
- Cost of running equipment, power supplies, and maintenance
- *Packaging and shipping costs*
- *Any royalty or license payments*
- *Scrap handling and loss*

These fixed and variable costs are included in the product cost and are plotted for a specific product as shown in Figure 11.3. The total cost of the product will be the addition of fixed cost and the variable costs, and depends on the volume produced. To obtain a profit, total sales volume should be greater than the break-even point. There are many things, including maximization of machine time utilization and proper allocation of machine and manpower, involved in minimizing product cost or shifting the break-even point to the left. These are discussed in Section 11.8.

11.5 Cost Estimating Techniques

There is no standard technique available for evaluating product costs. Moreover, due to the lack of a database and the proprietary nature of cost information, not much work has been done in this field. However, some work has been done in estimating the cost of production for aerospace parts. Northrop Corporation has developed the *Advanced Composites Cost Estimating Manual* (*ACCEM*) for the hand lay-up process.[1] MIT researchers have worked on developing a theoretical model for calculating the processing time for each sub-process and then adding them up to get the total time for fabrication of advanced composites.[2] The model is called the "First-Order

Cost Estimation

TABLE 11.1

Calculation of Cost per Housing

Factors	Material	Labor	Tooling	Equipment	Overhead	Total
Glass-filled nylon	$2.00					$2.00
Aluminum insert	$0.10					$0.10
Putting insert and core in the mold		$0.20			$0.60	$0.80
Running the machine		$0.30			$0.90	$1.20
Tooling cost per part			$0.14			$0.14
Machine cost per part				$0.70		$0.70
Total	$2.10	$0.50	$0.14	$0.70	$1.50	$4.94
% of Total	42.5	10	2.8	14.1	30.3	

Model." There are other methods, including the industrial engineering approach, technical cost model, and analogy approach, available for cost estimation. These methods are described in the following sections.

11.5.1 Industrial Engineering Approach (Methods Engineering)

In the industrial engineering approach, all the elements of work are listed in great detail, and the cost associated with each work item is determined and then summed to obtain the total cost. This method is widely used for the fabrication of metal components, as well as for composite fabrication.

A simplified example of this method is illustrated for the fabrication of a composite housing using an injection molding process. The material selected is 45 wt% long glass-filled nylon ($2.50/lb); the equipment consists of a 2000-ton (clamping force) injection molding machine ($700,000); the capacity of which is 100,000 parts per year; and the tool is a one-cavity steel tool ($70,000). The cost of each housing is calculated as per Table 11.1

These calculations are made assuming a tool life of 5 years and a machine life of 10 years. It is assumed that the machine is running at full capacity. The housing uses about 363 g (4/5 lb) of glass-filled nylon. The equivalent aluminum housing made by die casting requires several machining operations, such as turning, threading, and drilling. Injection molding provides net-shape components and eliminates machining operations.

For complex products in which several parts are joined and assembled, more rigorous work and computations are needed. Boothroyd Dewhurst Inc.[3] has developed software that determines optimum assembly and processing sequence for the economic production of a product.

11.5.2 ACCEM Cost Model

The ACCEM (Advanced Composites Cost Estimating Manual) cost model was developed by Northrop Corporation for the U.S. Air Force and published in 1976.[1] This model is based on a primitive cost model, for which a manufacturing process is viewed as a collection of primitive steps that, when repeated again and again, gives the final part or a feature in the part. This model can be utilized for some composite manufacturing processes where processing is of an additive nature and machining operations where it is of a subtractive nature. This approach is well-developed for machining, mechanical assembly, hand lay-up, and automatic tow placement.[1,2] In hand lay-up or tape laying processes, a single prepreg sheet is laid over and over again to get the final part. In manual and automated tape laying, the primitive step is the laying of a single prepreg tape strip or the laying of a fabric cloth. In a machining operation, a primitive step is a single cutting pass. For a well-defined manufacturing process, these primitive steps are written in power-law form, or in tabular form, to obtain the total time for an operation. This method requires empirical data to derive equations, and therefore the accuracy of the estimate will depend on the quality of data. When sufficient data is available, the time for each sub-process step is plotted against some design parameter such as length, area, volume, or weight on log-log paper. A best-fit curve is plotted from the data, and a power law relationship of the form of Equation (11.1) is developed between the variables.

$$t = Ay^B \qquad (11.1)$$

where A and B are constants based on the best-fit curve.

Figure 11.4 shows the total time (hours) required as a function of tape length for laying a 3- and 12-in.-wide tape during a hand lay-up process. Power law equations are also shown for 3- and 12-in. wide tape in Figure 11.4. The data shown is for a manual operation, but a similar relationship exists for several machining and equipment operations. ACCEM is in the public domain and provides similar equations for several advanced composite operations. Table 11.2 gives the calculation made by the ACCEM model to fabricate a 30-ft (9.15-m) "J" stringer. It can be seen that the model breaks the entire process of manufacturing into several step-by-step tasks. For each task, the time is calculated to complete the sub-process. Once the time is known for each sub-process, it is multiplied by a cost factor to get the cost.

11.5.3 First-Order Model

MIT researchers have developed this theoretical model for estimating the processing time for the fabrication of a composite part. They found, after reviewing considerable data for composites fabrication processes as well as for machining operations, that manufacturing operations (human or machines) follow a first-order velocity response and thus can be modeled as having first-order dynamics.[2] According to this method, the total task of

Cost Estimation

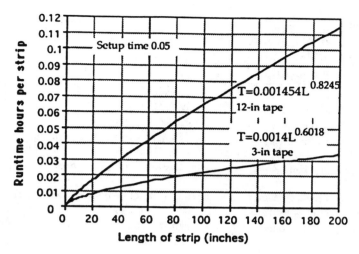

FIGURE 11.4
Lay-up hours as a function of strip length for hand lay-up of 3-in. (76.2-mm) and 12-in. (305-mm) wide tape. (From Northrop Corporation[1] and Gutowski et al.[2])

fabrication is divided into several sub-process steps. Each subprocess step is then modeled as the following:

$$v = v_0\left(1 - e^{-t/\tau}\right) \quad (11.2)$$

where τ is the dynamic time constant with the unit of time; and v_0 is the steady-state velocity or rate and has the dimension (y/t), where y is the appropriate extensive variable such as length, area, or weight for the task. Integration of Equation (11.2) leads to:

$$y = v_0\left[t - \tau\left(1 - e^{-t/\tau}\right)\right] \quad (11.3)$$

The advantage of the relationship between the extensive variable y and time t is that it utilizes two physical-based parameters (v_0 and τ) to define a sub-process or primitive task; τ represents the delay in attaining full speed and is related to the physical process or complexity of the task. In the ACCEM model, constants A and B (Equation (11.1)) do not have any physical significance.

This first-order model was used to represent various manufacturing operations and found to fit very well with the experimental data. Figures 11.5 through 11.8 compare first-order model with the ACCEM model for hand lay-up, automatic tape lay-up, debulking and lay-up for woven material. The fit in the published trend and first-order model is found to be remarkable. The first-order model is also found to fit very well with abrasion time estimates, as shown in Figure 11.9. The values of constants τ and v_0 are

TABLE 11.2
Steps Required to Make a 30-ft. (9.15-m) "J" Stringer with Associated Time Estimation Standards

No.	Procedure	Time
1	Clean lay-up tool surface	$= 0.000006 \times \text{area}$
2	Apply release agent to surface	$= 0.000009 \times \text{area}$
3	Position template and tape down	$= 0.000107 \times \text{area}^{0.77006}$
4	12 in. (305 mm) manual ply deposition	$= \text{IF}(\text{manual} = 1, 0.05 + \text{plies} \times (0.001454 \times \text{length}^{0.8245}), 0)$
5	12 in. (305 mm) hand-assist deposition	$= \text{IF}(\text{assist} = 1, 0.1 + \text{plies} \times (0.001585 \times \text{length}^{0.558}), 0)$
6	Tape layer (720 in. (18.3 m) min^{-1})	$= \text{IF}(\text{tape} = 1, 0.15 + \text{plies} \times (0.00063 \times \text{length}^{0.4942}), 0)$
7	Transfer from plate to stack	$= \text{plies} \times (0.000145 \times \text{area}^{0.6711})$
8	Transfer from stack to tool	$= 0.000145 \times \text{area}^{0.6711}$
9	Clean curing tool	$= 0.000006 \times \text{area}$
10	Apply release agent to curing tool	$= 0.000009 \times \text{area}$
11	Transfer lay-up to curing tool	$= 0.000145 \times \text{area}^{0.6711}$
12	Debulking (disposable bag)	$= 0.02 + 0.00175 \times \text{area}^{0.6911}$
13	Sharp male bend	$= \text{plies} \times (0.00007 \times \text{length}) \times \text{males}$
14	Sharp female bend	$= \text{plies} \times (0.00016 \times \text{length}) \times \text{females}$
15	Male radial	$= \text{IF}(\text{radius} > 2.0, \text{IF}(\text{radius} = 0, 0, \text{plies} \times (0.00007 \times \text{length}) \times \text{males}))$
16	Female radial	$= \text{IF}(\text{radius} = 0, 0, \text{IF}(\text{radius} > 2, \text{plies} \times (\text{length} \times 0.00047 \times \text{radius}^{-1.3585}) \times \text{females}, \text{plies} \times (0.00016 \times \text{length}) \times \text{females}))$
17	Stretch flange	$= \text{IF}(\text{flange} = 0, 0, \text{plies} \times (\text{length} \times 0.015 \times \text{radius}^{-0.5532} \times \text{flange}^{0.7456}))$
18	Shrink flange	$= \text{IF}(\text{flange} = 0, 0, \text{plies} \times (\text{length} \times 0.064 \times \text{radius}^{-0.5379} \times \text{flange}^{0.5178}))$
19	Set up	$= 0.07$
20	Gather details, prefit, disassemble, clean	$= 0.001326 \times \text{area}^{0.5252}$
21	Apply adhesive	$= 0.000055 \times \text{length}$
22	Assemble detail parts	$= 0.000145 \times \text{area}^{0.6711}$
23	Trim part	$= 0.00011 \times \text{perimeter}$
24	Apply porous separator film	$= 0.000009 \times (1.5 \times \text{area})$
25	Apply bleeder plies	$= \text{plies} \times (0.00002 \times \text{area})$
26	Apply non-porous separator film	$= 0.000009 \times (1.5 \times \text{area})$
27	Apply vent cloth	$= 0.00002 \times (1.5 \times \text{area})$
28	Install vacuum fittings	$= \text{IF}(\text{area} > 288, 0.0062 \times 2, 0.0062)$
29	Install thermocouples	$= \text{IF}(\text{area} > 288, 0.0162 \times 2, 0.0162)$
30	Apply seal strips	$= 0.00016 \times \text{perimeter}$
31	Apply disposable bag	$= 0.000006 \times (1.5 \times \text{area})$
32	Seal edges	$= 0.00054 \times \text{perimeter}$
33	Connect vacuum lines, apply vacuum	$= 0.0061$

TABLE 11.2 (continued)
Steps Required to Make a 30-ft. (9.15-m) "J" Stringer with Associated Time Estimation Standards

No.	Procedure	Time
34	Smooth down	$= 0.000006 \times (1.5 \times \text{area})$
35	Check seals	$= 0.000017 \times \text{perimeter}$
36	Disconnect vacuum lines	$= 0.0031$
37	Check autoclave interior	$= 0.03$
38	Load lay-up in tray	$= 0.000145 \times \text{area}^{0.6711}$
39	Roll tray in	$= 0.025$
40	Connect thermocouple	$= \text{IF}(\text{area} > 288, 0.0092 \times 2, 0.0092)$
41	Connect vacuum lines, apply vacuum	$= \text{IF}(\text{area} > 288, 0.0061 \times 2, 0.0061)$
42	Check bag, seal and fittings	$= 0.000006 \times (1.5 \times \text{area}) + 0.00027 \times \text{perimeter} + \text{IF}(\text{area} > 288, 0.0088 \times 2, 0.0088)$
43	Close autoclave	$= 0.0192$
44	Set recorders	$= 0.056$
45	Start cure cycle	—
46	Cycle check	$= 0.008$
47	Shut down	$= 0.00332$
48	Remove charts	$= 0.00332$
49	Open autoclave door	$= 0.0192$
50	Disconnect thermocouple leads	$= \text{IF}(\text{area} > 288, 0.0035 \times 2, 0.0035)$
51	Disconnect vacuum lines	$= \text{IF}(\text{area} > 288, 0.0031 \times 2, 0.0031)$
52	Roll tray out of autoclave	$= 0.012$
53	Remove lay-up from tray	$= 0.000145 \times \text{area}^{0.6711}$
54	Remove disposable bags	$= 0.000008 \times (1.5 \times \text{area})$
55	Remove thermocouples	$= \text{IF}(\text{area} > 288, 0.0095 \times 2, 0.0095)$
56	Remove vacuum fittings	$= \text{IF}(\text{area} > 288, 0.0029 \times 2, 0.0029)$
57	Remove vent cloth	$= 0.00007 \times (1.5 \times \text{area})$
58	Remove non-porous separator film	$= 0.00007 \times (1.5 \times \text{area})$
59	Remove bleeder plies	$= 0.00007 \times (1.5 \times \text{area})$
60	Remove porous separator film	$= 0.00007 \times (1.5 \times \text{area})$
61	Put used material aside	$= 0.00005 \times (1.5 \times \text{area})$
62	Remove lay-up	$= 0.00006 \times (1.5 \times \text{area})$
63	Clean tool	$= 0.00006 \times (1.5 \times \text{area})$
	Total time (h)	TOTAL

Source: See Northrop Corporation.[1]

obtained by fitting curves to the published data, as shown in Figures 11.5 to 11.9, and are plotted in Figures 11.10 to 11.12 for various manufacturing operations.

Equation (11.3) can be simplified to represent time t in terms of the extensive parameter y. Two simple approximations are as follows:

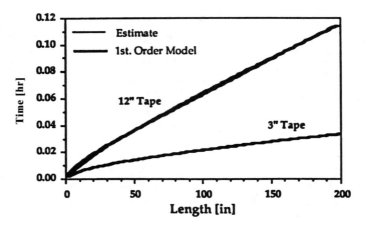

FIGURE 11.5
Comparison of first-order model and ACCEM for hand lay-up. (From Gutowski et al.[2])

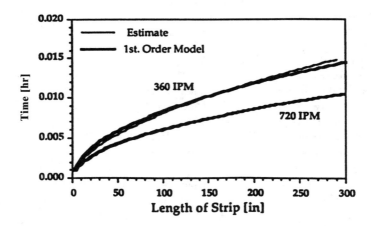

FIGURE 11.6
Comparison of first-order model and ACCEM for automatic tape lay-up. (From Gutowski et al.[2])

When $t < \tau$:

$$t = (2\tau y/v_0)^{1/2} \qquad (11.4)$$

When $t > \tau$:

$$t = \tau + (y/v_0) \qquad (11.5)$$

The linear approximation in Equation (11.5) is good for long times and makes conservative errors for short times. Equation (11.5) can be summed for the various operations to fabricate a composite part. For example, if n strips of

Cost Estimation

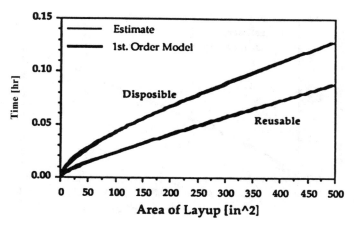

FIGURE 11.7
Comparison of first-order model and ACCEM for debulking. (From Gutowski et al.[2])

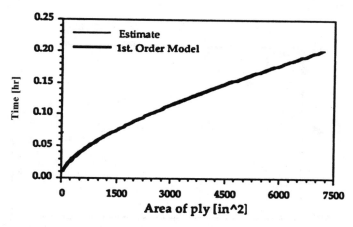

FIGURE 11.8
Comparison of first-order model and ACCEM for lay-up of woven material. (From Gutowski et al.[2])

width w are required to make a lamina or layer of area A, then the linear sum in the plane gives:

$$\sum_{1}^{n} t = n\tau + \frac{A}{wv_0} \quad (11.6)$$

Now, summing through N individual layers of thickness h yields:

$$t_i \sum_{1}^{N} \sum_{1}^{n} t = nN\tau + \frac{V}{whv_0} \quad (11.7)$$

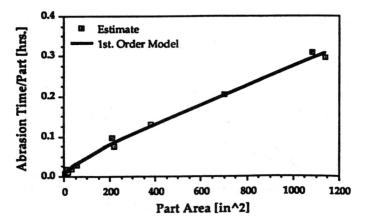

FIGURE 11.9
Verification of first-order model through industry estimates for abrasion operations. (From Gutowski et al.[2])

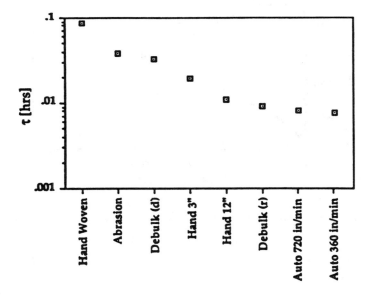

FIGURE 11.10
τ for various lay-up techniques (* indicates CONRAC automatic tape laying machine). (From Gutowski et al.[2])

where t_i represents the time required for ith step to fabricate the part. Other steps could be vacuum bagging preparation, debulking, cutting of plies, and other similar operations.

Equations (11.4) through (11.7) developed for first-order model are applicable for making simple shapes such as flat plates. For complex shapes, when

Cost Estimation

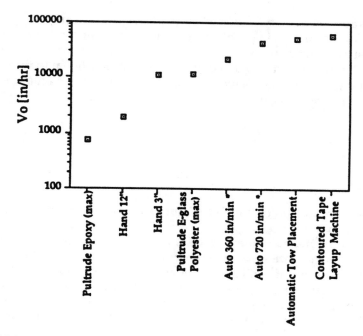

FIGURE 11.11
v_o for various processing techniques (* indicates CONRAC automatic tape laying machine). (From Gutowski et al.[2])

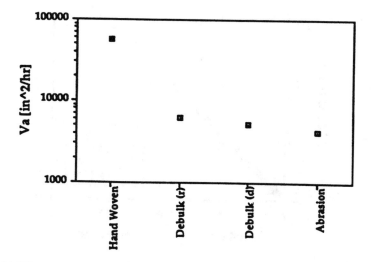

FIGURE 11.12
Steady-state velocities (v_a) for various operations. Subscript a denotes area as the extensive parameter. (From Gutowski et al.[2])

the operation is done on a curved surface, the complex path will have the effect of either increasing the dynamic time constant τ or decreasing the steady-state velocity v_0, or both. Using information theory, one can show that the information stored in a curved fiber is proportional to the enclosed angle of the fiber θ.[2] This shape complexity can be represented in the model using information theory as:

$$\tau = \tau_0 + b\theta \tag{11.8}$$

$$1/v = 1/v_0 + c\theta \tag{11.9}$$

The linear correlation of θ with the fabrication time in terms of τ and v_0 is addressed in Reference 3.

The time estimate outlined above is for simple and complex shapes and represents an individual step in a process. In actual fabrication, the total time consists of a series of similar operations, such as tool cleaning, application of release agent, bagging operation, and autoclave operation. This operational time can be summed to obtain the total time. Figures 11.13 and 11.14 show the comparison of total time using ACCEM and the first-order model.

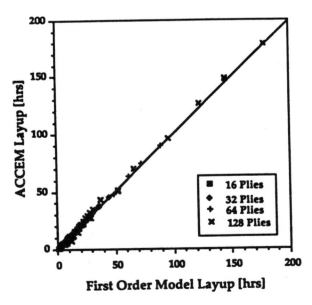

FIGURE 11.13
Comparisons of lay-up times estimated by the ACCEM model and first-order model. (From Gutowski et al.[2])

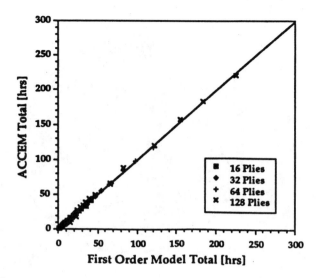

FIGURE 11.14
Comparison of total times estimated by the ACCEM model and the first-order model. (From Gutowski et al.[2])

11.5.4 Cost Estimating by Analogy

According to this method, future costs of a product or project are determined by past experience on a similar product or project. The past cost is used as a baseline and then allowances for cost escalation and size differences are added to the past cost. For example, a filament winding company can use its past cost estimating experience of making 2-in. diameter shafts to predict the cost of 4-in. diameter shafts of equal weight. For this, old records on cost data are required to make predictions for similar new products.

11.6 Cost Analysis for Composite Manufacturing Processes

The cost of manufacturing a product is different for different manufacturing processes. For example, a tube can be manufactured by filament winding, pultrusion, roll wrapping, autoclave processing, or resin transfer molding. The cost of manufacturing a tube is different for these various manufacturing processes. For small-volume production, roll wrapping is cheaper; whereas for larger-volume production, pultrusion is cheaper. The costs of labor, material, tooling, and equipment are different for these manufacturing processes and strongly depend on the production volume. This section analyzes the various manufacturing processes based on percentage amounts of material,

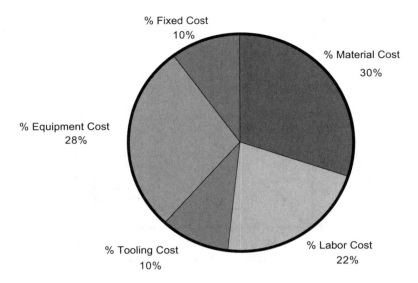

FIGURE 11.15
Percentage production costs for a typical manufacturing process at a production volume of 1000 parts per year.

labor, tooling, equipment, and fixed costs. For a typical production method, these percentages can be plotted as shown in Figure 11.15.

The product cost strongly depends on the production volume rate. Therefore, it is important to know the effect of production volume on product cost need. Figure 11.16 shows the effect of production volume on the above cost drivers.

In the following sections, various manufacturing processes will be compared based on materials, labor, tooling, equipment, and other cost factors.

11.6.1 Hand Lay-up Technique for Aerospace Parts

Hand lay-up is very common in aerospace industries. It is a highly labor-intensive process. This process is comprised of four main operations: prepreg cutting, prepreg laying on tool, bagging preparation, and autoclave curing. The material used in this process is commonly a graphite/epoxy (Gr/Ep) prepreg system. Various cost factors for making aerospace parts using the hand lay-up process are listed below.

- Part cost = $100 to $1000/lb
- Material cost (Gr/Ep prepreg) = $15 to $100/lb
- Labor rate = $15 to $25/hr, excluding fringe benefits
- Labor hour/lb = 1 to 250 hr
- Overhead = 2 to 4 times the labor cost

Cost Estimation

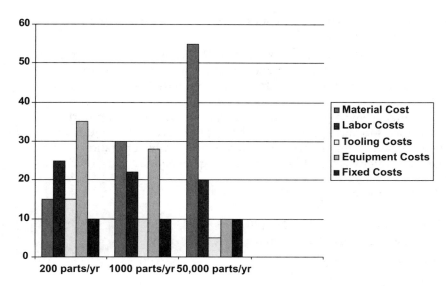

FIGURE 11.16
Percentage material, labor, tooling, equipment, and fixed costs for a manufacturing process for various production volumes.

- Equipment cost for autoclave = $500,000 to $1,000,000
- Autoclave production rate = 4 to 6 hr/part
- Tooling cost = $500 to $1000/ft^2

Although part costs for composites used in aerospace structures are expensive, it may be worth using composites in terms of life-cycle cost. It provides increased fuel economy with increased payload and reduced maintenance costs due to improved fatigue and corrosion resistance. There is good potential for lowering these costs to make composite structures more competitive.

MIT researchers[4] conducted a cost study survey of more than 50 parts fabricated by three major aerospace firms. All parts were hand lay-up of prepreg material, followed by autoclave cure. The parts weights ranged from 0.1 to 300 lb and included various types of constructions, from simple skins and channels to more complex parts made by co-curing mechanical fasteners. Full-depth honeycomb and panelized/ribbed construction were also included. The parts investigated were well down the learning curve, exceeding 100 or 200 units. The following conclusions can be drawn from their study.

1. Despite very high direct materials cost, labor costs were the highest cost source. Labor costs can be divided into hand lay-up of prepreg, bagging preparation, assembly of inserts and fasteners, etc. For simple structures, hand lay-up contributed the maximum cost. For

assembled parts, the cost of assembly was often equal to all other labor costs.

2. The buy-to-fly ratio was typically 2 because of the high scrap rate. Direct materials costs (cost of prepreg and inserts) are much higher than the cost of materials used to make the part.

Cost breakdowns for two hypothetical but typical Gr/Ep parts are listed below for 10-lb and 100-lb parts.[4] The smaller part (10 lb) requires no assembly, whereas the larger part (100 lb) requires assembly operation for fasteners. The labor rate was $22/h + $9/h for fringe benefits.

For the 10-lb part:

Material cost = 10 lb × 2 × $40/lb = $800.00
Labor cost = 10 lb × 3 h/lb × $31/h = $930.00
Overhead = Labor × 2 = $1860.00
Total = $3590.00

For the 100-lb. part:

Material cost = 100 lb × 2 × $40/lb = $8000.00
Labor cost = 100 lb × 4 h/lb × $31/h = $12,400.00
Overhead = Labor × 2 = $24,800.00
Total = $45,200.00

For parts that do not require assembly, labor costs decrease as the weight of the part increases, as shown in Figure 11.17. The data in the figure represents labor hour per pound of 55 parts which are in production at three major aerospace companies. A log-log curve of the form

$$Y = aX + b \tag{11.10}$$

was fitted with a very low correlation coefficient R^2 of only 0.320. Here, $Y = \log_{10} y$, and $X = \log_{10} x$. y and x are labor hours per pound and weight (lb), respectively. The overall slope "a" of the curve is −0.249.

Interestingly, when the data is grouped by firm, very good correlation coefficients (R^2 = 0.826, 0.883, and 0.658) are found, with slopes much steeper although significantly different (a = −0.724, −0.462, and −0.429). This shows that there are differences from firm to firm, but that a good trend exists within a firm. The data shows that larger parts provide more cost-effectiveness than smaller parts.

Boeing conducted a study of the cost of fabrication for various aerospace components such as body section, fuselage panel, wing box, etc. for thermoset- and thermoplastic-based composites. Table 11.3 shows fabrication costs

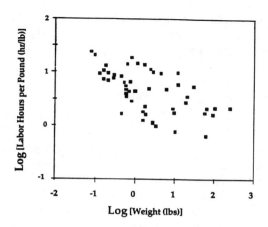

FIGURE 11.17
Relationship between labor hours per pound and part weight for composite aerospace parts (no assembly). (From Gutowski et al.[4])

TABLE 11.3

Fabrication Costs for Aircraft Fuselage Panel and Compass Cope Horizontal Stabilizer for Graphite/Epoxy and Graphite/Polysulfone Composites. Example 1 is for Aircraft Fuselage Panel and Example 2 is for Compass Cope Horizontal Stabilizer

Material	No. of Units	Production Hours	Material Cost ($)	Tooling Hours	Total Cost ($)	Total Cost (Relative)
Example 1						
Graphite/epoxy	1	1,425	2,452	1,310	84,502	1
($25/lb)	10	10,140	24,520	1,310	386,020	4.57
	100	62,349	245,200	1,310	2,154,970	25.5
Graphite/polysulfone	1	961	2,099	2,184	96,449	1.14
($25/lb)	10	6,838	20,990	2,184	291,650	3.45
	100	42,048	209,900	2,184	1,536,860	18.2
Example 2						
Glass/epoxy	1	1,068	600	810	56,940	1
($12/lb)	10	7,600	6,000	810	258,296	4.54
	100	46,730	60,000	810	1,486,178	26.1
Graphite/polysulfone	1	775	2,760	1,690	76,710	1.35
($65/lb)	10	5,515	27,600	1,690	243,747	4.28
	100	33,909	276,000	1,690	1,343,980	23.6

Source: See Hoggatt.[5,6]

for aircraft fuselage panels and compass cope horizontal stabilizers for two matrix material systems (polysulfone [thermoplastic] and epoxy [thermoset]). Thermoplastic composites provided a lower cost of fabrication compared to thermoset composites (Table 11.3), although the tooling cost was considered 70% higher for thermoplastic processing than for thermoset processing. The cost savings in thermoplastic composites derive from various factors, including faster processing cycle time, simpler storage, reduced scrap rate, and increased potential for repair.

11.6.2 Filament Winding for Consumer Goods

Filament winding is used for the fabrication of tubes, shafts, electric poles, pressure vessels, golf shafts, and aerospace parts such as fuselage and rocket launch tubes. For consumer goods, filament winding utilizes inexpensive fiber and resin systems such as E-glass fiber with epoxy. Filament winding can be highly automated and labor costs can be as low as 10 to 25% of total product cost. Many times, for consumer goods, autoclave curing is not done. For curing, the wound material with the mandrel is passed through a cabin where steam heating or electric heating is done to lower the cost of curing. Various cost elements in filament wound parts are shown below.

- Part cost = $4 to $30/lb
- Material cost = $2 to $10/lb
- Labor rate = $15 to $25/hr, excluding fringe benefits
- Labor hour/lb = 0.05 to 0.5 hr
- Overhead = 2 to 4 times the labor cost
- Equipment cost for filament winding = $250,000 to $500,000
- Mandrel cost = $50 to $500/ft^2

11.6.3 Compression Molded SMC Parts for Automotive Applications

Compression molded sheet molding compound (SMC) is currently used in a wide range of structural and nonstructural automotive applications, such as radiator supports, instrument panels, grill opening panels, instrument panels, cross-car beams, hoods, spoilers, decklids, fenders, and door panels. According to the SMC Automotive Alliance, the production of SMC for automotive applications has grown from 40 million pounds per year in 1970 to 240 million pounds in 1996, and was expected to climb to nearly 400 million pounds by 2000. This method is suitable for high-volume production rates and thus meets automotive requirements. Based on Dieffenbach's study,[7] this section provides a cost analysis of SMC parts for exterior body panel applications and compares it with existing steel body panels. SMC parts provide many advantages, including mass saving, corrosion resistance, and formability, over steel body panels. The exterior body panels for a midsize four-door sedan contain 17 individual pieces as listed below:

- Hood (outer and inner)
- Decklid (outer and inner)
- Fender (right and left)

- Front door (right and left, inner and outer)
- Rear door (right and left, inner and outer)
- Quarter panel (right and left)
- Roof

The mass, area, and piece count of the above components are shown in Tables 11.4 and 11.5 for steel and SMC, respectively. An increased area in SMC inner parts (Table 11.5) provides more contour and stiffness and is

TABLE 11.4

Mass, Area, and Piece Count for Automobile Exterior Body Panel for Steel

Steel Exterior Panel Set Design	Piece Count per Vehicle	Surface Area (cm²)	Material Scrap (% of blank)	Piece Mass (kg)	Percent of Steel Weight
Hood, outer	1	17,024	14	10.7	100
Roof	1	16,204	10	9.5	100
Decklid, outer	1	9,406	13	5.5	100
Fender (Left (L) & Right (R))	2	5,000	48	2.9	100
Front door, outer (L & R)	2	7,955	43	5.0	100
Rear door, outer (L & R)	2	7,103	47	4.5	100
Quarter panel (L&R)	2	4,175	70	2.5	100
Hood, inner	1	12,844	35	6.6	100
Decklid, inner	1	7,455	29	3.8	100
Front door, inner (L & R)	2	6,097	56	3.4	100
Rear door, inner (L & R)	2	5,684	57	3.3	100

Source: See Dieffenbach.[7]

TABLE 11.5

Mass, Area, and Piece Count for Automobile Exterior Body Panel for SMC

SMC Exterior Panel Set Design	Piece Count per Vehicle	Surface Area (cm²)	Material Scrap (% of blank)	Piece Mass (kg)	Percent of Steel Weight
Hood, outer	1	17,024	2.5–5	8.4	79
Roof	1	16,204	2.5–5	8.0	84
Decklid, outer	1	9,406	2.5–5	4.6	84
Fender (Left (L) & Right (R))	2	5,000	2.5–5	2.5	84
Front door, outer (L & R)	2	7,955	2.5–5	3.9	79
Rear door, outer (L & R)	2	7,103	2.5–5	3.5	79
Quarter panel (L&R)	2	4,175	2.5–5	2.1	84
Hood, inner	1	14,770	2.5–5	6.5	98
Decklid, inner	1	8,574	2.5–5	3.7	98
Front door, inner (L & R)	2	7,011	2.5–5	3.1	91
Rear door, inner (L & R)	2	6,536	2.5–5	2.9	85

Source: See Dieffenbach.[7]

TABLE 11.6
Common Input Assumptions for Cost Analysis of Exterior Body Panel Set

Common Input Assumptions	All
Annual production volume	200,000 vehicles/yr
Length of production run	4 yr
Sheet steel price	$0.75/kg
Sheet steel scrap credit	$0.10/kg
SMC charge price	$1.75/kg
SMC charge scrap credit	$0.00/kg
Direct labor wage	$21.60/h, with benefits
Working days per year	250
Working hours per day	16
Fabrication equipment life	15 yr
Assembly equipment life	8 yr
Building life	20 yr

Source: See Dieffenbach.[7]

assumed to compensate for the reduced "baseline" stiffness that SMC offers relative to the steel. The assumptions made for cost analysis are shown in Table 11.6. To show the impact of improvements in SMC technology over the past 10 years, two SMC cases — in 1986 and 1996 — are studied. The four inputs that change in going from the 1986 to the 1996 scenario are:

1. Cycle time
2. Material scrap
3. Process yield
4. Capacity utilization

It is assumed that no significant changes occurred in stamping technology in these 10 years. The fabrication and assembly assumptions for these three cases are given in Table 11.7. Using this information and technical cost modeling,[7] costs per panel were determined as a function of annual production volume. Cost comparisons for SMC 1986 and steel body panel are given in Figure 11.18, and those for SMC 1996 and steel body panel in Figure 11.19. It can be seen that SMC 1986 body panels are cheaper until 70,000 panels per year, whereas SMC 1996 is cost-effective until a production volume of 150,000. Figure 11.20 combines the previous two figures and presents the comparison in a single graph. A cost breakdown in terms of material, labor, tooling, equipment, and other costs (utilities, building, maintenance, and cost of capital, or interest) is shown in Figure 11.21 for all three cases at

TABLE 11.7
Fabrication and Assembly Assumptions for Exterior Body Panel Set

Case-Specific Assumptions	Steel Baseline	1986 SMC	1996 SMC
Fabrication assumptions			
Laborers per station	6	2	Same as 1986
Cycle time (s)	6–9	135	90
Equipment investment cost ($000)	$12,000	$600	Same as 1986
Total cost per set ($000)	$1000–$2800	$275–$500	Same as 1986
Assembly assumptions			
Laborers per station	2	2	Same as 1986
Cycle time (s)	60	120	Same as 1986
Equipment investment cost ($000)	$100	$150	Same as 1986
Total cost per set ($000)	$100	$100	Same as 1986

Source: See Dieffenbach.[7]

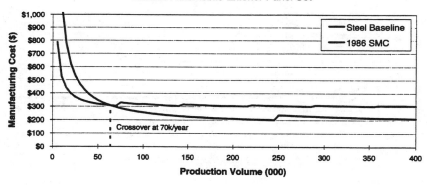

FIGURE 11.18
Comparison of manufacturing costs for SMC (1986) and steel midsize automobile exterior body panel set. (From Dieffenbach.[7])

production volumes of 50,000, 100,000, and 200,000 panel sets per year. With SMC parts, material cost is the most significant cost contributor; whereas with steel, tooling is the major cost contributor at 50,000 volume and decreases with increasing production volume. The use of SMC can be justified on a cost basis for vehicle models that require less than 150,000 units per year. Vehicles such as the Corvette, Lincoln Town Car, Dodge Viper, BMW Z3, Mazda RX-7, and more than 40 other models come in this category (less than 150,000 units per year). This number would increase significantly if light trucks are added.

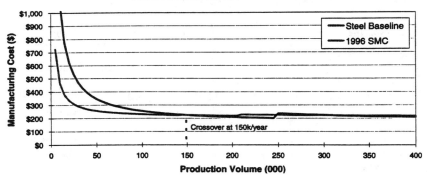

FIGURE 11.19
Comparison of manufacturing costs for SMC (1996) and steel midsize automobile exterior body panel set. (From Dieffenbach.[7])

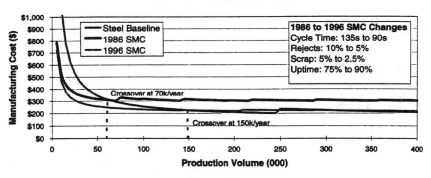

FIGURE 11.20
Cost vs. annual volume for 1986 SMC, 1996 SMC, and steel midsize automobile exterior body panel set. (From Dieffenbach.[7])

11.7 Learning Curve

It is human nature to perform better and better as the number of repetitions increases for the same task. That is, the time required to perform a task decreases with increasing repetition. There is also the saying that practice makes a man perfect. It is true in a manufacturing environment that workers produce more product in a given time as their experience with the job increases. If the task is short, simple, and routine, a modest amount of

Cost Estimation

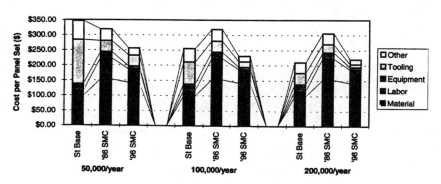

FIGURE 11.21
Cost breakdown by element vs. annual production volume. (From Dieffenbach.[7])

improvement takes place quickly. If the task is fairly complex and of longer duration, the amount of improvement occurs over a longer period of time. There are various reasons that contribute to the improvement in a worker's performance. These would include an increase in an employee's skill level, improved production methods, better management practices involving scheduling and production planning, and implementation of quality policies such as ISO 9000 and QS9000. Several other factors, such as ergonomics, production run time, standardization of the product and process, worker and management relationship, and nature of the job, contribute to the improvement of output. Generally, the production process that is dominated by people shows better improvement than a machine-dominated production process such as a chemical process plants for making resins.

A typical learning curve is shown in Figure 11.22. The learning curve is represented by:

$$y = ax^n \tag{11.11}$$

where y is the production time (hours/unit), x is the cumulative production volume, a is the time required to complete the first unit, and n represents the negative slope of the curve. On a log-log scale, the learning curve is represented by a straight line. The value of intercept on the y axis is the time for making the first unit.

Example 11.2

A worker takes 10 hours to make the first unit of an aerospace part by the hand lay-up technique. If this activity is identified as a 90% learning curve, predict the time required to make the second, fourth, and sixteenth units of the same part.

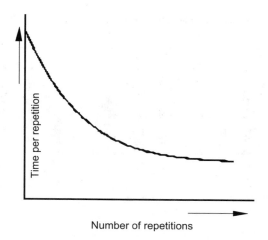

FIGURE 11.22
Representation of a learning curve. The time required per repetition decreases as the number of repetitions increases.

SOLUTION:
For a 90% learning curve, the time required to make the above units is as follows:

Unit	Unit Time (h)
1	10
2	0.9 × 10 = 9
4	0.9 × 9 = 8.1
8	0.9 × 8.1 = 7.29
16	0.9 × 7.29 = 6.561

It is interesting to note that the time required to make successive product decreases constantly. Management uses this information to predict the cost and production volume. In the above example, the time to make the third unit or the fifth unit is not determined. To determine the time required for those events, use Equation (11.11).

11.8 Guidelines for Minimization of Production Cost

To be successful in the competitive market, all possible consideration should be given to reduce the product cost. The principal sources of product cost derive from material, labor, equipment, tooling, and fixed costs. These costs need to be minimized by all possible means to reduce the overall cost. The following suggestions will help minimize the product cost.

1. Design products that are easy and cost-effective to manufacture. Because product cost depends on the product design, sufficient amounts of time and staff must be allocated to utilize the design for manufacturing (DFM) strategy for the reduction of product cost. DFM is discussed in detail in Chapter 5.
2. Build quality into the product — not by inspection, but by proper process design.
3. Minimize number of parts and thus minimize assembly time and cost.
4. Maximize the use of machine time. An idle machine increases product cost by doing nothing.
5. Minimize process cycle time. Because time is directly related to the cost, every consideration should be given to minimize the processing time. This can be achieved by selecting the fastest cure cycle resin system, by increasing the laminating speed or winding speed, and by automation.
6. Allocate machine power and manpower prudently.
7. Design the shop floor for good material flow.
8. Minimize scrap.
9. Minimize labor-intensive processes.
10. Minimize processing requirements. Decrease the high-pressure requirement or high-temperature requirement for processing, as these add costs in building and running the equipment.

References

1. Northrop Corporation, Advanced Composites Cost Estimating Manual (ACCEM), AFFDL-TR-76-87, August 1976.
2. Gutowski, T., Hoult, D., Dillon, G., Neoh, E., Muter, S., Kim, E., and Tse, M., Development of a theoretical cost model for advanced composite fabrication, *Composites Manufact.*, 5(4), 231, 1994.
3. Boothroyd, G. and Dewhurst, P., *Product Design for Assembly*, Boothroyd Dewhurst, Inc. 1991.
4. Gutowski, T., Henderson, R., and Shipp, C., Manufacturing costs for advanced composites aerospace parts, *SAMPE J.*, 1991.
5. Hoggatt, J.T., Advanced Fiber Reinforced Thermoplastics Program, Contract F33615-76-C-3048, December 1976.
6. Hoggatt, J.T., Advanced Fiber Reinforced Thermoplastics Program, Contract F33615-76-C-3048, April 1977.
7. Dieffenbach, J.R., Compression molded sheet molding compound (SMC) for automotive exterior body panels: a cost and market assessment, *SAE Int. Congr. Exposition*, Technical Paper No. 970246, February 1997.

Bibliography

1. Ostward, P.F., *Cost Estimating*, 2nd ed., Prentice-Hall, Englewood Cliffs, NJ, 1984.
2. Vernon, I.R., *Realistic Cost Estimating for Manufacturing*, Society of Manufacturing Engineers, Dearborn, MI, 1968.
3. Malstorm, E.M., *What Every Engineer Should Know about Manufacturing Cost Estimating*, Marcel Dekker, New York, 1981.

Questions

1. Why should the manufacturing engineer play a key role in the preparation of a manufacturing cost estimate?
2. What are the objectives of preparing cost estimates?
3. Why is it that product design essentially determines manufacturing cost?
4. What minimum basic information is required to prepare a valid manufacturing cost estimate?
5. What is the most commonly used approach to cost estimating, and how does it work?
6. What is the difference between recurring and nonrecurring costs?
7. Which composites manufacturing process has a higher labor cost?
8. Which manufacturing processes require higher initial investment for equipment and tooling? Create a ranking on five composites manufacturing processes based on higher equipment and tooling costs.
9. Which manufacturing process would you recommend for prototyping purposes based on cost?
10. Perform a comparative study between roll wrapping and filament winding processes for making golf shafts. Compare these processes based on material, tooling, equipment, and labor costs and provide an estimate on which manufacturing process is the best choice for making 100, 1000, and 5000 golf shafts. Fabrication of golf shafts by the roll wrapping process is explained in Chapter 6.
11. Why is compression molding (SMC) a process of choice for making less than 150,000 automotive parts per year?

12
Recycling of Composites

12.1 Introduction

With the increase in the use of composite materials in various industrial sectors, the scrap materials and composite waste parts cannot just be landfilled; instead, these need to be recycled for a better environment. Currently in many business sectors, composite wastes are landfilled with little regard for recovering fibers and plastics for future use. Governments and customers are becoming aware of the environmental pollution created by these materials and passing strict regulations for recycling of plastics and composites waste. Germany, England, France, Italy, and other European countries have mandated that plastics and composites waste must be recycled. Japan is running a similar program for waste disposal. Japan currently incinerates more than 70% of its 47 billion kg total wastes.[1] In the United States, plastic constitutes 7% of the total municipal solid waste (MSW) by weight.[2] According to the EPA (Environmental Pollution Agency), 13.2 billion kg of plastics were disposed to MSW in 1990, out of which only about 150 million kg (1%) was recycled.[3] The major method of MSW disposal is landfilling, which accounts for 73%. Other methods, such as recycling (11%), incineration (14%), and composting (2%), also contribute to MSW disposal.[4] In the United States, the annual growth rate for plastics production in the past three decades was about 10.3%, as compared to a 3.2% national product growth rate. The annual growth rate for composite materials is 4% in the United States. Composite material production amounted to 3.9 billion lb in the United States in 2000 (as discussed in Chapter 1). In 1980, plastics production exceeded 58.3 billion lb (26.5 billion kg). The Society of Plastic Industries estimated sales of 75.7 billion lb (34.4 billion kg) of plastics by the end of 2000.[5] Because plastics constitute one of the major ingredients in composites, recycling of plastics is also discussed in this chapter to some extent.

12.2 Categories of Dealing with Wastes

There are several ways of handling municipal and industrial solid wastes. These wastes can be burned, buried, or reused.[2] For some people, reuse is recycling, whereas others include burning in recycling, provided the energy or by-products during the burning process are utilized. Few consider burying to be recycling. These techniques of handling wastes are described below.

12.2.1 Landfilling or Burying

Landfilling has been the most common way of handling waste. In this process, the waste is carried to a specific place and unloaded there. Because plastics and composites are not biodegradable, they cause environmental pollution. This prevalent method of disposal in landfills is becoming prohibitive due to the shortage of space, environmental concern, negative public opinion, and legislation. This method is becoming increasingly restricted by governments.

12.2.2 Incineration or Burning

Incineration is an option for dealing with waste composite materials, but it destroys valuable materials in the process and can be a source of pollution. In this method, waste is burned in a conventional municipal incinerator. This method is popular in Japan because of a shortage of space; Japan incinerates more than 70% of its total waste stream. In Japan every year, 5 million kg of plastics enter the waste stream, of which 65% is incinerated, 23% is landfilled, and 5% is recycled. If the energy generated during burning is used or saved as a gas, then this process comes under the category of a quaternary recycling process, which is described in Section 12.2.3.

12.2.3 Recycling

Recycling is becoming popular around the world for the disposal of plastics and composites because of its characteristics to preserve the natural resources as well as to render a better environment. Rathje[6] has divided recycling and burning into four categories: primary, secondary, tertiary, and quaternary. Primary recycling involves reprocessing the waste to obtain the original or a comparable product. An example of this is the reprocessing of scrap plastic buckets into new buckets or mugs. Products made from unreinforced thermoplastics fall in this category. In secondary recycling, the waste material is transformed into products that do not require virgin material properties. The reprocessing of scrap polyethylene terephthalate (PET) bevarage containers into products such as carpet backing falls in this category.[1] During continuous

TABLE 12.1

Recycling Processes for Composites

Recycling Product/Input	Recycling Method	Category[a]	Output
Post-consumer SMC	Pyrolysis	2,3,4	Fuel gas, oil, inorganic solid
SMC	Pyrolysis and milling	2,3,4	Fuel gas, oil, inorganic solid
Polyurethane foams	Pyrolysis	2,3,4	Gas, oil, solid waste
Auto shredder residue	Pyrolysis	2,3,4	Gas, oil, solid waste
Mixed polymer waste, ASR	Incineration	4	Heat, solid and gaseous wastes
Post-consumer RIM	Regrinding	2	Ground particles for fillers
Phenolic scrap parts	Regrinding	2	Ground particles for fillers
SMC	Regrinding	2	Ground particles for fillers
Thermoplastic composites	Regrinding	2	Flake for compression molding
Thermoplastic composites	Regrinding	2	Injection molding pellets

[a] 1 = Primary (reprocessing into similar product); 2 = secondary (reprocessing into degraded service); 3 = tertiary (reprocessing into chemicals); 4 = quaternary (energy recovery).

Source: Data adapted from Henshaw et al.[1]

use of a product, the material properties may become degraded and therefore some properties cannot be achieved. The tertiary recycling process breaks the polymers used in composites into their chemical building blocks. These lower-chain hydrocarbons can be used to make monomers, polymers, fuels, or chemicals, thus promoting conservation of petroleum resources. The fibers and fillers separated during this process are reused as molding compounds. Tertiary recycling may fall into either primary or secondary recycling, depending on the final use of these chemicals. In a quaternary process, the waste is burned off and the energy (in terms of fuel and gas generated by this process) is used for other applications. Based on these categories, recycling methods such as pyrolysis and regrinding have been developed; these are described in Section 12.3.

Monolithic materials are easier to recycle than composite materials.[7] The reinforcing fibers in composites offer unique properties but create complications in recycling. Thermoplastic composites have the potential of primary or secondary recycling, whereas thermoset composites usually fall into secondary, tertiary, or quaternary recycling. Table 12.1 shows applicable recycling methods for various types of composites. The final product and wastes (output) generated by the recycling process are also shown.

12.3 Recycling Methods

With the increase in composites usage, the concern for recycling these materials has also increased. Engineers and researchers are developing ways to recycle these materials. New recycling methods are required to overcome the technical hurdles associated with recycling of plastics and composites. At the heart of the issue are the costs of recycling these materials and their resulting values.[8] Currently available techniques include regrinding, pyrolysis, incineration, and acid digestion. Regrinding and pyrolysis are described below. Incineration was described in Section 12.2.2, but is impractical in terms of environmental pollution. It destroys the valuable carbon and aramid fibers during the burning process and creates pollution. Acid digestion uses harsh chemicals and conditions to dissolve the polymer. During this process, a mixture of hydrocarbons and acid is formed that needs further processing. This method is also impractical from an environmental point of view.

12.3.1 Regrinding

Regrinding is a secondary recycling process in which composite waste is ground to suitable sizes to be reused as fillers. The type of application of these fillers depends on the type of polymer used in the composite material. For a thermoplastic matrix, the resulting materials are used in injection molding and compression molding processes. For thermoset composites, ground materials are used as fillers for SMC, bulk molding, or reinforced concrete.

Developmental work on the recycling of reaction injection molded (RIM) polyurethane automotive scrap parts shows that ground RIM parts can be used as a filler in other molding processes.[9] Compression molding with these fillers requires high pressure and is suitable for simple shapes. Phenolic suppliers have reported that scrap phenolic composites containing glass or carbon fibers are capable of being pulverized into particulates and can be used as fillers and extenders in molding compounds.[10] The properties of compounds containing recycled phenolics gave similar properties to those of virgin materials.[11] A study by Owens-Corning Fiberglass on the recycling of SMC composites demonstrates that the milled and ground SMC powder can be used as fillers and reinforcements in BMC and thermoplastic polyolefin molding compounds.[12] Another study on SMC recycling by Union Carbide Chemicals and Plastics Company shows that the properties of thermoplastic composites containing recycled SMC as fillers is as good as or better than those of the virgin materials properties.[13]

12.3.2 Pyrolysis

Pyrolysis is a tertiary recycling process in which polymer is thermally decomposed at elevated temperatures in the absence of oxygen. This process breaks the polymer into reusable hydrocarbon fractions as monomers, fuels, and chemicals and thus preserves scarce petroleum resources. During this process, fibers are separated from polymers and reused as fillers or reinforcements. A schematic diagram of a pyrolysis process is shown in Figure 12.1. A typical pyrolysis process includes a shredder or grinder, reaction chamber,

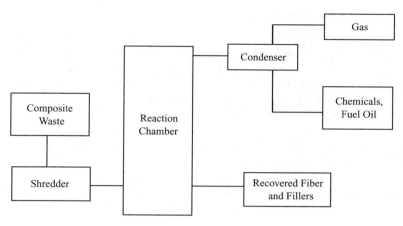

FIGURE 12.1
Schematic flow diagram for a pyrolysis process.

furnace, condenser, and storage tanks. The polymer matrix gets converted to low-molecular-weight hydrocarbons under the action of heat and catalysts and is removed from the fibers as a gas. These hydrocarbons are refined and used as fuels or feedstocks in the petrochemical industry. The fibers can be reused as reinforcements or fillers for new applications. Results show that most types of matrix materials — thermosets as well as thermoplastics — can be converted to valuable hydrocarbon products.[14]

Attempts have been made to pyrolyze cured and uncured SMC scraps at a tire pyrolysis facility.[15] The pyrolysis resulted in 30% gas and oil, and 70% solid by-product. The solid residue, which contains mostly fibers, is then milled to be used as filler. These fillers are then reused to make SMC parts. The results show that this is a technically feasible recycling technology.[16] There are other articles written describing material properties of recycled SMC parts. An insignificant loss in strength is observed due to the addition of a small percentage (<10%) of recycled materials as fillers.[17,18] Phoenix Fiberglass (Toronto, Canada) has a patented process for recovering fiber glass from SMC waste. The recycled material sells for $0.60/kg ($0.27/lb).[19]

12.4 Existing Infrastructure for Recycling

At present, no well-known recycling infrastructure for handling composites waste is available. The purpose of this section is to make the reader aware of current practices in the automotive and aerospace industries regarding the handling of scrap products. Researchers need to identify how and at what stage of the existing infrastructure composite recycling methods can be incorporated to make composite recycling a profitable business.

12.4.1 Automotive Recycling Infrastructure

Automotive recycling is a profitable business; today, the automobile is the most recycled product in the world. This was not the case when scrap automobiles were discarded in junkyards and on roadsides. The first automotive recycling challenge was resolved with the development of automotive shredders, which efficiently separate ferrous metals for sale to electric arc furnace minimills. The development of these minimills was also critical because it had to handle impure automobile iron scrap, which was not possible with the established oxygen furnace technology.[8] Once again, the automotive industry is facing the challenge of recycling, but now for plastic and composite parts, in a cost-effective manner. For simplicity, the current automobile recycling infrastructure can be divided into three main categories[8]:

- Dismantling
- Shredding/ferrous separation
- Nonferrous separation

In the dismantling phase, good and bad parts are separated. The good parts include radio, transmission, engine block and anything else that has value. The bad parts are those that have no value and that can damage the recycling equipment or can affect the efficiency of recycling. The bad parts include tires, radiators, fuel tanks, batteries, air bag cannisters, etc. After the separation of good and bad parts, a dismantler flattens the vehicle into a hulk and sells it to a ferrous separator. The dismantler makes money by selling good parts.

The ferrous separator buys the hulks from the dismantler and feeds them into a shredder, where hulks are pulverized into small chunks using a set of spinning hammers. The ground ferrous metals are then magnetically separated and sold to the scrap metal market. The remaining materials, called automotive shredder residue (ASR), contain two streams of materials: a heavy stream consisting of nonferrous metals and plastics, and a lightweight stream containing mostly plastics (including foams), glasses, and other waste. The lightweight stream predominantly ends up in the landfill, but does have some recycling potential. The heavy stream is sold to nonferrous separators. The money earned from selling recovered materials is more than the cost of recovering these materials.

In the third stage, the heavy stream supplied by the ferrous separator is separated into aluminum, zinc, magnesium, and other metals through gravity action, hand-sorting, and some proprietary technology. The resulting plastic-intensive waste goes for landfilling. The recovered metals are sold to the scrap metal market. These materials are sold at greater cost than that of recovering them.

The present recycling infrastructure is profitable but plastics are not recovered to any significant extent. The increase in plastics and composites use

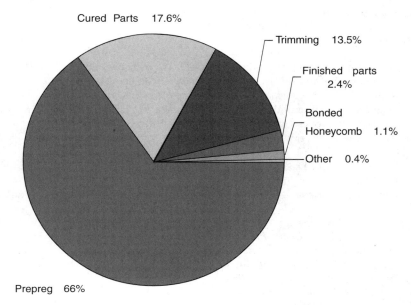

FIGURE 12.2
Average distribution of composite waste in aerospace industries. (Adapted from Unser et al.[19])

in automobiles is causing environmental concerns and thus there is a need for improvement in the present recycling infrastructure for these new material systems.

12.4.2 Aerospace Recycling Infrastructure

Aerospace industries commonly use prepreg materials for part fabrication and a typical buy-to-fly ratio is 1.7 to 1.0.[20] It means that for every 1.7 kg of prepreg purchased, 0.7 kg is unused and scrapped. In 1996, Environmental Technical Services (ETS), under a contract by Adherent Technologies Inc., conducted a survey among composite industries in the aerospace market.[19] They sent almost 450 questionnaires to a wide variety of companies and found that carbon/epoxy is the most dominant material system. At present, the most common method of scrap disposal is landfilling.[19] McDonnell Douglas Aerospace estimates that it annually sends approximately 49,000 lb of waste composite materials for landfilling. Figure 12.2 shows the average distribution of composite wastes in aerospace industries. Several programs have started to enhance the recycling activities in aerospace industries. Because graphite fibers are expensive, their recovery shows promise for profitable business.

References

1. Henshaw, J.M., Han, W. and Owens, A.D., An overview of recycling issues for composite materials, *J. Thermoplastic Composite Materials*, 9, 4, 1996.
2. Rathje, W. and Murphy, C., *Rubbish,* Harper Collins, 1992.
3. Hegberg, B.A., Brenniman, G.R., and Hallenbeck, W.H., *Mixed Plastics Recycling Technologies,* Noyes Data Corporation, Park Ridge, NJ, 1992, 11.
4. Anonymous, Plastics makers recycle for new growth, *Chemicalweek,* p. 28, December 18–25, 1991.
5. Curlee, R.T. and Das, S., *Plastic Wastes,* Noyes Data Corporation, Park Ridge, NJ, 1991, 15.
6. Leindner, J., *Plastic — Recovery of Economic Value,* Marcel Dekker, New York, 1981, 22.
7. Henshaw, J.M., Design for recycling: new paradigm or just the latest 'design for X' fad?, *Int. J. Mater. Prod. Technol.*, 9, 125, 1994.
8. Dieffenbach, J.R. and Mascarin, A.E., Cost Simulation of the Automotive Recycling Infrastructure: The Impact of Plastics Recovery, Society of Automotive Engineers, March 1993, 930557.
9. Wood, S.A., *Modern Plastics,* June, 62, 1988.
10. Reiss, R., The German approach to plastics recycling, *Metals and Mater.*, 8(11), 592, 1992.
11. Anonymous, German recycling targets, *European Plastics News,* 20(2), 20, 1993.
12. Jutte, R.B. and Graham, W.D., Recycling SMC scrap as a reinforcement, *Plastics Engineering,* p.13, May 1991.
13. Godlewski, R. and Herdle, W., Recycling of ground SMC into polypropylene and recovery of scrap Azdel to produce quality molded parts, Advanced Composites: Design, Materials and Processing Technologies, ASM International, Chicago, IL, November 1992, 239.
14. Allred, R.E., Recycling process for scrap composites and prepregs, *SAMPE J.*, 32(5), 46, 1996.
15. Norris, D.R., 1990, *Proc. of ASM/ESD Advanced Composites Conf.*, 6, 277, 1990.
16. Graham, W.D., SMC composites recycling, *Proc. 9th Annu. ASM/ESD Advanced Composites Conference,* Detroit, MI, 1993, 609.
17. Patterson, J. and Nilson, P., Recycling of SMC and BMC in standard process equipment, *J. Thermoplastic Composite Mater.,* 7, January 1994.
18. Inoh, T. et al., SMC recycling technology, *J. Thermoplastic Composite Mater.,* 7, January 1994.
19. Unser, J.F., Staley, T., and Larsen, D., Advanced composites recycling, *SAMPE J.,* 32(5), 52, 1996.
20. Composite Industry Survey, report prepared by McDonnell Douglas Aerospace, dated November 27, 1991.

Questions

1. What are the common recycling methods for composite materials?
2. What are the challenges in recycling of composite materials?
3. What are the methods of dealing with wastes?
4. What is a pyrolysis process?

Index

A

Abrasion, 321, 353, 359
ACCEM, 350-352
Acid etching, 320
Adherend, 310
Adhesive, 12, 62, 90, 180, 291, 292, 310, 311, 312, 315
    Acrylic, 316
    Anaerobic, 317
    Cyanoacrylates, 317
    Epoxy, 315, 317
    Hot melt 317
    Mix, 315
    No-mix, 316
    Polyurethane, 316, 317
    Selection guidelines, 319
    Types, 315
    Water based, 318
Adhesive bonding, 12, 62, 90, 309, 310, 318
    Advantages, 318
    Basic science, 313
    Design guidelines, 322
    Disadvantages, 319
    Failure modes, 313
Adhesive failure, 313
Adsorption, 314
Aerospace, 14, 58, 71, 92, 119, 200, 362, 371
    Applications, 14, 15
    Industry, 14, 331
    Market, 58
Aircraft, 14, 15, 290, 292, 309
Alignment, 90
Aluminum, 1, 2, 25, 53, 54, 107
Amorphous, 33
Anodizing, 321
Applications, 13, 14, 15, 17, 18, 58
    Aerospace, 14, 15
    Automotive, 17, 18
    Construction, 19
    Consumer goods, 19
    Industrial, 19
    Marine, 18

Aramid fiber, 24, 54, 333
    Manufacturing, 28
Assembly, 12, 85, 87, 90, 94, 95, 325
Autoclave, *see also* Autoclave processing, 12, 24, 120, 125, 126, 127, 128, 134, 217, 218, 219, 221, 222, 225, 236, 238, 241, 270
    Process, 12, 24, 120, 125, 126, 241, 295, 306, 307
Autoclave processing, 218
    Advantages, 221
    Applications, 219
    Limitations, 222
    Making of the part, 219
    Methods of applying heat and pressure, 221
    Processing steps, 221
    Raw materials, 219
    Tooling, 219
Autohesion, 220, 279
Automotive, 17, 18, 58, 68, 71, 92, 119, 200, 214, 226
    Industry, 17, 71, 214
    Market, 17, 58

B

BCC, 235
BMC, 24, 51
BMI, 32
Barrier film, 124, 127
Bathtub, 136, 137, 138
Barriers, 20
Bevel joint, 311
Bicycle, 18, 88, 160, 161
Bill of materials, 290, 305
Bi-ply, 39
Bismaleimide, 32
Black & Decker, 93
Bladder molding, 12
Bleeder, 124, 127
Blow molding, 12
BMC molding, 12, 24

385

Boat, 128, 129, 132, 136
Body-centered cubic, 235
Boeing, 15
Bolt, 89
Bolted joint, 310, 324
 Design parameters, 326
 Failure mode, 325
 Preparation, 327
Bolting, 323, 341
Bond thickness, 312
Bonded joints, 323
Boron fiber, 24, 25
Boundary conditions, 28, 244
Braiding, 44, 45, 86
Breather, 124, 12
Bumper beam, 92, 211
Butt joint, 311, 324

C

CAD, 105, 141, 147
CNC, 107, 148
CVD, 334, 335
CVDD, 334
Calcium carbonate, 48, 136, 162
Capacity planning, 289, 290, 304
Carbon / epoxy, 8, 122, 325
Carbon fiber, 24, 25, 333
 Manufacturing, 27
Carbon / PEEK, 202, 219, 223, 245, 252
Carriage unit, 141, 144
Ceramics, 1-4
Chassis, 92
Chemical resistance, 68, 320
Chemical treatment, 322
Chemical vapor deposition, 334
Chopped fiber, 26, 47
Chromic acid, 320
Cleavage stress, 310
Clip, 90
CO_2 laser, 340
Coefficient of thermal expansion, 7, 104, 333, 266
Cohesive failure, 313
Collapsible mandrel, 144
Commercial aircraft, 15
Compaction, 209, 237, 253, 254
Compliant, 86
Compression molding, 12, 79, 101, 179, 180, 188, 210, 211, 366
 Advantages, 188, 214
 Applications, 180, 210
 Limitations, 188, 215
 Making of the part, 181

Methods of applying heat and pressure, 187, 214
 Mold design, 186, 213
 Processing steps, 187
Raw materials, 181, 211
Computer-aided design, 105
Computer numerical control, 148
Concept development, 74
Concept feasibility phase, 77
Concurrent engineering, 74, 75, 85
Consolidation, 116, 204, 206, 207, 209, 237, 264, 270, 275
Continuous fiber, 5, 8, 26
Continuous fiber composites, 5, 8
Continuous improvement, 80
Copper, 1, 2
Core, 53, 54, 71, 130, 134, 164, 291-293, 305
 Balsa, 54, 134
 Foam, 54, 134
 Honeycomb, 53, 54, 134
Corner radii, 105
Corona treatment, 321
Corrosion resistance, 67
Cost, 58, 60, 62, 63, 66, 79, 80, 85, 87, 94, 100
Cost estimating, 345
 Reasons, 346
 Requirements, 348
 Techniques, 350
Cost breakdown, 347
Cradle, 107
Crystalline, 33, 34
Customer-driven, 73
Cutting of composites, 331
Cutting tools, 334
Cynate ester, 32, 122

D

DFA, 7, 86, 94
 Benefits, 95
 Guidelines, 95
DFM, 7, 86, 87, 88, 92, 93
 Implementation guidelines, 88
 When to apply, 93
Debulking, 353
Deburring, 305
Decklid, 180, 366
Defect, 86, 89, 95, 96, 97
 Material-related, 97
 Process-related, 97
 Product-related, 97
Degree of compaction, 237
Delamination, 342
Delivery point, 141

Index

Demand planning, 289
Depreciation cost, 348
Design evaluation method, 93
Design flow, 88
Design for assembly, 7, 85, 94
Design for life cycle, 85
Design for manufacturing, 7, 85
Design for no assembly, 95
Design for quality, 85
Design problem, 86, 87
Design review, 80
Detailed design phase, 78
Diaphragm forming, 12, 222-225
    Advantages, 225
    Applications, 222
    Limitations, 222
    Making of the part, 223
    Methods of applying heat and pressure, 225
    Raw materials, 223
    Tooling, 223
Die, 154-156, 208, 223
Difficult, 87
    To assemble, 87
    To manufacture, 87
Diffusion, 315
Doctor blade, 145
Double lap joint, 311, 324
Draft, 105
Drilling, 305, 341
Driveshaft, 61, 142

E

EPA, 375
EDM, 105
E-glass, 24, 25
Ejector pin, 186
Electrical discharge machining, 105
End mills, 335
Engine components, 16
Environmental pollution agency, 375
Epoxy, 2, 30, 52, 71, 117, 143, 310, 315
Error-proof, 97
Expert system, 68

F

FAA, 9
FEA, 62, 169, 238
FMEA, 81, 82
FRP, 106, 131, 336
Fabric, 26, 36, 100, 107, 112, 131, 132
    Noncrimp, 37
    Stitched, 39, 132
    Woven, 26, 36
Fabrication, 71, 79, 91
Failure mode, 313, 325, 333
Failure mode and effects analysis, 81, 82
Fairing, 112, 121
Fastener, 89
Fastening, 12, 90
Fender, 366
Fiberglass sheet, 291
Filament winding, 12, 13, 14, 23, 26, 44, 61, 86, 102, 104, 105, 116, 118, 120, 140, 141, 142, 143, 144, 145, 146, 149, 200, 222, 238, 242, 252, 263, 264, 267, 349, 366
    Advantages, 149
    Applications, 141
    Limitations, 150
    Making of the part, 144
    Methods of applying heat and pressure, 146
    Processing steps, 148
Raw materials, 143
    Tooling, 143
Filler, 48
Film adhesive, 291
Finishing, 13
Finite element analysis, 62, 169, 238
Fishing rod, 141
Fixed cost, 348, 350
Flame treatment, 320
Flap, 292-294, 296, 302
Floor panel, 290
Forming, 12
Fracture toughness, 68
Freezer, 295, 308
Friction, 67
Full-scale production, 80
Fusion bonding, 12

G

GMT, 210, 211, 213-215
Galvanic corrosion, 325
Gate location, 115
Glass / epoxy, 8, 122, 312, 338
Glass fiber, 23, 24, 25, 48, 61, 143
    Manufacturing, 27
Glass transition temperature, 32
Go / no-go, 77
Golf, 18, 141, 189, 190, 191, 304
Graphite, 25
Graphite / epoxy, 261, 294, 338

H

Hand lay-up, 12, 24, 40, 100, 106, 116, 119, 128, 350, 353, 362, 371
Handrail, 151, 152
Helicopter, 112, 113, 121
High-speed steel, 334
Honeycomb, 52, 53, 134, 164, 290, 291
    Aluminum, 52, 53
    Methods for the manufacture, 54
    Nomex, 52, 53, 290, 291
    Thermoplastic, 52, 53
Hood, 366
Hot gas, 245, 250
Hot press, 24, 215, 217, 241, 270
    Advantages, 218
    Applications, 215
    Limitations, 218
    Making of the part, 216
    Methods of applying heat and pressure, 217
    Processing steps, 217
Raw materials, 215
    Tooling, 215
Hull, 130, 131

I

Impact properties, 8, 9
Impregnation, 116
Incineration, 376, 377
Initial condition, 238
Injection molding, 12, 23, 24, 197, 226-229, 236
    Advantages, 229
    Applications, 197, 226
    Limitations, 229
    Making of the part, 198, 227
Processing steps, 228
Raw materials, 197, 226
    Tooling, 198, 226
In-service monitoring, 6
Inserts, 335
Interdiffusion, 204
Interference-fit, 90
Intimate contact, 217, 270, 271, 272, 275
Inventory control, 95
Isotropic, 86

J

Joining, 12, 309
Joint, *see also* Mechanical joint, 309
Just-in-time, 306

K

Kevlar, 25
Kevlar / epoxy, 8, 122, 338
Kite, 18

L

Ladder, 151
Land filling, 375, 376
Landing gear, 121
Lap joint, 310-312
Lap length, 312
Laser, 207, 245, 250, 251, 339, 340
Laser-assisted processing, 245
Laser cutting, 339, 340
Laser printer, 92
Layout planning, 289
Leaf spring, 68, 150
Learning curve, 370-372
Life cycle, 85
Liquid board, 107, 108
Low Earth orbit (LEO), 15

M

MSDS, 291
Machinability, 66, 67
Machining, 12, 99, 296, 304, 331
    Challenges, 332
    Failure mode, 333
    Purpose, 351
    Types, 336
Maintainability, 75, 90
Make-and-break approach, 62
Make or buy, 290, 346
Mandrel, 113, 143, 144, 145, 202, 205, 366
Manufacturing, 61, 72, 86, 99, 100
    Processes, 61, 72, 86
Manufacturing instructions, 289, 290, 294
Marine, 18, 119
    Applications, 18
Markets, 13
Master model, 106, 107, 108
Mat, 39, 128, 136, 145, 210
Material selection, 57, 58, 86
    Methods, 63
    Need, 57
    Reasons, 58
    Steps, 60
Matrix, 5
    Functions, 5, 6
    Materials, 28
Mechanical fastening, 12

Index

Mechanical joint, *see also* Bolted joint, 309, 323, 325
   Advantages, 325
   Disadvantages, 325
Melt time, 250
Metals, 1, 2, 4, 88
Methyl acrylate, 310
Microstructural, 86
Military aircraft, 14, 16
Model, 235
Modular design, 89, 91, 92
Mold, 112, 131, 135-138, 158, 162, 178, 179, 213, 214
Mold making, 104
   Design criteria, 104
   Methods, 105
Molding compound, 47
   Injection moldable compound, 52
   Sheet molding compound, 48
   Thick molding compound, 49
Monocoque, 88
Multifunctional, 91
Multiple cavities, 110
Multiple ports, 110

N

Near-net-shape, 7, 99
Net-shape, 7, 99
No-assembly, 86
Noise, vibration, and harshness (NVH), 8,
Nomex, 53, 290
Non-recurring cost, 348
Noncrimp, 37
Normal stress, 310
Nut, 89, 90
Nylon, 2, 34, 43

O

O-ring, 90
One-component, 86
Online process monitoring, 6
Order entry, 80
Oscilloscope, 248
Overhead cost, 349

P

PEEK, 35, 43, 202, 203, 219
PPS, 35, 43, 202, 203
Packaging, 290, 295, 348, 350
Part count, 88

Part integration, 6, 88
Parting line, 199, 228
Payout eye, 148
Peel stress, 311
Phenolic, 29, 31, 52
Pickup box, 92
Pilot-scale, 79
Pin joint, 323
Pipe, 145
Plasma treatment, 320
Plastics, 1, 2, 3, 88
Plunger, 199, 227
Polyamide, 124
Polycrystalline diamond (PCD), 334
Polyester, 29, 31, 48, 52, 61, 117, 136, 143, 162
Polyetheretherketone (PEEK), 35, 43, 202
Polyimide, 32
Polyphenylene sulfide (PPS), 35, 43, 202
Polypropylene, 2, 35
Polyurethane, 32, 310, 316
Preform, 26, 44, 46, 47, 71, 169, 178
Preliminary design, 80
Prepreg, 24, 40, 42, 43, 59, 71, 96, 102, 107, 110, 116, 119, 120, 122, 123, 124, 125, 127, 128, 143, 216-219, 261, 291, 293, 294, 305, 349
Prepreg lay-up, 119, 120, 122, 124, 125, 127, 128
   Advantages, 128
   Applications, 120
   Limitations, 128
   Making of the part, 122
   Manufacturing challenges, 127
   Methods of applying heat and pressure, 126
   Processing steps, 127
   Raw materials, 122
   Tooling requirement, 122
Preproduction, 79
Process design, 74, 75
Process flow, 79, 295
Process models, 235
   Importance, 235
Processing, 10, 12, 100
Processing conditions, 100
Processing parameters, 100
Product design, 73, 74, 75
Product defect, 89
Product development, 71, 75, 79, 80, 85, 95
   Cycle time, 71, 85
   Process, 71, 72, 77
   Reasons, 72
   Team (PDT), 72, 75, 80, 81
Product life cycle, 76
Product quality, 73, 87

Production planning, 289, 290
Production rate, 60, 100, 197, 305, 319, 348
Property, 2, 8, 9, 59, 63
    Nonquantitative, 67
Prototype development, 78
Pultrusion, 12, 23, 26, 61, 86, 91, 101, 102, 104, 105, 120, 200, 208-210, 222
    Advantages, 158, 210
    Applications, 150, 208
    Limitations, 158, 210
    Making of the part, 154, 209
    Methods of applying heat and pressure, 157, 209
    Processing steps, 158
    Raw materials, 152, 208
    Tooling, 154, 208
Pyrolysis, 377, 378

Q

Quality, 73, 80, 85, 86, 87, 88, 95
Quality assurance, 80
Quality control, 132, 140, 290, 296
Quaternary, 376

R

RTM, 7, 12, 14, 24, 26, 44, 45, 61, 86, 102, 104, 105, 108, 109, 110, 112, 113, 116, 132, 143, 235-238, 252, 284, 305
Rack, 306, 307
Radiation, 146
Radomes, 120
Recurring cost, 349
Recycling, 375, 376
    Infrastructure, 379
Methods, 377
Regrinding, 377, 378
Release film, 124, 127
Reliability, 75, 94
Repairability, 67
Residual stress, 127, 236, 237, 283
Resin, 29, 71
    Epoxy, 30
    Properties, 29
    Thermoplastic, 33
    Thermoset, 29
Resin transfer molding, *see also* RTM, 7, 159
    Advantages, 173
    Applications, 159
    Limitations, 174
    Making of the part, 164
    Methods of applying heat and pressure, 169

    Processing steps, 170
    Raw materials, 161
    Tooling, 162
Resistance, 67
    Chemical, 68
    Corrosion, 67
    Heating, 206
    Wear, 67
Rigid body, 86
Riveting, 323, 341
Roll wrapping, 12, 24, 61, 188, 304
    Advantages, 196
    Applications, 188
    Limitations, 196
    Making of the part, 189
    Methods of applying heat and pressure, 194
    Processing steps, 194
    Raw materials, 189
    Tooling, 189
Roving, 24, 26, 27

S

S-Glass, 24, 25
SCRIMP, 12, 175
SMC 24, 48, 49, 79, 101, 366, 368, 378
    Molding, 12, 24, 48, 79
SRIM, *see also* Structural reaction injection molding, 12, 24, 44, 79, 103, 105, 108
Sand blasting, 320
Sandwich panel, 130, 290
Satellite components, 17
Scaling, 67
Scarf joint, 311
Scrap, 99
Screw, 89, 90
Selection criteria, 100
Semi-crystalline, 33
Serial approach, 74
Serviceability, 66, 75, 87, 90
Shear edge, 186
Sheet molding compound, 48, 180
Shipping, 290, 292
Short fiber composites, 2, 5
Short shots, 228
Shredding, 380
Shrinkage, 185, 219, 228, 237
Shrink tape, 192, 196
Silicon carbide, 1
Single lap joint, 310, 312, 323, 324
Single piece, 86
Slip-fit, 90

Index

Snap-fit, 89
Solidification, 117
Solvent resistance, 10
Specific strength, 7
Specific stiffness, 7
Sporting goods, 18
    Industry, 18
    Market, 18
Spot weld, 216, 219, 223
Spraygun, 136, 137, 139
Spray-up, 12, 26, 106, 107, 132, 135 - 140
    Advantages, 140
    Applications, 135
    Limitations, 140
    Making of the part, 139
    Methods of applying heat and pressure, 139
    Processing steps, 139
Raw materials, 136
    Tooling requirement, 136
Spring, 90
Standardize, 88
Stapler, 86, 95
Step joint, 311
Steel, 2, 107
Stiffness, 7, 67, 71, 86
Stop block, 186
Storage tank, 128, 136
Strap joint, 311
Strength, 7, 67, 68, 71
Structural reaction injection molding, *see also* SRIM, 175
    Advantages, 179
    Applications, 176
    Limitations, 179
    Making of the part, 178
    Methods of applying heat and pressure, 178
    Processing steps, 179
Raw materials, 176
    Tooling, 178
Surface preparation, 320
Swimming pool, 128, 136

T

Tail rotor, 112, 113
Tape lay-up, 353
Tape winding, 12, 24, 201, 236, 238, 242, 244, 252, 270, 275
    Advantages, 207
    Applications, 202
    Limitations, 207
    Making of the part, 203
    Methods of applying heat and pressure, 206
    Raw materials, 202
    Tooling, 202
Teach-in-programming, 147
Tennis racquet, 110, 111
Thermochemical, 237, 238, 243, 245
Thermocouple, 245, 306
Thermoforming, 223
Thermosets, 11, 29, 60, 117, 118, 339
    Cutting, 339
    Processing, 117, 118
    Prepreg, 43
    Resin, 29
Thermoplastics, 11, 33, 43, 44, 60, 117, 118, 339
    Compression molding, 210
    Cutting, 339
    Prepreg, 43, 44
    Processing, 117, 118
    Pultrusion, 208
    Resin, 33
    Tape winding, 201
Threaded joint, 310
Tool making, 104
    Design criteria, 104
    FRP, 106
    Guidelines, 108
    Methods, 105
Tooling, *see also* Tooling section for various manufacturing processes, 346, 347
Tooling panel, 295-297
Trimming, 305
Two-component, 315
Two-plate mold, 109

U

UV, 146
    Curing, 146
Ultraviolet, 146
Unidirectional, 2, 8, 42
Composites, 2, 8
    Prepreg, 41

V

VARTM, 132, 175, 225
Vacuum bagging, 124, 127, 133, 219, 296, 349
Variable cost, 349, 350
Vinylester, 32, 61, 117, 136, 143

W

Warehousing, 80
Warp, 37
Warpage, 127
Waste, 376
Waterjet cutting, 337, 338
Wear resistance, 67
Weft
Weighted property, 66
Welding, 216, 223
Weld line, 114, 115, 228
Wet lay-up, 100, 104, 106, 116, 119, 128, 131, 132, 134-136, 139
    Advantages, 135
    Applications, 128
    Limitations, 135
    Making of the part, 131
    Methods of applying heat and pressure, 134
    Processing steps, 134
    Raw materials, 131
    Tooling requirement, 131
Wetting, 314
Windmill blades, 128
Woven fabric, 26, 36, 59

Y

Yacht, 128, 130
Yarns, 26, 37